CAMBRIDGE TRACTS IN MATHEMATICS

---

189 Nonlinear Perron–Frobenius Theory

CAMBRIDGE TRACTS IN MATHEMATICS

GENERAL EDITORS

A complete list of books in the series can be found at www.cambridge.org/mathematics. Recent titles include the following:

155. The Direct Method in Soliton Theory. By R. HIROTA. Edited and translated by A. NAGAI, J. NIMMO, and C. GILSON
156. Harmonic Mappings in the Plane. By P. DUREN
157. Affine Hecke Algebras and Orthogonal Polynomials. By I. G. MACDONALD
158. Quasi-Frobenius Rings. By W. K. NICHOLSON and M. F. YOUSIF
159. The Geometry of Total Curvature on Complete Open Surfaces. By K. SHIOHAMA, T. SHIOYA, and M. TANAKA
160. Approximation by Algebraic Numbers. By Y. BUGEAUD
161. Equivalence and Duality for Module Categories. By R. R. COLBY and K. R. FULLER
162. Lévy Processes in Lie Groups. By M. LIAO
163. Linear and Projective Representations of Symmetric Groups. By A. KLESHCHEV
164. The Covering Property Axiom, CPA. By K. CIESIELSKI and J. PAWLIKOWSKI
165. Projective Differential Geometry Old and New. By V. OVSIENKO and S. TABACHNIKOV
166. The Lévy Laplacian. By M. N. FELLER
167. Poincaré Duality Algebras, Macaulay's Dual Systems, and Steenrod Operations. By D. MEYER and L. SMITH
168. The Cube-A Window to Convex and Discrete Geometry. By C. ZONG
169. Quantum Stochastic Processes and Noncommutative Geometry. By K. B. SINHA and D. GOSWAMI
170. Polynomials and Vanishing Cycles. By M. TIBĂR
171. Orbifolds and Stringy Topology. By A. ADEM, J. LEIDA, and Y. RUAN
172. Rigid Cohomology. By B. LE STUM
173. Enumeration of Finite Groups. By S. R. BLACKBURN, P. M. NEUMANN, and G. VENKATARAMAN
174. Forcing Idealized. By J. ZAPLETAL
175. The Large Sieve and its Applications. By E. KOWALSKI
176. The Monster Group and Majorana Involutions. By A. A. IVANOV
177. A Higher-Dimensional Sieve Method. By H. G. DIAMOND, H. HALBERSTAM, and W. F. GALWAY
178. Analysis in Positive Characteristic. By A. N. KOCHUBEI
179. Dynamics of Linear Operators. By F. BAYART and É. MATHERON
180. Synthetic Geometry of Manifolds. By A. KOCK
181. Totally Positive Matrices. By A. PINKUS
182. Nonlinear Markov Processes and Kinetic Equations. By V. N. KOLOKOLTSOV
183. Period Domains over Finite and p-adic Fields. By J.-F. DAT, S. ORLIK, and M. RAPOPORT
184. Algebraic Theories. By J. ADÁMEK, J. ROSICKÝ, and E. M. VITALE
185. Rigidity in Higher Rank abelian Group Actions I: Introduction and Cocycle Problem. By A. KATOK and V. NIŢICĂ
186. Dimensions, Embeddings, and Attractors. By J. C. ROBINSON
187. Convexity: An Analytic Viewpoint. By B. SIMON
188. Modern Approaches to the Invariant Subspace Problem. By I. CHALENDAR and J. R. PARTINGTON
189. Nonlinear Perron–Frobenius Theory. By B. LEMMENS and R. NUSSBAUM
190. Jordan Structures in Geometry and Analysis. By C.-H. CHU
191. Malliavin Calculus for Lévy Processes and Infinite-Dimensional Brownian Motion. By H. OSSWALD
192. Normal Approximations with Malliavin Calculus. By I. NOURDIN and G. PECCATI

Nonlinear Perron–Frobenius Theory

BAS LEMMENS

University of Kent, Canterbury

ROGER NUSSBAUM

Rutgers University, New Jersey

CAMBRIDGE
UNIVERSITY PRESS

CAMBRIDGE UNIVERSITY PRESS
Cambridge, New York, Melbourne, Madrid, Cape Town,
Singapore, São Paulo, Delhi, Mexico City

Cambridge University Press
The Edinburgh Building, Cambridge CB2 8RU, UK

Published in the United States of America by Cambridge University Press, New York

www.cambridge.org
Information on this title: www.cambridge.org/9780521898812

First published 2012

Printed in the United Kingdom at the University Press, Cambridge

A catalogue record for this publication is available from the British Library

Library of Congress Cataloguing in Publication data
Lemmens, Bas.
Nonlinear Perron-Frobenius theory / Bas Lemmens, Roger Nussbaum.
p. cm. – (Cambridge tracts in mathematics ; 189)
Includes bibliographical references and index.
ISBN 978-0-521-89881-2 (hardback)
1. Non-negative matrices. 2. Eigenvalues. 3. Eigenvectors.
4. Algebras, Linear. I. Nussbaum, Roger D., 1944– II. Title.
QA188.L456 2012
512′.5–dc23
2011053268

ISBN 978-0-521-89881-2 Hardback

Contents

Preface

Sometimes in mathematics a simple-looking observation opens up a new road to a fertile field. Such an observation was made independently by Garrett Birkhoff [25] and Hans Samelson [192], who remarked that one can use Hilbert's (projective) metric and the contraction mapping principle to prove some of the theorems of Perron and Frobenius concerning eigenvectors and eigenvalues of nonnegative matrices. This idea has been pivotal for the development of nonlinear Perron–Frobenius theory.

In the past few decades a number of strikingly detailed nonlinear extensions of Perron–Frobenius theory have been obtained. These results provide an extensive analysis of the eigenvectors and eigenvalues of various classes of order-preserving (monotone) nonlinear maps and give information about their iterative behavior and periodic orbits. Particular classes of order-preserving maps for which there exist nonlinear Perron–Frobenius theorems include subhomogeneous maps, topical maps, and integral-preserving maps. The latter class of order-preserving maps can be regarded as a nonlinear generalization of column stochastic matrices, whereas topical maps generalize row stochastic matrices

The main purpose of this book is to give a systematic, self-contained introduction to nonlinear Perron–Frobenius theory and to provide a guide to various challenging open problems. We hope that it will be a stimulating source for non-experts to learn and appreciate this subject. To keep our task manageable, we restrict ourselves to **finite-dimensional** vector spaces, which allows us to avoid the use of sophisticated fixed-point theorems, the fixed-point index, and topological degree theory. The material in this book requires familiarity only with basic real analysis and topology, and is accessible to graduate students.

Classical Perron–Frobenius theory was developed in the early 1900s. In fact, Perron published his work [179, 180] on eigenvalues and eigenvectors of matrices with positive coefficients in 1907. His results were generalized by

Frobenius, in a series of papers [70–72] a few years later, to irreducible matrices with nonnegative coefficients. Their collective results and subsequent work are known today as Perron–Frobenius theory and are considered one of the most beautiful topics in linear algebra. The theory has numerous applications in probability theory, game theory, information theory, dynamical systems theory, mathematical biology, mathematical economics, and computer science.

In a seminal paper, Kreĭn and Rutman [117] placed the Perron–Frobenius theorem in the more general context of linear operators that leave a cone in a Banach space invariant. Their work has led to numerous studies of the spectral theory of positive linear operators on Banach spaces, including work by Bonsall [29–32] and Schaefer [193–195]. A detailed account can be found in Schaefer's book [196]. In their work Kreĭn and Rutman combined analytic methods with geometric ones. Among other methods they applied the Brouwer fixed-point theorem to find a positive eigenvector of a nonsingular matrix that leaves a cone invariant. Their geometric ideas are another source of inspiration for nonlinear Perron–Frobenius theory.

The book contains nine chapters and two appendices. The first four chapters contain preliminaries to Chapters 5, 6, 7, 8, and 9, which form the core of the book. The main objective of Chapter 1 is to introduce the reader to a variety of questions in nonlinear Perron–Frobenius theory. To this end we recall the classical results from Perron–Frobenius theory. Some of their proofs are given in Appendix B and are nonlinear in spirit. We also introduce various classes of nonlinear order-preserving maps for which there exist nonlinear Perron–Frobenius theorems and provide motivating examples.

Chapter 2 develops the relation between order-preserving maps and non-expansive maps. It shows how various classes of order-preserving maps are non-expansive under Hilbert's metric, Thompson's metric, or a polyhedral norm. At the heart of these results lies the observation by Birkhoff and Samelson. In addition, several results concerning the geometry and topology of Hilbert's and Thompson's metric spaces are collected.

Chapter 3 covers several useful topics on the iterative behavior of non-expansive maps including ω-limit sets, fixed-point theorems, horofunctions, and horoballs. It also contains a discussion of "Denjoy–Wolff type" theorems for fixed-point free non-expansive maps on metric spaces whose geometry resembles that of a hyperbolic space, which are due to Beardon [18, 19] and were further developed by Karlsson [99].

Chapter 4 focuses on the dynamics of sup-norm non-expansive maps. A characteristic property of sup-norm non-expansive maps is that either all orbits are unbounded or each orbit converges to a periodic orbit. Moreover, there exists an a-priori upper bound for the periods of their periodic points in terms of the dimension of the underlying space. These results are a key ingredient in the analysis of the iterative behavior of order-preserving subhomogeneous maps on polyhedral cones. In addition, the dynamics and periodic orbits of topical maps are discussed.

Chapter 5 deals with eigenvectors and the cone spectral radius of order-preserving homogeneous maps on closed cones in a finite-dimensional vector space. The cone spectrum and the cone spectral radius are analyzed. Among other results it is shown that there exists an eigenvector in the cone corresponding to the cone spectral radius. The chapter also treats the continuity problem of the cone spectral radius and discusses nonlinear generalizations of the classical Collatz–Wielandt formula for the spectral radius of nonnegative matrices.

Chapter 6 is mainly concerned with the question whether there exists an eigenvector in the interior of the cone for a given order-preserving homogeneous map. It appears that this problem is irreducibly difficult. Several general principles are discussed that are helpful when faced with this problem. These principles and their limitations are illustrated by analyzing particular order-preserving homogeneous maps involving means.

In Chapter 7 we illustrate how the results from Chapters 5 and 6 can be applied to finding solutions to various matrix scaling problems. We follow the fixed-point approach, as pioneered by Menon [143]. Among other matters we discuss the elegant solution, independently discovered by Sinkhorn and Knopp [206] and Brualdi, Parter, and Schneider [39], of the classic DAD problem: Given an $n \times n$ nonnegative matrix A, when do there exist positive diagonal matrices D and E such that DAE is doubly stochastic?

In Chapter 8 we derive a variety of results for order-preserving subhomogeneous maps on finite-dimensional cones. A central question is the behavior of the iterates of such maps. We provide a detailed analysis of the periodic orbits of order-preserving subhomogeneous maps on polyhedral cones. We also discuss "Denjoy–Wolff type" theorems for order-preserving homogeneous maps which do not have an eigenvector in the interior of the cone.

Chapter 9 is devoted to nonlinear Perron–Frobenius theorems for order-preserving integral-preserving maps on the standard positive cone. Such maps are non-expansive under the ℓ_1-norm. It is shown how the dynamics of order-preserving integral-preserving maps is related to the dynamics of lower semi-lattice homomorphisms. This connection yields a complete combinatorial characterization of the set of possible periods of periodic points of order-preserving integral-preserving maps in terms of periods of so-called admissible arrays. This characterization allows one to compute the set of possible periods of periodic points of order-preserving integral-preserving maps on the standard positive cone in finite time.

This book does not attempt to be an encyclopedic coverage of nonlinear Perron–Frobenius theory, even for finite-dimensional spaces. The expert reader may note that the following topics have been omitted: applications to the theory of ordinary differential equations [87,88,115,164,228], the cycle time problem [36, 51, 152], and the spectral theory of order-preserving convex maps [3, 4]. However, the authors believe that an understanding of the material in this book will leave the reader well equipped to master the existing literature on these topics.

Nonlinear Perron–Frobenius theory is related to monotone dynamical systems theory. In the theory of monotone dynamical systems one considers discrete and continuous dynamical systems that are strongly order-preserving or strongly monotone. Pioneering work in this field was done by Hirsch [87] who showed, among other results, that in a continuous time strongly monotone dynamical system almost all pre-compact orbits converge to the set of equilibrium points. For discrete time strongly monotone dynamical systems one has generic convergence to periodic orbits under appropriate conditions; see [182]. An extensive overview of these results was given by Hirsch and Smith [89]; see also Smith [207]. In monotone dynamical systems theory, emphasis is placed on strong monotonicity. If, however, one only assumes the dynamical system to be monotone, most of the theory is not applicable. In nonlinear Perron–Frobenius theory one usually considers discrete time dynamical systems that are only monotone (order-preserving), but satisfy an additional assumption such as subhomogeneity, additive homogeneity, or the integral-preserving condition. Another notable difference between the two theories is that in monotone dynamical systems theory one usually assumes the system to be differentiable. In nonlinear Perron–Frobenius theory, the discrete dynamical system is often only continuous. These different assumptions on the dynamical system require different methods and give the two theories a very different character. We hope that this book will also be a valuable addition to the existing literature on monotone dynamical systems theory.

Acknowledgment

We are grateful to our wives Elizabeth and Joyce for their patience and continued support during the long process of writing this book.

1

What is nonlinear Perron–Frobenius theory?

To get an impression of the contents of nonlinear Perron–Frobenius theory, it is useful to first recall the basics of classical Perron–Frobenius theory. Classical Perron–Frobenius theory concerns nonnegative matrices, their eigenvalues and corresponding eigenvectors. The fundamental theorems of this classical theory were discovered at the beginning of the twentieth century by Perron [179, 180], who investigated eigenvalues and eigenvectors of matrices with strictly positive entries, and by Frobenius [70–72], who extended Perron's results to irreducible nonnegative matrices. In the first section we discuss the theorems of Perron and Frobenius and some of their generalizations to linear maps that leave a cone in a finite-dimensional vector space invariant. The proofs of these classical results can be found in many books on matrix analysis, e.g., [15, 22, 73, 148, 202]. Nevertheless, in Appendix B we prove some of them once more using a combination of analytic, geometric, and algebraic methods. The geometric methods originate from work of Alexandroff and Hopf [8], Birkhoff [25], Kreĭn and Rutman [117], and Samelson [192] and underpin much of nonlinear Perron–Frobenius theory. Readers who are not familiar with these methods might prefer to first read Chapters 1 and 2 and Appendix B. Besides recalling the classical Perron–Frobenius theorems, we use this chapter to introduce some basic concepts and terminology that will be used throughout the exposition, and provide some motivating examples of classes of nonlinear maps to which the theory applies.

We emphasize that throughout the book we will always be working in a **finite-dimensional** real vector space V, unless we explicitly say otherwise.

1.1 Classical Perron–Frobenius theory

An $n \times n$ matrix $A = (a_{ij})$ is said to be *nonnegative* if $a_{ij} \geq 0$ for all i and j. It is called *positive* if $a_{ij} > 0$ for all i and j. Similarly, we call a vector $x \in \mathbb{R}^n$

nonnegative (or *positive*) if all its coordinates are nonnegative (or positive). The *spectrum* of A is given by

$$\sigma(A) = \{\lambda \in \mathbb{C} : Ax = \lambda x \text{ for some } x \in \mathbb{C}^n \setminus \{0\}\}.$$

Recall also that the *spectral radius* of A is given by

$$r(A) = \max\{|\lambda| : \lambda \in \sigma(A)\},$$

and satisfies the equality $r(A) = \lim_{k\to\infty} \|A^k\|^{1/k}$. Notice that the limit is independent of the choice of the matrix norm, or norm on $\mathbb{R}^{n^2}$, as they are all equivalent; see Rudin [190]. The following result is due to Perron [180].

Theorem 1.1.1 (Perron) *If A is a positive matrix, then the following assertions hold:*

(i) $r(A) > 0$, and $r(A)$ is an algebraically simple eigenvalue of A and the corresponding normalized eigenvector v is unique and positive.
(ii) Any nonnegative eigenvector of A is a multiple of v.
(iii) For each eigenvalue $\lambda \in \sigma(A)$ with $\lambda \neq r(A)$ we have that $|\lambda| < r(A)$.

In a series of papers Frobenius [70–72] extended Perron's theorem to so-called irreducible matrices. An $n \times n$ nonnegative matrix $A = (a_{ij})$ is called *reducible* if $\{1, \ldots, n\}$ can be partitioned into two non-empty sets I and J such that $a_{ij} = 0$ for all $i \in I$ and $j \in J$. In other words, A is reducible if and only if there exists a permutation matrix P such that

$$P^T AP = \begin{bmatrix} B & C \\ O & D \end{bmatrix},$$

where B and D are square matrices and O is the zero matrix. A nonnegative matrix is said to be *irreducible* if it is not reducible. In particular, any nonnegative 1×1 matrix is irreducible. Frobenius proved the following generalization of Perron's result.

Theorem 1.1.2 (Perron–Frobenius) *If A is a nonnegative irreducible $n \times n$ matrix, then the following assertions hold:*

(i) $r(A)$ is an algebraically simple eigenvalue of A and the corresponding normalized eigenvector v is unique and positive. Moreover $r(A) > 0$, if $A \neq [0]$.
(ii) Any nonnegative eigenvector of A is a multiple of v.
(iii) If, in addition, A has exactly q eigenvalues λ with $|\lambda| = r(A)$, then these eigenvalues are given by $r(A)e^{2\pi ik/q}$ for $0 \leq k < q$.

The integer q in Theorem 1.1.2 is called the *index of cyclicity* of A.

Nonnegative matrices leave the cone of nonnegative vectors in $\mathbb{R}^n$ invariant. This is a crucial property of nonnegative matrices. In fact, a major part of the Perron–Frobenius theory can be generalized to linear maps that leave a cone in a vector space invariant. We will discuss this important fact in greater detail now. Let V be a finite-dimensional real vector space. A subset K of V is called a *cone* if it is convex, $\mu K \subseteq K$ for all $\mu \geq 0$, and $K \cap (-K) = \{0\}$. It is said to be a *closed* cone if it is a closed set in V with respect to the standard topology. Given a subset $S \subseteq V$, the interior, closure, and boundary of S with respect to the standard topology on V are denoted, respectively by $\mathrm{int}(S)$, $\mathrm{cl}(S)$, and ∂S. A cone is said to be *solid* if it has a non-empty interior. Basic examples of solid closed cones include the *standard positive cone*,

$$\mathbb{R}^n_+ = \{(x_1, \ldots, x_n) \in \mathbb{R}^n : x_i \geq 0 \text{ for } 1 \leq i \leq n\},$$

and the *Lorentz cone*,

$$\Lambda_{n+1} = \{(x_1, \ldots, x_{n+1}) \in \mathbb{R}^{n+1} : x_1^2 - x_2^2 - \ldots - x_{n+1}^2 \geq 0 \text{ and } x_1 \geq 0\}.$$

Other interesting examples arise in spaces of matrices. For example, if V is the space of real $n \times n$ symmetric matrices, then the set of positive-semidefinite matrices, $\Pi_n(\mathbb{R})$, is a solid closed cone in V.

We write V^* to denote the dual space of V and define the *dual cone* of K by

$$K^* = \{\varphi \in V^* : \varphi(x) \geq 0 \text{ for all } x \in K\}.$$

A *face* of a closed cone K in V is a non-empty convex subset F of K such that if $x, y \in K$ and $\lambda x + (1-\lambda)y \in F$ for some $0 < \lambda < 1$, then $x, y \in F$. Note that K itself and $\{0\}$ are faces. These two faces of K are called *improper faces*. All other faces of K are said to be *proper*. The relative interior of a convex set $C \subseteq V$, denoted relint (C), is the interior of C, regarded as a subset of the affine hull of C in V. It is known (see [186, theorem 6.2]) that $\mathrm{cl}(C)$ and relint (C) are non-empty and convex if C is a non-empty convex subset of V. We can use this to show that each face of a closed cone is closed. Indeed, given a face F of K, let z be a point in the relative interior of F. Now note that, for each $x \in \mathrm{cl}(F)$ with $x \neq z$, there exists $y \in \mathrm{cl}(F)$ such that z is in the relative interior of the straight-line segment from x to y, and hence $x \in F$.

A face F of K is called an *exposed face* if there exists $\varphi \in K^*$ such that $F = K \cap \{x \in V : \varphi(x) = 0\}$. In general not every face of a cone is exposed. However, each face of a polyhedral cone is exposed. A cone K in V is *polyhedral* if it is the intersection of finitely many closed half-spaces, i.e., there exists $\varphi_1, \ldots, \varphi_m \in V^*$ such that

$$K = \{x \in V : \varphi_i(x) \geq 0 \text{ for all } 1 \leq i \leq m\}.$$

A face F of a polyhedral cone K is called a *facet* if $\dim(F) = \dim(K) - 1$. Here $\dim(F)$ denotes the dimension of the linear span of F. The following basic result from polyhedral geometry [201, section 8.4] will be useful in the sequel.

Lemma 1.1.3 *If $K \subseteq V$ is a polyhedral cone with N facets, then there exist N linear functionals $\psi_1, \ldots, \psi_N$ such that*

$$K = \{x \in V : \psi_i(x) \geq 0 \text{ for all } 1 \leq i \leq N\} \cap \mathrm{span}(K)$$

and each linear functional ψ_i corresponds to a unique facet of K.

The functionals $\psi_1, \ldots, \psi_N$ in Lemma 1.1.3 are called *facet-defining functionals* of the polyhedral cone. For example, $\mathbb{R}^n_+$ has 2^n faces, $F_I = \{x \in \mathbb{R}^n_+ : x_i > 0 \text{ if and only if } i \in I\}$ for $I \subseteq \{1, \ldots, n\}$, and n facets corresponding to those I with $|I| = n - 1$.

A cone K in V induces a partial ordering $\leq_K$ on V by

$$x \leq_K y \text{ if } y - x \in K.$$

If K has a non-empty interior, then we write $x \ll_K y$ if $y - x \in \mathrm{int}(K)$. We also write $x <_K y$ if $x \leq_K y$ and $x \neq y$. We simply write $x \leq y$, $x < y$, and $x \ll y$ if K is clear from the context. In the special case where $K = \mathbb{R}^n_+$ we note that $x \leq y$ if and only if $x_i \leq y_i$ for all $1 \leq i \leq n$. Note that a linear map $A : V \to V$ leaves a cone K invariant if and only if $0 \leq_K Ax$ for all $0 \leq_K x$ in V.

There is a natural way to generalize the concept of irreducibility to linear maps that leave a solid closed cone invariant.

Definition 1.1.4 A linear map $A : V \to V$ that leaves a solid closed cone K invariant is said to be *irreducible* if no proper face of K is left invariant by A.

It is a simple exercise to show that a nonnegative matrix A is irreducible in the sense of Definition 1.1.4 if and only if it is irreducible in the usual sense. Another equivalent way to define the notion of irreducibility is given in the following proposition.

Proposition 1.1.5 *A linear map $A : V \to V$ that leaves a solid closed cone K invariant is irreducible if and only if $(\lambda I - A)^{-1}(K \setminus \{0\}) \subseteq \mathrm{int}(K)$ for some $\lambda > r(A)$.*

Proof Suppose, for the sake of contradiction, that A is irreducible and there exists $z \in K \setminus \{0\}$ such that $(\lambda I - A)^{-1} z \notin \mathrm{int}(K)$ for all $\lambda > r(A)$. Write $u = (\lambda I - A)^{-1} z$. By the Hahn–Banach separation theorem [186, Theorem 11.6] there exists $\varphi \in K^* \setminus \{0\}$ such that $\varphi(u) = 0$ and $\varphi(x) > 0$ for all $x \in \mathrm{int}(K)$. Define $\psi \in V^*$ by

$$\psi(x) = \varphi((\lambda I - A)^{-1}x) = \sum_{k=0}^{\infty} \lambda^{-k-1}\varphi(A^k x),$$

and remark that $\psi(x) > 0$ for all $x \in \text{int}(K)$. Obviously $\psi(z) = 0$ and hence $F = \{x \in K : \psi(x) = 0\}$ is a proper (exposed) face of K. Moreover, for each $x \in F$ we have that

$$0 \leq \psi(Ax) = \varphi((\lambda I - A)^{-1}(Ax)) = \sum_{k=0}^{\infty} \lambda^{-k-1}\varphi(A^{k+1}x)$$

$$= \lambda \sum_{k=1}^{\infty} \lambda^{-k-1}\varphi(A^k x) \leq \lambda\psi(x) = 0.$$

This implies that $\psi(Ax) = 0$ for all $x \in F$. Thus, A leaves the proper face F invariant, which is impossible.

To establish the equivalence, suppose that for some $\lambda > r(A)$ we have that $(\lambda I - A)^{-1}z \in \text{int}(K)$ for all $z \in K \setminus \{0\}$. Now if there exists a proper face F of K such that $A(F) \subseteq F$, then it follows from the identity, $(\lambda I - A)^{-1} = \sum_{k=0}^{\infty} \lambda^{-k-1}A^k$, that $(\lambda I - A)^{-1}(F) \subseteq F$, which is impossible. □

Note that the proof of Proposition 1.1.5 also shows that a linear map A is irreducible if and only if it leaves no proper exposed face of K invariant.

As mentioned earlier the theorems of Perron and Frobenius can be generalized to linear maps that leave a cone invariant. This important observation was made by Kreĭn and Rutman in their pioneering work [117], in which they studied linear operators that leave a cone in a possibly infinite-dimensional normed space invariant. As we are only concerned with finite-dimensional spaces in this book, we give a finite-dimensional version of their theorem here.

Theorem 1.1.6 (Kreĭn–Rutman) *If $A : V \to V$ is a linear map that leaves a solid closed cone K invariant, then $r(A)$ is an eigenvalue of A and $r(A)$ has an eigenvector in K.*

Just as for the Perron–Frobenius Theorem 1.1.2 we have uniqueness of the eigenvector if A is irreducible.

Theorem 1.1.7 *If $A : V \to V$ is an irreducible linear map that leaves a solid closed cone K invariant, then the following assertions hold:*

(i) $r(A)$ is an algebraically simple eigenvalue of A and the corresponding normalized eigenvector v is unique and lies in $\text{int}(K)$. *Furthermore, $r(A) > 0$, if $A \neq [0]$.*
(ii) Any nonnegative eigenvector of A is a multiple of v.

The third part of the Perron–Frobenius Theorem 1.1.2 cannot be extended to arbitrary closed cones. Simply consider the Lorentz cone Λ_3 and the linear map $A : \mathbb{R}^3 \to \mathbb{R}^3$ given by

$$A\begin{pmatrix} x_1 \\ x_2 \\ x_3 \end{pmatrix} = \begin{pmatrix} 1 & 0 & 0 \\ 0 & \cos\vartheta & -\sin\vartheta \\ 0 & \sin\vartheta & \cos\vartheta \end{pmatrix}\begin{pmatrix} x_1 \\ x_2 \\ x_3 \end{pmatrix}.$$

Clearly $A(\Lambda_3) \subseteq \Lambda_3$ and $\sigma(A) = \{1, e^{\pm i\vartheta}\}$. If ϑ is an irrational multiple of 2π, then $e^{i\vartheta}$ is not a root of unity. The third part of the Perron–Frobenius Theorem 1.1.2 can, however, be generalized to polyhedral cones (see [117, section 8]).

Theorem 1.1.8 *If $A : V \to V$ is a linear map that leaves a solid polyhedral cone K with N facets invariant, then for each $\lambda \in \sigma(A)$ with $|\lambda| = r(A)$ there exists $1 \leq q \leq N$ such that*

$$\lambda^q = r(A)^q.$$

Theorem 1.1.8 can be used to prove the following result concerning the iterative behavior of linear maps that leave a polyhedral cone invariant; see Appendix B for details.

Theorem 1.1.9 *If $A : V \to V$ is a linear map that leaves a solid polyhedral cone K with N facets invariant, then there exists an integer $p \geq 1$ such that*

$$\lim_{k\to\infty} A^{kp}x$$

exists for each $x \in K$ with $(\|A^k x\|)_k$ bounded. Moreover, p is the order of a permutation on N letters.

For linear maps $A : V \to V$ the condition that $A(K) \subseteq K$ is equivalent to $x \leq_K y$ implies $Ax \leq_K Ay$ for $x, y \in V$. Given cones $K \subseteq V$ and $K' \subseteq V'$, a map $f : X \to X'$, with $X \subseteq V$ and $X' \subseteq V'$, is said to be *order-preserving* if $x \leq_K y$ implies $f(x) \leq_{K'} f(y)$ for $x, y \in X$. It is said to be *strongly order-preserving* if $x <_K y$ implies $f(x) \ll_{K'} f(y)$. Moreover, we say that $f : X \to X'$ is *order-reversing* if $x \leq_K y$ implies $f(y) \leq_{K'} f(x)$ for $x, y \in X$.

Nonlinear Perron–Frobenius theory is primarily concerned with order-preserving maps and treats questions like:

- Is there a sensible definition of the spectral radius for order-preserving maps $f : K \to K$, and does there exist a corresponding eigenvector?
- When does an order-preserving map have an eigenvector in the interior of the cone, and when is it unique?

- How do the orbits

$$x, f(x), f^2(x) = f(f(x)), f^3(x) = f(f(f(x))), \ldots$$

of an order-preserving map f behave in the long term?
- When does an order-preserving map have the property that every bounded orbit converges to a periodic orbit?

These questions arise in numerous applications and lie at the heart of nonlinear Perron–Frobenius theory. They motivate much of the material discussed in this book. As we shall see, strikingly detailed answers exist for a variety of classes of order-preserving map.

1.2 Cones and partial orderings

Given a cone $K \subseteq V$ and $x, y \in V$ it is often useful to use the following result to decide whether $x \leq_K y$.

Lemma 1.2.1 *If $K \subseteq V$ is a closed cone, then $x \leq_K y$ if and only if $\varphi(x) \leq \varphi(y)$ for all $\varphi \in K^*$.*

Proof Note that if $x \not\leq_K y$, then $y - x \notin K$. By the Hahn–Banach separation theorem [186, theorem 11.4] there exist $\alpha \in \mathbb{R}$ and $\psi \in V^*$ such that $\psi(y-x) < \alpha$ and $\psi(v) > \alpha$ for all $v \in K$. Remark that, as $\lambda v \in K$ for all $\lambda \geq 0$ and $v \in K$, $\psi(v) \geq 0$ for all $v \in K$, and hence $\psi \in K^*$. Also note that $\psi(0) = 0$, so that $\alpha < 0$. Thus, $\psi(y) < \psi(x)$. The other implication is trivial. □

We say that $y \in K$ *dominates* $x \in V$ if there exist $\alpha, \beta \in \mathbb{R}$ such that $\alpha y \leq x \leq \beta y$. This notion yields an equivalence relation, $\sim_K$, on K by $x \sim_K y$ if x dominates y and y dominates x. It is easy to verify that $x \sim_K y$ if and only if there exist $0 < \alpha \leq \beta$ such that $\alpha y \leq x \leq \beta y$. The equivalence classes in K are called *parts* of the cone, and we write $[x]$ to denote the part of K containing x. The set of all parts of K is denoted by $\mathcal{P}(K)$. For instance, the parts of $\mathbb{R}^2_+$ are $\{0\}$, $\mathrm{int}(\mathbb{R}^2_+)$, $\{(x_1, 0) \in \mathbb{R}^2_+ : x_1 > 0\}$ and $\{(0, x_2) \in \mathbb{R}^2_+ : x_2 > 0\}$. The parts of Λ_3 are given by $\{0\}$, $\mathrm{int}(\Lambda_3)$, and $\{\lambda(1, \cos\vartheta, \sin\vartheta) : \lambda > 0\}$ for $0 \leq \vartheta < 2\pi$.

Lemma 1.2.2 *If $K \subseteq V$ is a closed cone, then the parts of K are precisely the relative interiors of the faces of K.*

Proof It is known (see [186, theorem 8.2]) that the relative interiors of the faces of K partition K. Let $[x]$ be the part of x in K and F_x be the face of K with x in its relative interior. Note that F_x is itself a closed cone. Suppose that y is in the relative interior of F_x. Then there exists $\alpha > 0$ such that $y - \alpha x \in F_x$ and $x - \alpha y \in F_x$. As $F_x \subseteq K$, we deduce that $\alpha x \leq y \leq x/\alpha$, and hence $x \sim_K y$.

On the other hand, if $y \in K$ is such that $x \sim_K y$, then for $\delta > 0$ sufficiently small, $x - \delta y \sim_K y$ and $y - \delta x \sim_K x$. As $x \sim_K y$ implies $\lambda x \sim_K \mu y$ for all $\lambda, \mu > 0$, we deduce for $\varepsilon > 0$ sufficiently small that

$$x_\varepsilon = (1+\varepsilon)x - \varepsilon y \sim_K y \quad \text{and} \quad y_\varepsilon = (1+\varepsilon)y - \varepsilon x \sim_K x.$$

Note that $x = ax_\varepsilon + by_\varepsilon$ for $a = (1+\varepsilon)/(1+2\varepsilon)$ and $b = \varepsilon/(1+2\varepsilon)$. As $x \in F_x$, we deduce that x_ε and y_ε in F_x, since x is in the relative interior of F_x. As $y = \lambda y_\varepsilon + (1-\lambda)x$ for $\lambda = 1/(1+\varepsilon)$, we conclude that y is in the relative interior of F_x. □

It follows from Lemma 1.2.2 that a polyhedral cone has finitely many parts. The combinatorics of the parts of a polyhedral cone will play an important role in the sequel, especially in Chapter 8. Given a polyhedral cone K with N facets and facet-defining functionals $\psi_1, \ldots, \psi_N$ define for each $x \in K$ the set $I_x = \{i : \psi_i(x) > 0\}$. Likewise, if P is a part of K we let $I(P) = \{i : \psi_i(x) > 0 \text{ for some } x \in P\}$.

The set of parts $\mathcal{P}(K)$ has a natural partial ordering $\trianglelefteq$ given by $P \trianglelefteq Q$ if there exist $x \in P$ and $y \in Q$ such that y dominates x. Equivalently, $P \trianglelefteq Q$ if for each $x \in P$ and each $y \in Q$ we have that y dominates x.

Lemma 1.2.3 *If $K \subseteq V$ is a polyhedral cone with N facets, then the following assertions hold:*

(i) $I_x = I_y$ if and only if $x \sim_K y$.
(ii) For $P \in \mathcal{P}(K)$ we have

$$P = \{x \in K : \psi_i(x) > 0 \text{ if and only if } i \in I(P)\}.$$

(iii) $P \trianglelefteq Q$ if and only if $I(P) \subseteq I(Q)$.
(iv) $|\mathcal{P}(K)| \leq 2^N$.

Proof Let $\psi_1, \ldots, \psi_N \in K^*$ be the facet-defining functionals of K. By Lemma 1.1.3 we know that $x \leq_K y$ is equivalent to $\psi_i(x) \leq \psi_i(y)$ for all i. This implies that $I_x \subseteq I_y$ if and only if y dominates x. Therefore $x \sim_K y$ is equivalent to $I_x = I_y$. It also shows that $x \in P$ if and only if $I_x = I(P)$, which proves the second assertion.

Now suppose that $P \trianglelefteq Q$. Then there exist $x \in P$ and $y \in Q$ such that y dominates x. So, $I(P) = I_x \subseteq I_y = I(Q)$. On the other hand, if $I(P) \subseteq I(Q)$, then for $x \in P$ and $y \in Q$ we know that there exists $\beta > 0$ such that $\psi_i(x) \leq \beta\psi_i(y)$ for all i. Thus, by Lemma 1.1.3 we find that y dominates x and hence $P \trianglelefteq Q$. The final assertion is a direct consequence of the third one. □

The shape of the cone plays a fundamental role in nonlinear Perron–Frobenius theory. A distinguished class of cones is the set of so-called strictly convex cones. A closed cone $K \subseteq V$ is called *strictly convex* if, for all $x, y \in \partial K \setminus \{0\}$, with $x \neq \alpha y$ for all $\alpha > 0$, we have that

$$\{\lambda x + (1-\lambda)y : 0 < \lambda < 1\} \subseteq \mathrm{int}(K).$$

The Lorentz cone, Λ_{n+1}, is a prime example of a strictly convex cone for $n \geq 2$. In the sequel it will become clear that there is a marked contrast in the theory between polyhedral cones, strictly convex cones, and cones that are neither polyhedral nor strictly convex. An important example of the latter type is the cone of positive-semidefinite matrices $\Pi_n(\mathbb{R})$ for $n \geq 3$.

The partial ordering induced by the standard positive cone is particularly nice. To fully appreciate this we need a few more definitions. Let K be a cone in V. If $S \subseteq V$ and $u \in V$ is such that $s \leq_K u$ for all $s \in S$, we say that u is an *upper bound* of S. If, in addition, each upper bound $v \in V$ of S satisfies $u \leq_K v$, then we call u the *supremum* of S and we write $u = \sup(S)$. In a similar way lower bounds and the *infimum* of S can be defined. Note that if $K = \mathbb{R}^n_+$, then $\sup(x, y)$ exists and satisfies $\sup(x, y)_i = \max\{x_i, y_i\}$ for all i. Likewise, $\inf(x, y)$ exists and satisfies $\inf(x, y)_i = \min\{x_i, y_i\}$ for all i. In that case we write $x \wedge y = \inf(x, y)$ and $x \vee y = \sup(x, y)$, so $(x \wedge y)_i = \min\{x_i, y_i\}$ and $(x \vee y)_i = \max\{x_i, y_i\}$ for $1 \leq i \leq n$. The operations $\wedge$ and $\vee$ turn $\mathbb{R}^n$ into a vector lattice.

In general cones $\sup(x, y)$ and $\inf(x, y)$ need not exist for all $x, y \in V$. Cones for which $\sup(x, y)$ and $\inf(x, y)$ exist for all $x, y \in V$ are called *minihedral*. It is known that a solid closed minihedral cone is essentially equal to $\mathbb{R}^n_+$. More precisely, it was shown in [117,151] that a solid closed cone K in an n-dimensional real vector space V is minihedral if and only if K is *simplicial*, i.e., there exist n linearly independent vectors $x^1, \ldots, x^n \in V$ such that

$$K = \{x \in V : x = \sum_i \alpha_i x^i \text{ where } \alpha_i \geq 0 \text{ for } 1 \leq i \leq n\}.$$

We conclude this section with some basic facts about dual cones and the normality property. Recall that the dual cone of $K \subseteq V$ is given by $K^* = \{\varphi \in V^* : \varphi(x) \geq 0 \text{ for all } x \in K\}$. In general K^* need not be a cone. However, the following is true.

Lemma 1.2.4 *If $K \subseteq V$ is a solid closed cone, then K^* is a solid closed cone. Moreover, if $\varphi \in \mathrm{int}(K^*)$, then $\varphi(x) > 0$ for all $x \in K$ with $x \neq 0$, and*

$$\Sigma_\varphi = \{x \in K : \varphi(x) = 1\}$$

is a compact convex subset of V.

Proof Clearly K^* is convex and $\lambda K^* \subseteq K^*$ for all $\lambda \geq 0$. Suppose that φ and $-\varphi$ are both in K^*, and $\varphi \neq 0$. Then $\varphi(x) \geq 0$ and $\varphi(x) \leq 0$ for all $x \in K$. This implies that $K \subseteq \{x \in V : \varphi(x) = 0\}$, which contradicts $\mathrm{int}(K) \neq \emptyset$.

Let $K^* - K^* = \{\varphi - \psi : \varphi, \psi \in K^*\}$, which is the smallest subspace of V^* containing K^*. For the sake of contradiction, suppose that $\mathrm{int}(K^*) = \emptyset$. We claim that $K^* - K^* \neq V^*$ in that case. Indeed, if $K^* - K^* = V^*$, then there exist linearly independent functionals in $\varphi_1, \ldots, \varphi_n \in K^*$, where $n = \dim(V^*)$. This implies that the convex hull of $\{0, \varphi_1, \ldots, \varphi_n\}$ is an n-dimensional simplex in K^* and hence $\mathrm{int}(K^*) \neq \emptyset$. Thus $K^* - K^* \neq V^*$, and hence there exists $z \in V \setminus \{0\}$ such that $K^* - K^* \subseteq \{\varphi \in V^* : \varphi(z) = 0\}$. It now follows from Lemma 1.2.1 that $z \in K$ and $-z \in K$, which is impossible, as K is a cone.

Let $\varphi \in \mathrm{int}(K^*)$ and suppose that $\varphi(x) = 0$ for some $x \in K$ with $x \neq 0$. As K is a cone, $-x \notin K$. By the Hahn–Banach separation theorem [186, theorem 11.4] there exists $\alpha \in \mathbb{R}$ and $\psi_x \in K^*$ such that $\psi_x(-x) < \alpha$ and $\psi_x(v) > \alpha$ for all $v \in K$. Since $\lambda v \in K$ for all $\lambda \geq 0$ and $v \in K$, $\psi_x(v) \geq 0$, so that $\psi_x \in K^*$ and $\alpha < 0$. This implies that $-\psi_x(x) < \alpha < 0$. As $\varphi \in \mathrm{int}(K^*)$ there exists $\varepsilon > 0$ such that $\varphi - \varepsilon\psi_x \in \mathrm{int}(K^*)$. So, $\varphi(x) - \varepsilon\psi_x(x) < 0$, which contradicts Lemma 1.2.1.

Let $\|\cdot\|$ be any norm on V and $S = \{x \in K : \|x\| = 1\}$, which is a compact set, as K is closed. Since $\varphi : K \to \mathbb{R}$ is continuous and strictly positive on S, there exists $\delta > 0$ such that $\varphi(y) \geq \delta$ for all $y \in S$. Thus, for each $x \in \Sigma_\varphi$ we have that $\|x\| \leq \delta^{-1}$, which shows that Σ_φ is bounded. Obviously Σ_φ is closed and convex, and hence compact, since V is finite-dimensional. □

If $K \subseteq V$ is a cone and V is equipped with a norm $\|\cdot\|$, we say that K is *normal* if there exists a constant $\delta > 0$ such that $0 \leq x \leq y$ implies $\|x\| \leq \delta\|y\|$. The infimum of all such $\delta > 0$ is called the *normality constant*. We call $\|\cdot\|$ a *monotone norm* for K if the normality constant is equal to 1. It is a basic fact that every closed cone in a finite-dimensional normed space is normal.

Lemma 1.2.5 *Every closed cone K in $(V, \|\cdot\|)$ is normal.*

Proof For the sake of contradiction suppose that there exist sequences (x_k) and (y_k) in K such that $x_k \leq y_k$ for all $k \geq 1$ and

$$\lim_{k\to\infty} \frac{\|y_k\|}{\|x_k\|} = 0.$$

By rescaling x_k and y_k we may assume $\|x_k\| = 1$ and $x_k \leq y_k$ for all $k \geq 1$. This implies that $y_k \to 0$ as $k \to \infty$. Since V is finite-dimensional, we may further assume, after taking a subsequence, that (x_k) converges to some $x \in K$

with $\|x\| = 1$. But this implies that $y_k - x_k \to -x$ as $k \to \infty$, which is impossible, since $-x \notin K$. □

1.3 Order-preserving maps

Central objects in nonlinear Perron–Frobenius theory are order-preserving maps $f : K \to K$, where K is a solid closed cone in a finite-dimensional vector space V. To decide whether a map $f : K \to K$ is order-preserving with respect to K one can use, besides the definition, the Fréchet derivative, $Df(x)$, at $x \in \text{int}(K)$ of f. Recall that if $U \subseteq V$ is open, then $f : U \to V$ is said to be *Fréchet differentiable at* $x \in U$ if there exists a linear map $L : V \to V$ such that

$$\lim_{h \to 0} \frac{\|f(x+h) - f(x) - Lh\|}{\|h\|} = 0,$$

and we denote the *Fréchet derivative* by $Df(x) = L$.

The proof of the next theorem uses some more advanced results from real analysis. In particular, it uses Rademacher's theorem [184], which asserts that if $f : U \to V$ is locally Lipschitz, then f is Fréchet differentiable at $x \in U$ for Lebesgue almost all x, and $\|Df(x)\|$ is uniformly bounded on compact subsets of U. Moreover, the map $x \mapsto Df(x)$ is Lebesgue measurable. At first reading the reader may wish to assume that f is Fréchet differentiable on K, which simplifies the proof considerably.

Theorem 1.3.1 *Let $K \subseteq V$ be a solid closed cone and $U \subseteq \text{int}(K)$ convex and open. If $f : U \to K$ is locally Lipschitz, then $Df(x)$ exists for Lebesgue almost all $x \in U$, and f is order-preserving with respect to K if and only if $Df(x)(K) \subseteq K$ for all $x \in U$ for which $Df(x)$ exists.*

Proof If $x \in U$ is such that $Df(x)$ exists and $y \in K$, then there exists $\delta > 0$ such that $x + ty \in U$ for all $0 < t < \delta$ and

$$\lim_{t \to 0} \frac{f(x+ty) - f(x)}{t} = Df(x)y.$$

If f is order-preserving, the left-hand side of the previous equation is an element of K; so, $Df(x)(K) \subseteq K$.

By Rademacher's theorem [184] we know that $Df(x)$ exists for Lebesgue almost all $x \in U$. Assume for all such $x \in U$ that $Df(x)$ maps K into itself. Under these assumptions we must prove that f is order-preserving. As f is continuous, it suffices to show that $w, z \in U$ and $w \ll_K z$ implies $f(w) \leq_K f(z)$. By Lemma 1.2.4 there exists $\varphi \in \text{int}(K^*)$ such that $\varphi(z) = 1$. Write

$\Sigma_\varphi = \{x \in K : \varphi(x) = 1\}$ and recall that Σ_φ is compact by Lemma 1.2.4. Define $B_\varepsilon(z) = \{\eta \in U : \|z-\eta\| < \varepsilon\}$, where $\varepsilon > 0$ is such that $w \notin \mathrm{cl}(B_\varepsilon(z))$ and $\mathrm{cl}(B_\varepsilon(z)) \subseteq U$. Let $Z = \{x \in V : \varphi(x) = 1\}$ be the affine hyperplane in V containing Σ_φ and define $Z_\varepsilon = \{(1-t)w+t\eta : 0 < t < 1 \text{ and } \eta \in Z \cap B_\varepsilon(z)\}$. Remark that $\mathrm{cl}(Z_\varepsilon)$ is a compact subset of U, as U is convex.

Define $U_\varepsilon = (0, 1) \times \big(Z \cap B_\varepsilon(z)\big)$ and let $\vartheta : U_\varepsilon \to Z_\varepsilon$ be given by

$$\vartheta(t, \eta) = (1 - t)w + t\eta \quad \text{for } (t, \eta) \in U_\varepsilon.$$

We claim that ϑ is one-to-one. Indeed, if $\vartheta(t, \eta) = \vartheta(s, \zeta)$ for $(t, \eta), (s, \zeta) \in U_\varepsilon$, then

$$(1 - t)w + t\eta = (1 - s)w + s\zeta.$$

Applying φ to both sides gives

$$\varphi(w)(s - t) = (s - t) = \varphi(z)(s - t).$$

Recall that $w \ll_K z$, so that $\varphi(w) < \varphi(z)$, and hence $s = t$ and $\eta = \zeta$.

If $v = (1 - t)w + t\eta \in Z_\varepsilon$, then $\varphi(v) = \varphi(w) + t(1 - \varphi(w))$, so that $\varphi(w) < \varphi(v) < 1$ and, moreover,

$$t = \frac{\varphi(v) - \varphi(w)}{1 - \varphi(w)} \quad \text{and} \quad \eta = \frac{(1 - \varphi(w))v - (1 - \varphi(v))w}{\varphi(v) - \varphi(w)}.$$

It follows that ϑ^{-1} is locally Lipschitz. Since locally Lipschitz maps take sets of Lebesgue measure zero to Lebesgue measure zero, the set

$$\{(t, \eta) \in U_\varepsilon : Df((1 - t)w + t\eta) \text{ does not exist}\}$$

has Lebesgue measure zero. Moreover there exists a constant $M > 0$ such that $\|Df((1 - t)w + t\eta\| \leq M$ for all $(t, \eta) \in U_\varepsilon$ for which $Df((1 - t)w + t\eta)$ exists.

Let $E = \{(t, \eta) \in U_\varepsilon : Df((1-t)w+t\eta) \text{ exists}\}$. By Rademacher's theorem [184], E is Lebesgue measurable and $U_\varepsilon \setminus E$ has Lebesgue measure zero. Fubini's theorem implies for Lebesgue almost all $\eta \in Z \cap B_\varepsilon(z)$ that $(t, \eta) \in E$ for Lebesgue almost all $t \in (0, 1)$. Notice that for each $\eta \in Z \cap B_\varepsilon(z)$ the map $t \mapsto f((1 - t)w + t\eta)$ is Lipschitz and hence absolutely continuous for $0 \leq t \leq 1$. Thus,

$$\frac{\mathrm{d}}{\mathrm{d}t} f((1 - t)w + t\eta)$$

exists for Lebesgue almost all $t \in (0, 1)$ and

$$f(\eta) - f(w) = \int_0^1 \frac{\mathrm{d}}{\mathrm{d}t} f((1 - t)w + t\eta)\mathrm{d}t.$$

If $\eta \in Z \cap B_\varepsilon(z)$ is such that $(t, \eta) \in E$ for almost all $t \in (0, 1)$, then

$$\frac{\mathrm{d}}{\mathrm{d}t} f((1-t)w + t\eta) = Df((1-t)w + t\eta)(\eta - w)$$

for Lebesgue almost all $t \in (0, 1)$, and

$$f(\eta) - f(w) = \int_0^1 Df((1-t)w + t\eta)(\eta - w)\mathrm{d}t.$$

Since $w \ll_K z$, we know that, for all η close enough to z, $w \ll_K \eta$. As $Df(x)(K) \subseteq K$ for those x for which $Df(x)$ is defined, we see that $f(\eta) - f(w) \in K$ for all η sufficiently close to z. As f is continuous, it follows that $f(z) - f(w) \in K$, and we are done. □

1.4 Subhomogeneous maps

In this book we focus on several special classes of order-preserving maps. In particular, we consider the class of order-preserving subhomogeneous maps. Let $K \subseteq V$ and $K' \subseteq V'$ be cones. A map $f : K \to K'$ is called *subhomogeneous* if $\lambda f(x) \leq f(\lambda x)$ for all $x \in K$ and $0 < \lambda < 1$. It is said to be *strictly subhomogeneous* if $\lambda f(x) \ll f(\lambda x)$ for all $x \in K \setminus \{0\}$ and $0 < \lambda < 1$. The map $f : K \to K'$ is called *homogeneous of degree* p if $\lambda^p f(x) = f(\lambda x)$ for all $x \in K$ and $\lambda > 0$. We simply say that f is *homogeneous* if it is homogeneous of degree 1.

Means and their iterates. Order-preserving subhomogeneous maps arise naturally in the study of means. A basic example is the *Gauss arithmetic-geometric mean*, $f : \mathbb{R}^2_+ \to \mathbb{R}^2_+$,

$$f(x) = \left(\frac{x_1 + x_2}{2}, \sqrt{x_1 x_2}\right) \quad \text{for } x \in \mathbb{R}^2_+.$$

In this case it is not hard to show that for each $x \in \mathrm{int}(\mathbb{R}^2_+)$ there exists $\lambda > 0$ such that

$$\lim_{k\to\infty} f^k(x) = (\lambda, \lambda). \tag{1.1}$$

A much deeper fact, which was proved by Landen, and later independently by Lagrange and Gauss, is the following identity:

$$\int_0^{\pi/2} (x_1^2 \cos^2 \varphi + x_2^2 \sin^2 \varphi)^{-1/2} \mathrm{d}\varphi = \frac{\pi}{2\lambda}. \tag{1.2}$$

For a historic account of arithmetic-geometric mean and its applications in the study of elliptic integrals and the computation of the digits of π see [33, 53].

There are numerous variants on means and their iterates for which it is of interest to prove analogues of (1.1). Formulas like (1.2) are, however, known in very few cases; see [49].

For $r \in \mathbb{R} \setminus \{0\}$ and a probability vector $\sigma \in \mathbb{R}^n_+$, so $\sum_i \sigma_i = 1$, we can define the *(r, σ)-mean of $x \in \text{int}(\mathbb{R}^n_+)$* by

$$M_{r\sigma}(x) = \Big(\sum_i \sigma_i x_i^r\Big)^{1/r} \quad \text{for } x \in \text{int}(\mathbb{R}^n_+). \tag{1.3}$$

One can prove (see [82]) that

$$\lim_{r \to 0} M_{r\sigma}(x) = \prod_{i \in \text{supp}(\sigma)} x_i^{\sigma_i} \quad \text{for } x \in \text{int}(\mathbb{R}^n_+),$$

where $\text{supp}(\sigma) = \{i : \sigma_i > 0\}$ is the *support* of σ.

We define

$$M_{0\sigma}(x) = x^\sigma = \prod_{i \in \text{supp}(\sigma)} x_i^{\sigma_i} \quad \text{for } x \in \text{int}(\mathbb{R}^n_+). \tag{1.4}$$

For $r = 1, 0$, and -1 the (r, σ)-mean is, respectively, an *arithmetic mean*, a *geometric mean*, and a *harmonic mean*.

The map $x \mapsto M_{r\sigma}$ is order-preserving and homogeneous on the interior of $\mathbb{R}^n_+$. It is easy to verify that it has a continuous order-preserving homogeneous extension to $\mathbb{R}^n_+$. For $r \geq 0$ the continuous extension is given by equations (1.3) and (1.4). For $r < 0$ it is given by

$$M_{r\sigma}(x) = \begin{cases} \Big(\sum_{i \in \text{supp}(\sigma)} \sigma_i x_i^r\Big)^{1/r} & \text{if } x_i > 0 \text{ for all } i \in \text{supp}(\sigma), \\ 0 & \text{otherwise.} \end{cases}$$

It is known (see [82]) that the map $x \mapsto M_{r\sigma}(x)$ is super-additive if $r \leq 1$, and sub-additive if $r \geq 1$. For fixed σ and $x \in \mathbb{R}^n_+$, the map $r \mapsto M_{r\sigma}(x)$ is increasing, so $M_{r\sigma}(x) \leq M_{s\sigma}(x)$ for $r \leq s$. Furthermore,

$$\lim_{r \to -\infty} M_{r\sigma}(x) = \min_{i \in \text{supp}(\sigma)} x_i \quad \text{and} \quad \lim_{r \to \infty} M_{r\sigma}(x) = \max_{i \in \text{supp}(\sigma)} x_i.$$

Thus, it is natural to define

$$M_{-\infty\sigma}(x) = \min_{i \in \text{supp}(\sigma)} x_i \quad \text{and} \quad M_{\infty\sigma}(x) = \max_{i \in \text{supp}(\sigma)} x_i$$

for $x \in \mathbb{R}^n_+$ and a probability vector $\sigma \in \mathbb{R}^n_+$.

By using the means $M_{r\sigma}$ we can define a rich class of continuous order-preserving homogeneous maps $f : \mathbb{R}^n_+ \to \mathbb{R}^n_+$ as follows. For $1 \leq i \leq n$ let Γ_i be a non-empty finite set of pairs (r, σ), where $r \in [-\infty, \infty]$ and $\sigma \in \mathbb{R}^n_+$ is a probability vector. For each $(r, \sigma) \in \Gamma_i$ let $c_{ir\sigma} > 0$ be a positive real. Now let $f : \mathbb{R}^n_+ \to \mathbb{R}^n_+$ be given by

$$f_i(x) = \sum_{(r,\sigma)\in\Gamma_i} c_{ir\sigma} M_{r\sigma}(x) \tag{1.5}$$

for $1 \leq i \leq n$ and $x \in \mathbb{R}^n_+$.

We shall write M to denote the class of continuous order-preserving homogeneous maps f on $\mathbb{R}^n_+$ where each coordinate function of f is of the form (1.5) with $-\infty < r < \infty$.

For maps $f \in M$ it is interesting to understand whether there exists $u \in \operatorname{int}(\mathbb{R}^n_+)$ such that for each $x \in \operatorname{int}(\mathbb{R}^n_+)$ we have

$$\lim_{k\to\infty} f^k(x) = \lambda_x u,$$

for some $\lambda_x > 0$. In Chapter 6 we will discuss this problem in detail.

Matrix scaling problems. The following matrix scaling problem in combinatorial matrix theory [15,39] is known as a DAD problem. Given a nonnegative $m \times n$ matrix A and positive vectors $r \in \mathbb{R}^m$ and $c \in \mathbb{R}^n$, can we find positive diagonal matrices D and E such that DAE has row sums r_i for $1 \leq i \leq m$, and column sums c_j for $1 \leq j \leq n$?

This problem arises in information theory and statistics [35, 50, 205]. For example, suppose that we have observed a finite number of transitions of a Markov chain with n states, but we do not know the transition probabilities p_{ij}. The only information we have is that $P = (p_{ij})$ is a doubly stochastic matrix. In that case a reasonable approximation for P is the solution of the DAD problem where $r = c = (1, \dots, 1) \in \mathbb{R}^n_+$ and A has entries a_{ij} equal to the relative frequency of observed transitions from state i to state j.

Menon [143] associated with a DAD problem a nonlinear order-preserving homogeneous map $T : \operatorname{int}(\mathbb{R}^n_+) \to \operatorname{int}(\mathbb{R}^n_+)$ and showed that the DAD problem has a solution if and only if T has an eigenvector in $\operatorname{int}(\mathbb{R}^n_+)$. The map T is of the form

$$T(x) = (A^T \circ R_c \circ A \circ R_r)(x),$$

where $R_b(x) = (b_1/x_1, \dots, b_n/x_n)$ for $x \in \operatorname{int}(\mathbb{R}^n_+)$. In Chapter 7 we will see how results from nonlinear Perron–Frobenius theory can be used to solve a variety of DAD problems.

Nonlinear matrix equations. A basic problem in matrix analysis is to understand whether for given functions f and g the equation

$$f(X) = g(X)$$

has a solution in a set of matrices. A simple example is *Bushell's equation* [45],

$$X^2 = BXB^*,$$

where B is an invertible $n \times n$ matrix and X is an $n \times n$ positive-definite Hermitian matrix. Like ordinary equations, matrix equations are often hard to solve. In special cases, however, nonlinear Perron–Frobenius theory can be of help.

Let $\mathrm{Herm}(n)$ denote the space of all $n \times n$ Hermitian matrices. We view $\mathrm{Herm}(n)$ as a real vector space. Note that $\mathrm{Herm}(n)$ contains the solid closed cone, $\Pi_n(\mathbb{C})$, consisting of all positive-semidefinite Hermitian matrices. Recall that if $X \in \mathrm{Herm}(n)$, then there exists a unitary matrix U such that

$$U\Lambda U^* = X,$$

where $\Lambda = \mathrm{diag}(\lambda_1, \ldots, \lambda_n)$ and $\lambda_1 \geq \ldots \geq \lambda_n$ are the eigenvalues of X. This allows us to give meaning to expressions like $\log X$ and X^r for $r \geq 0$. Indeed, given a continuous function $f : \sigma(X) \subseteq \mathbb{R} \to \mathbb{C}$, one defines

$$f(X) = U \begin{pmatrix} f(\lambda_1) & & 0 \\ & \ddots & \\ 0 & & f(\lambda_n) \end{pmatrix} U^*$$

for $X \in \mathrm{Herm}(n)$. Basic results concerning this functional calculus can be found in [58, 215].

It is of interest to us to understand the functions $f : (0, \infty) \to \mathbb{R}$ that yield an order-preserving map. The following result by Löwner [135] provides an answer. A detailed exposition of Löwner's theory can be found in [24, 57].

Theorem 1.4.1 (Löwner) *If $f : (0, \infty) \to \mathbb{R}$ is a continuous function with an analytic extension to $\mathbb{C} \setminus \{z \in \mathbb{C} : \mathrm{Im}(z) = 0$ and $\mathrm{Re}(z) \leq 0\}$, and $f(H_+) \subseteq H_+$, where $H_+ = \{z \in \mathbb{C} : \mathrm{Im}(z) > 0\}$, then $f : \mathrm{int}(\Pi_n(\mathbb{C})) \to \mathrm{Herm}(n)$ is order-preserving.*

Basic examples include $f(x) = \log x$ and $f(x) = x^r$ for $0 < r < 1$. For $f(x) = x^r$ with $0 < r < 1$ there exists the following direct argument, which illustrates the workings of the partial ordering induced by $\Pi_n(\mathbb{C})$ on $\mathrm{Herm}(n)$. We need the following lemma.

Lemma 1.4.2 *The map $f : \mathrm{int}(\Pi_n(\mathbb{C})) \to \mathrm{int}(\Pi_n(\mathbb{C}))$ with $f(X) = X^{-1}$ for $X \in \mathrm{int}(\Pi_n(\mathbb{C}))$ is order-reversing.*

Proof Note that if $A, B \in \mathrm{int}(\Pi_n(\mathbb{C}))$, then $A \leq B$ if and only if $I \leq A^{-1/2}BA^{-1/2}$, which is equivalent to $\sigma(A^{-1/2}BA^{-1/2}) \subseteq [1, \infty)$. Thus, by the spectral mapping theorem, $\sigma(A^{1/2}B^{-1}A^{1/2}) \subseteq (0, 1]$. This implies that $A^{1/2}B^{-1}A^{1/2} \leq I$, and hence $B^{-1} \leq A^{-1}$. □

There exists the following integral representation for A^r (see [102, p. 286]):

$$A^r = \frac{\sin(\pi r)}{\pi} \int_0^\infty \lambda^{r-1} A(\lambda I + A)^{-1} \mathrm{d}\lambda, \tag{1.6}$$

for $A \in \mathrm{int}(\Pi_n(\mathbb{C}))$ and $0 < r < 1$. As $A(\lambda I + A)^{-1} = (\lambda A^{-1} + I)^{-1}$, it follows from Lemma 1.4.2 that the map $A \mapsto A(\lambda I + A)^{-1}$ is order-preserving for all $\lambda > 0$. Thus by (1.6), $A \leq B$ implies $A^r \leq B^r$ for $0 < r < 1$.

Also remark that $f(X) = X^r$ is subhomogeneous for $0 < r < 1$, as

$$\lambda f(X) \leq \lambda^r f(X) = f(\lambda X)$$

for $0 < \lambda < 1$ and $X \in \mathrm{int}(\Pi_n(\mathbb{C}))$.

In particular, we see that positive-definite solutions of Bushell's equation, $X^2 = BXB^*$, correspond to fixed points of the order-preserving subhomogneous map $X \mapsto (BXB^*)^{1/2}$, where $X \in \mathrm{int}(\Pi_n(\mathbb{C}))$. Other interesting equations of this type related to Riccati equations are *Stein's equation*,

$$X = A + BXB^*,$$

and the *discrete algebraic Riccati equation*,

$$X = A + B(C + X^{-1})^{-1}B^*,$$

where $A, C \in \Pi_n(\mathbb{C})$, B is an invertible matrix, and $X \in \mathrm{int}(\Pi_n(\mathbb{C}))$.

1.5 Topical maps

A map $f : \mathbb{R}^n \to \mathbb{R}^n$ is said to be *additively homogeneous* if

$$f(x + \lambda \mathbb{1}) = f(x) + \lambda \mathbb{1} \quad \text{for all } x \in \mathbb{R}^n \text{ and } \lambda \in \mathbb{R}.$$

Here $\mathbb{1} = (1, \ldots, 1) \in \mathbb{R}^n$. It is called *additively subhomogeneous* if

$$f(x + \lambda \mathbb{1}) \leq f(x) + \lambda \mathbb{1} \quad \text{for all } x \in \mathbb{R}^n \text{ and } \lambda \geq 0,$$

where the ordering is induced by $\mathbb{R}^n_+$. If $f : \mathbb{R}^n \to \mathbb{R}^n$ is additively homogeneous, respectively additively subhomogeneous, and f order-preserving with respect to $\mathbb{R}^n_+$, then f is called a *topical map*, respectively a *sub-topical map*. Topical maps are closely related to order-preserving homogeneous maps on $\mathbb{R}^n_+$. To see this, consider the map $E : \mathbb{R}^n \to \mathrm{int}(\mathbb{R}^n_+)$ with

$$E(x) = (e^{x_1}, \ldots, e^{x_n}) \quad \text{for } x = (x_1, \ldots, x_n) \in \mathbb{R}^n.$$

The inverse, $L : \mathrm{int}(\mathbb{R}^n_+) \to \mathbb{R}^n$, of E is the coordinatewise logarithm, so

$$L(x) = (\log x_1, \dots, \log x_n) \quad \text{for } x = (x_1, \dots, x_n) \in \mathrm{int}(\mathbb{R}^n_+).$$

It is easy to verify that if $f : \mathbb{R}^n \to \mathbb{R}^n$ is a (sub-)topical map, then the *log-exp transform* $g : \mathrm{int}(\mathbb{R}^n_+) \to \mathrm{int}(\mathbb{R}^n_+)$ of f given by

$$g(z) = (E \circ f \circ L)(z) \quad \text{for } z \in \mathrm{int}(\mathbb{R}^n_+) \tag{1.7}$$

is an order-preserving (sub)homogeneous map.

Max-plus maps. Important examples of topical maps are so-called max-plus maps, which arise in optimal control and scheduling theory. To introduce these maps let $\mathbb{R}_{\max} = \mathbb{R} \cup \{-\infty\}$ denote the *max-plus semi-ring*, meaning that the addition operation $\oplus$ is taking maximum, and the multiplication operation $\otimes$ is addition. For instance, $2 \oplus 3 = 3$ and $2 \otimes 3 = 5$. Let $A = (a_{ij})$ be an $n \times n$ matrix with entries from $\mathbb{R}_{\max}$ and suppose that for each i there exists j such that $a_{ij} \neq -\infty$. A *max-plus map* $f_A : \mathbb{R}^n \to \mathbb{R}^n$ is defined by

$$f_A(x)_i = (A \otimes x)_i = \max_j \{a_{ij} + x_j\} \quad \text{for } x \in \mathbb{R}^n \text{ and } 1 \leq i \leq n. \tag{1.8}$$

It easy to verify that max-plus maps are topical. Max-plus maps are linear maps over the max-plus semi-ring. The analogy with linear algebra is very powerful and gives rise to a rich theory for max-plus maps; see [11,47,162].

To see how max-plus maps naturally arise in scheduling theory we imagine a network of n interconnected machines, each of which has a single task to perform. Typically a machine cannot start a fresh cycle of activity until certain that other machines have completed their current cycle. This happens, for instance, when a machine requires inputs from certain other machines. Let a_{ij} be the number of time units machine i has to wait before it can start its k-th cycle, after machine j has started its $(k-1)$-th cycle. If i does not need to wait for j to finish, we put $a_{ij} = -\infty$. The number a_{ij} can be thought of as the time needed for machine j to produce the input for machine i. If we let x_i^k denote the earliest time at which machine i can start its k-th cycle, then

$$x_i^{k+1} = \max_j \{a_{ij} + x_j^k\} \quad \text{for } 1 \leq i \leq n \text{ and } k \geq 1.$$

Thus the evolution of cycles in the machine network can be modeled by a discrete dynamical system,

$$x^{k+1} = f_A(x^k),$$

where $f_A : \mathbb{R}^n \to \mathbb{R}^n$ is a max-plus map. In the context of machine scheduling it is particularly interesting to understand the *cycle time* vector,

$$\lim_{k\to\infty} \frac{f_A^k(x)}{k} = \lim_{k\to\infty} \frac{f_A^k(x) - f_A^{k-1}(x) + f_A^{k-1}(x) - \ldots - x}{k}, \tag{1.9}$$

which is a measure of the speed of the network.

The machine network in the previous paragraph is a typical example of a discrete event system. Roughly speaking a discrete event system is a system consisting of repeated events that depend on each other. For such systems it is often natural to assume that if the execution of an event is delayed, then the time of the next execution of any other event cannot be brought forward. In addition, if each event is delayed by the same amount of time λ, then the next execution of each event will also be delayed by λ. Such discrete event systems are usually modeled by topical maps; see [80].

Markov decision processes. Other interesting examples of topical maps arise in Markov decision theory [224], which provides a framework for studying decision-making in random processes whose outcomes can be partly controlled by the decision-maker. For example, imagine a system with finite states $S = \{1, \ldots, n\}$. Periodically the decision-maker observes the current state $i \in S$ of the system and selects an action α from a finite set A_i of possible actions. On performing the action two things happen:

- The decision-maker receives a reward $r_i^\alpha \in \mathbb{R}$.
- The system moves from the current state i to a new state j with probability p_{ij}^α.

After a predetermined number of stages, k, the decision-maker receives a terminal reward x_j^k depending on the final state j of the process. The goal of the decision-maker is to maximize the total expected reward.

For $i \in S$, let v_i^k denote the optimal expected reward of the process consisting of k stages and initial state i. Then the following recursion relation holds:

$$v_i^{k+1} = \max_{\alpha \in A_i} r_i^\alpha + \sum_{j=1}^{n} p_{ij}^\alpha v_j^k, \tag{1.10}$$

where $v^0 = x^k \in \mathbb{R}^n$ is the terminal pay-off vector. This is a backward recursion relation in the sense that if you know which decisions to make for a k-stage process, you can use it to determine the first action in a $(k+1)$-stage process.

The recursion relation (1.10) provides an effective way to compute the optimal expected reward. This dynamic programming method is known as *value iteration* and goes back to Bellman [21] and Howard [92]. Note that the recurrence relation (1.10) gives rise to a topical map $f : \mathbb{R}^n \to \mathbb{R}^n$, where

$$f_i(x) = \max_{\alpha \in A_i} r_i^\alpha + \sum_{j=1}^{n} p_{ij}^\alpha x_j \quad \text{for } 1 \leq i \leq n \text{ and } x \in \mathbb{R}^n, \tag{1.11}$$

since $\sum_j p_{ij}^\alpha = 1$ for all i.

Stochastic games. Stochastic games are two-player zero-sum games with finite states and multiple stages, and go back to Shapley [203]; see also [23, 152, 208]. At each stage both players select an action. The current state and the chosen actions determine a reward that one player receives from the other player and the transition probability from the current state to the next state. Stochastic games generalize Markov decision processes in the sense that a Markov decision process is a stochastic game, where one of the players is a "dummy" having only one action available in each state. The dynamic programming method for computing the value of a stochastic game also leads to a topical map.

Denote the two players by P_1 and P_2 and the finite set of states by $S = \{1, \ldots, n\}$. Suppose that the game terminates after k stages. At each stage each player P_m, $m = 1, 2$, selects an action from a finite set of possible actions $A_m(i)$, which may depend on the current state i of the game. Suppose that P_1 has selected α and P_2 has chosen β. Then the following happens:

- P_1 receives a reward $r_i(\alpha, \beta) \in \mathbb{R}$ from P_2. (Note that $r_i(\alpha, \beta)$ can be negative.)
- The game moves from the current state i to a new state j with probability $p_{ij}(\alpha, \beta)$.

After k stages P_1 receives a terminal pay-off $x_j^0 \in \mathbb{R}$ from P_2, which depends on the final state j of the game. The vector $x^0 \in \mathbb{R}^n$ is called the *pay-off vector*.

A fundamental problem in game theory is to understand how P_1 and P_2 should play to maximize the total expected reward and minimize the total expected loss, respectively. To analyze this problem further the notion of a strategy is needed. A *strategy* for P_1 in a k-stage stochastic game with pay-off vector $x^0 \in \mathbb{R}^n$ is a sequence $\sigma = (s_1, \ldots, s_k)$, where each s_m is a function which assigns to each $i \in S$ a probability distribution $\pi_m(i)$, on $A_1(i)$. In the same way strategies for player P_2 can be defined. Following the dynamic programming set-up players use their strategies as follows. If after q stages the game has reached state $i_q \in S$, then player P_1 uses probability distribution $\pi_{k-q}(i_q)$ to select an action from $A_1(i_q)$. In the same way player P_2 applies his strategy.

We denote by $V_i(\sigma, \tau)$ the total expected reward for player P_1 when the game starts in state $i \in S$ and player P_1 uses strategy σ and P_2 applies strategy τ. As this is a zero-sum game, $V_i(\sigma, \tau)$ is also the total expected loss for

P_2. A stochastic game with initial state $i \in S$ and k stages is said to have a *value* V_i^k if P_1 has a strategy σ^* and P_2 has a strategy τ^* such that

$$V_i(\sigma, \tau^*) \leq V_i^k \leq V_i(\sigma^*, \tau),$$

for all strategies σ of P_1 and τ of P_2. Thus, V_i^k is the expected amount P_1 can secure to win, whatever strategy P_2 may apply. On the other hand, it is also the minimum expected amount P_2 can secure to lose, whatever strategy P_1 may use.

As in Markov decision processes the value of a game can be determined using a dynamic programming method. Before we can give the dynamic programming operator, we need to introduce several more definitions. To begin we call a function r which assigns to each $i \in S$ a probability distribution on $A_m(i)$ a *policy* for player P_m, $m = 1, 2$. Given $i \in S$ and policies s for P_1 and t for P_2, define $L_i(s, t) : \mathbb{R}^n \to \mathbb{R}$ by

$$L_i(s,t)(x) = \sum_{\alpha \in A_1(i)} s(\alpha \mid i) \sum_{\beta \in A_2(i)} t(\beta \mid i)\Big[r_i(\alpha, \beta) + \sum_{j=1}^{n} p_{ij}(\alpha, \beta) x_j\Big],$$

for $x \in \mathbb{R}^n$. Here $s(\alpha \mid i)$ denotes the probability that α is chosen from $A_1(i)$ and $t(\beta \mid i)$ denotes the probability that β is selected from $A_2(i)$.

Note that $L_i(s, t)x$ is the expected reward for P_1 of the one-stage stochastic game with initial state $i \in S$ and pay-off vector $x \in \mathbb{R}^n$, when P_1 applies policy s and P_2 uses policy t. The *Shapley operator* $\Psi : \mathbb{R}^n \to \mathbb{R}^n$ is defined by

$$\Psi_i(x) = \max_s \min_t L_i(s, t)x$$

for $i \in S$ and $x \in \mathbb{R}^n$. Here the maximum is taken over all policies s of P_1, and the minimum is taken over all policies t of P_2. Note that $\Psi : \mathbb{R}^n \to \mathbb{R}^n$ is order-preserving with respect to $\mathbb{R}^n_+$ and satisfies $\Psi(x + \lambda \mathbb{1}) = \Psi(x) + \lambda \mathbb{1}$ for all $\lambda \in \mathbb{R}$ and $x \in \mathbb{R}^n$. Thus, Ψ is a topical map.

To see how the Shapley operator relates to the value of the k-stage stochastic game, we first note that for each $i \in S$ and $x \in \mathbb{R}^n$ we have an $|A_1(i)| \times |A_2(i)|$ matrix $A = (a_{\alpha\beta})$, where

$$a_{\alpha\beta} = r_i(\alpha, \beta) + \sum_j p_{ij}(\alpha, \beta) x_j.$$

If we think of A as the pay-off matrix of a zero-sum two-player game, then we know from von Neumann's minimax theorem (see for example [208, p. 155]) that there exist optimal strategies, i.e., probability distributions, p_i^* on $A_1(i)$ and q_i^* on $A_2(i)$ such that

$$\max_{p_i} \min_{q_i} p_i^T A q_i = (p_i^*)^T A q_i^* = \min_{q_i} \max_{p_i} p_i^T A q_i,$$

where the maximum is taken over all probability distributions p_i on $A_1(i)$ and the minimum is taken over all probability distributions q_i on $A_2(i)$. The probability distributions p_i^*, for $i \in S$, make up a policy s^* for P_1. Likewise, the probability distributions q_i^* give a policy t^* for P_2. So,

$$\Psi(x)_i = L_i(s^*, t^*)x.$$

Now for $m = 1, \ldots, k$ define $w^m = \Psi(w^{m-1})$ where $w^0 = x^0$ (the pay-off vector). Furthermore, let s_m^* be a policy for P_1 and t_m^* be a policy such that

$$\Psi_i(w^{m-1}) = L_i(s_m^*, t_m^*)w^{m-1}$$

for each $i \in S$. So, the i-th component of the policies s_m^* and t_m^* are optimal mixed strategies of the zero-sum two-player game where the entries of the pay-off matrix are given by

$$a_{\alpha\beta} = r_i(\alpha, \beta) + \sum_j p_{ij}(\alpha, \beta) w_j^{m-1}.$$

We can now prove the following result.

Theorem 1.5.1 *A stochastic game with pay-off vector $x^0 \in \mathbb{R}^n$ and k stages has value w^k and optimal strategies $\sigma^* = (s_k^*, \ldots, s_1^*)$ for player P_1 and $\tau^* = (t_k^*, \ldots, t_1^*)$ for player P_2.*

Proof By definition of σ^* and τ^* we have that $V(\sigma^*, \tau^*) = w^k$. We now show by induction on k that

$$V_i(\sigma, \tau^*) \leq V_i(\sigma^*, \tau^*)$$

for all strategies σ of P_1 and $i \in S$. The case $k = 1$ follows directly from the definition of σ^* and τ^*. Given a strategy $\sigma = (s_{k+1}, s_k, \ldots, s_1)$ we write $\overline{\sigma} = (s_k, \ldots, s_1)$. Note that

$$\begin{aligned} V_i(\sigma, \tau^*) = & \sum_{\alpha \in A_1(i)} s_{k+1}(\alpha \mid i) \sum_{\beta \in A_2(i)} t_{k+1}^*(\beta \mid i) \Big[r_i(\alpha, \beta) + \\ & \sum_{j=1}^{n} p_{ij}(\alpha, \beta) V_j(\overline{\sigma}, \overline{\tau}^*) \Big] \\ \leq & \sum_{\alpha \in A_1(i)} s_{k+1}(\alpha \mid i) \sum_{\beta \in A_2(i)} t_{k+1}^*(\beta \mid i) \Big[r_i(\alpha, \beta) + \\ & \sum_{j=1}^{n} p_{ij}(\alpha, \beta) V_j(\overline{\sigma}^*, \overline{\tau}^*) \Big] \end{aligned}$$

$$\leq \sum_{\alpha \in A_1(i)} s^*_{k+1}(\alpha \mid i) \sum_{\beta \in A_2(i)} t^*_{k+1}(\beta \mid i)\Big[r_i(\alpha, \beta) + \sum_{j=1}^{n} p_{ij}(\alpha, \beta) V_j(\overline{\sigma}^*, \overline{\tau}^*)\Big]$$
$$= V_i(\sigma^*, \tau^*).$$

The first inequality follows by induction, whereas the second inequality follows from the definition of σ^* and τ^*. In the same way it can be shown that

$$V_i(\sigma^*, \tau^*) \leq V_i(\sigma^*, \tau)$$

for all strategies τ of P_2 and $i \in S$. □

In the context of stochastic games it is of interest to understand the behavior of the value as the number of stages k becomes large, which amounts to analyzing the iterates of the Shapley operator.

1.6 Integral-preserving maps

An interesting class of nonlinear maps that preserve the ordering induced by $\mathbb{R}^n_+$ arises as generalizations of linear diffusion processes on finitely many states, or simply Markov chains. In a linear diffusion process with finite states $\{1, \ldots, n\}$ a fixed fraction $0 \leq p_{ij} \leq 1$ of the mass in state i is moved to state j after each time unit for all $1 \leq i, j \leq n$. The movement of mass is described by a linear map $P : \mathbb{R}^n_+ \to \mathbb{R}^n_+$, where

$$(Px)_j = \sum_{i=1}^{n} p_{ij} x_i \quad \text{for } 1 \leq j \leq n \text{ and } x \in \mathbb{R}^n_+.$$

Clearly if we start with more mass in one of the states, we end up with more mass in the states, and hence P is order-preserving with respect to $\mathbb{R}^n_+$. Note also that, as no mass is lost, the total amount of mass is preserved in the process, so

$$\sum_i (Px)_i = \sum_i x_i \quad \text{for all } x \in \mathbb{R}^n_+.$$

We say that $f : X \to \mathbb{R}^n$, with $X \subseteq \mathbb{R}^n$, is *integral-preserving* if

$$\sum_i f(x)_i = \sum_i x_i \quad \text{for all } x \in X.$$

Linear diffusion processes have other characteristic properties. First, the amount of mass that is moved from state i to state j is independent of the amount of mass in state j. Second, the mass at the top of the pile in state i is treated in the same way as the mass at the bottom of the pile. There are many natural examples of diffusion processes that lack one of these properties.

Nonlinear diffusion processes. Consider the following process, which one can imagine playing on the beach. Let $C_1, \ldots, C_n$ be n containers, each having an amount of sand x_i for $1 \leq i \leq n$ at time zero. With each container C_i a sequence of buckets $(b_{ij})_{j \in \mathbb{N}}$ is associated. Let a_{ij} denote the volume of bucket b_{ij}. Assume for convenience that $\sum_{j \in \mathbb{N}} a_{ij} = \infty$ for all $1 \leq i \leq n$ and that each container C_i has an infinite volume. After each time unit the following steps are performed. For each $1 \leq i \leq n$, sand from container C_i is poured into bucket b_{i1} until either b_{i1} is full or C_i is empty. If b_{i1} is full the remaining sand in C_i is poured into b_{i2} until either b_{i2} is full or C_i is empty. This process is repeated until C_i is empty and all the sand is distributed over the buckets. The amount of sand in bucket b_{ik} is given by

$$M_{ik}(x) = \min\Big(\max\{x_i - \sum_{j=1}^{k-1} a_{ij}, 0\}, a_{ik}\Big).$$

Subsequently the sand is poured back into the containers using a fixed rule $\gamma : \{1, \ldots, n\} \times \mathbb{N} \to \{1, \ldots, n\}$ in the following way. The contents of b_{ik} is poured into container $C_{\gamma(i,k)}$ for each $1 \leq i \leq n$ and $k \in \mathbb{N}$. The total amount of mass in container C_j after the procedure is given by

$$y_j = \sum_{\gamma(i,k)=j} M_{ik}(x). \tag{1.12}$$

This process is described by a nonlinear map $f : \mathbb{R}^n_+ \to \mathbb{R}^n_+$, where $f(x)_j = y_j$ and y_j is given by (1.12) for $1 \leq j \leq n$. The map f is called a *sand-shift map* with rule γ. In this diffusion process mass on top of the pile in a state (container) may be treated differently from mass at the bottom of the pile. It is easy to verify that every sand-shift map is order-preserving and integral-preserving.

There exist many other maps that possess these two properties. To give another set of examples, we let σ and τ be two one-to-one maps of $\{1, \ldots, n\}$ onto itself. Define $h : \mathbb{R}^n_+ \to \mathbb{R}^n_+$ by

$$h(x)_i = (x_{\sigma(i)} \vee 1) + (x_{\tau(i)} \wedge 1) - 1 \quad \text{for } 1 \leq i \leq n \text{ and } x \in \mathbb{R}^n_+.$$

The map h is order-preserving and integral-preserving. In addition, it satisfies

$$h(\lambda \mathbb{1}) = \lambda \mathbb{1} \quad \text{for all } \lambda \geq 0. \tag{1.13}$$

Maps that satisfy (1.13) are called *sup-norm decreasing*. The terminology is motivated by the following lemma. (Recall that the *sup-norm* is given by $\|x\|_\infty = \max_i |x_i|$ for $x \in \mathbb{R}^n$.)

Lemma 1.6.1 *If $f : \mathbb{R}^n_+ \to \mathbb{R}^n_+$ is order-preserving and integral-preserving, then f is sup-norm decreasing if and only if $\|f(x)\|_\infty \leq \|x\|_\infty$ for all $x \in \mathbb{R}^n_+$.*

Proof Clearly $x \leq (\max_i x_i)\mathbb{1}$ and hence $f(x) \leq (\max_i x_i)\mathbb{1}$, as f is sup-norm decreasing. This implies that $\|f(x)\|_\infty = \max_i f(x)_i \leq \max_i x_i = \|x\|_\infty$, as x and $f(x)$ are in $\mathbb{R}^n_+$. On the other hand, if $f : \mathbb{R}^n_+ \to \mathbb{R}^n_+$ is integral-preserving, then $n\lambda = \sum_i \lambda = \sum_i f(\lambda\mathbb{1})_i$ for all $\lambda \geq 0$. But also $f(\lambda\mathbb{1})_i \leq \lambda$ for all $\lambda \geq 0$ and $1 \leq i \leq n$, as $\|f(\lambda\mathbb{1})\|_\infty \leq \|\lambda\mathbb{1}\|_\infty = \lambda$. Therefore $f(\lambda\mathbb{1})_i = \lambda$ for all $\lambda \geq 0$ and $1 \leq i \leq n$, and we are done. □

Considering these nonlinear diffusion processes one might wonder how the iterates of order-preserving integral-preserving maps behave. In particular, one might ask if each initial distribution still converges to a periodic distribution as in the linear case.

2

Non-expansiveness and nonlinear Perron–Frobenius theory

Order-preserving maps possessing one of the additional properties discussed in Chapter 1 are non-expansive under some metric. As we shall see later the non-expansiveness property is a powerful tool in the analysis of nonlinear eigenvalue problems and in the study of the iterative behavior of order-preserving maps. It lies at the heart of many arguments in nonlinear Perron–Frobenius theory. The purpose of this chapter is to discuss the relation between order-preserving maps and non-expansive maps. In addition, we shall introduce Hilbert's metric and Thompson's metric, and review some of their geometric and topological properties that will be useful in the next chapter.

2.1 Hilbert's and Thompson's metrics

In this section we shall show how order-preserving (sub)homogeneous maps are related to maps that are non-expansive under Hilbert's (projective) metric and Thompson's (part) metric. We start with the definition of Hilbert's metric on a cone. Let K be a cone in a finite-dimensional real vector space V. For $x, y \in K$ with $x \sim_K y$ and $y \neq 0$, define

$$M(x/y; K) = \inf\{\beta > 0 : x \leq \beta y\}$$

and

$$m(x/y; K) = \sup\{\alpha > 0 : \alpha y \leq x\}.$$

When K is clear from the context we simply write $m(x/y)$ and $M(x/y)$. *Hilbert's (projective) metric* $d_H : K \times K \to [0, \infty]$ is defined by

$$d_H(x, y) = \log\left(\frac{M(x/y)}{m(x/y)}\right),$$

for all $x \sim_K y$ in K with $y \neq 0$. Furthermore, we put $d_H(0,0) = 0$ and define $d_H(x,y) = \infty$ otherwise.

Proposition 2.1.1 *If K is a cone in V, then (K, d_H) satisfies:*

(i) $d_H(x,y) \geq 0$ and $d_H(x,y) = d_H(y,x)$ for all $x, y \in K$,
(ii) $d_H(x,z) \leq d_H(x,y) + d_H(y,z)$ for all $x \sim_K y \sim_K z$, and
(iii) $d_H(\alpha x, \beta y) = d_H(x,y)$ for all $\alpha, \beta > 0$ and $x, y \in K$.

If, in addition, K is closed, then $d_H(x,y) = 0$ if and only if $x = \lambda y$ for some $\lambda > 0$. In that case, if $X \subseteq K$ has the property that for each $x \in K \setminus \{0\}$ there exists a unique $\lambda > 0$ such that $\lambda x \in X$ and P is a part of K, then (Σ, d_H), where $\Sigma = P \cap X$, is a genuine metric space.

Proof To prove the first assertion it suffices to consider $x \sim_K y$ and $y \neq 0$, as the other cases are clear from the definition. In that case $x \neq 0$. For each $0 < \alpha < m(x/y)$ and $0 < M(x/y) < \beta$ we have that $\alpha y \leq x \leq \beta y$, so that $y \leq (\beta/\alpha)y$. This implies that $\beta/\alpha \geq 1$ for all $0 < \alpha < m(x/y)$ and $0 < M(x/y) < \beta$. Thus, $M(x/y)/m(x/y) \geq 1$ and hence $d_H(x,y) \geq 0$.

Remark that $M(x/y) = \inf\{\alpha > 0 : \alpha^{-1}x \leq y\} = m(y/x)^{-1}$. A similar argument shows that $m(x/y) = M(y/x)^{-1}$. From this we deduce that $d_H(x,y) = d_H(y,x)$.

To prove the triangle inequality, we take $x \sim_K y \sim_K z$ in K and $z \neq 0$. For each $0 < \alpha < m(x/y)$ and $0 < \gamma < m(y/z)$ we have that $\alpha y \leq x$ and $\gamma z \leq y$, so that $\alpha\gamma z \leq x$. This implies that $m(x/z) \geq m(x/y)m(y/z)$. A similar argument shows that $M(x/z) \leq M(x/y)M(y/z)$. From these inequalities we get that

$$\frac{M(x/z)}{m(x/z)} \leq \frac{M(x/y)M(y/z)}{m(x/y)m(y/z)},$$

so that $d_H(x,z) \leq d_H(x,y) + d_H(y,z)$.

The third assertion follows from the fact that

$$M(\alpha x/\beta y) = \frac{\alpha}{\beta}M(x/y) \quad \text{and} \quad m(\alpha x/\beta y) = \frac{\alpha}{\beta}m(x/y)$$

for all $\alpha, \beta > 0$.

If K is closed and $x \sim_K y$ in K with $y \neq 0$, then

$$m(x/y)y \leq x \leq M(x/y)y, \tag{2.1}$$

Thus, $d_H(x,y) = 0$ implies that $y \leq m(x/y)^{-1}x \leq M(x/y)m(x/y)^{-1}y = y$, so that $x = \lambda y$ for some $\lambda > 0$. The reverse implication follows directly from assertion (iii).

The final assertion can be deduced from the previous observations and the fact that $d_H(x,y) < \infty$ if x and y are in the same part of K. □

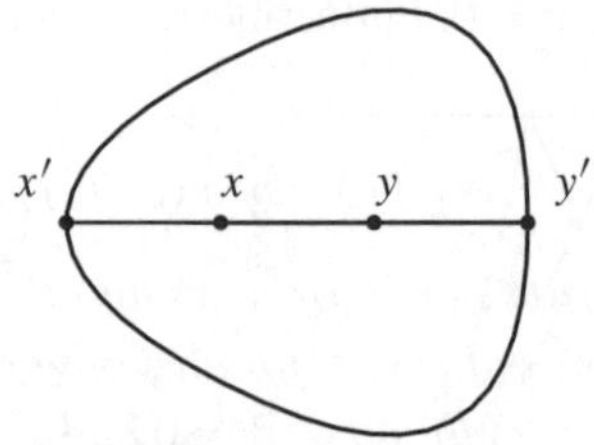

Figure 2.1 Cross-ratio.

Note that in the second part of Proposition 2.1.1 it is necessary to assume that K is closed. Indeed, if $K = \{x \in \mathbb{R}^2 : x_1 > 0\} \cup \{(0,0)\}$, then $x \sim_K y$ for all $x, y \in K \setminus \{0\}$. Moreover, $M(x/y) = x_1/y_1 = m(x/y)$, so that $d_H(x,y) = 0$.

Hilbert's metric on cones was introduced by Birkhoff [25] and Samelson [192], and is closely related to a metric discovered by Hilbert [86]. To introduce Hilbert's original metric, we consider a non-empty, open, bounded, convex set X in a finite-dimensional real affine space A. For $x, y \in X$ let $\ell_{x,y}$ be the affine line through x and y in A. Then $\ell_{x,y}$ intersects ∂X at two points x' and y', where we assume that x is between x' and y, and y is between y' and x; see Figure 2.1.

For $x', x, y, y' \in A$ the *cross-ratio* is defined by

$$[x', x, y, y'] = \frac{t}{s} \cdot \frac{1-s}{1-t},$$

where $x = (1-s)x' + sy'$ and $y = (1-t)x' + ty'$. Note that as $0 < s < t < 1$, we have that $[x', x, y, y'] > 1$. On X Hilbert defined the *cross-ratio metric*, $\kappa : X \times X \to [0, \infty)$, by

$$\kappa(x, y) = \log[x', x, y, y'] \quad \text{for } x \neq y \in X,$$

and $\kappa(x, x) = 0$ for all $x \in X$. As Hilbert [86] pointed out in a letter to Klein, the cross-ratio metric space is a natural generalization of Klein's model for the hyperbolic plane, which is obtained when X is an open disc.

To fully appreciate the relation between the two metrics κ and d_H, let $K \subseteq V$ be a solid closed cone, $\varphi \in \text{int}(K^*)$, and $\Sigma^\circ = \{x \in \text{int}(K) : \varphi(x) = 1\}$. Suppose that $x \neq y$ in Σ°, and write $\alpha = m(x/y)$ and $\beta = M(x/y)$. Remark that $a = x - \alpha y \in \partial K$ and $b = y - \beta^{-1}x \in \partial K$, as K is closed. Thus, $x' = a/\varphi(a)$ and $y' = b/\varphi(b)$, where x' and y' are the points of intersection of $\ell_{x,y}$ and $\partial\Sigma^\circ$; see Figure 2.2. Note that there exist $\sigma > 1$ and $\tau > 1$ such that $x' = y + \sigma(x - y)$ and $y' = x + \tau(y - x)$. So,

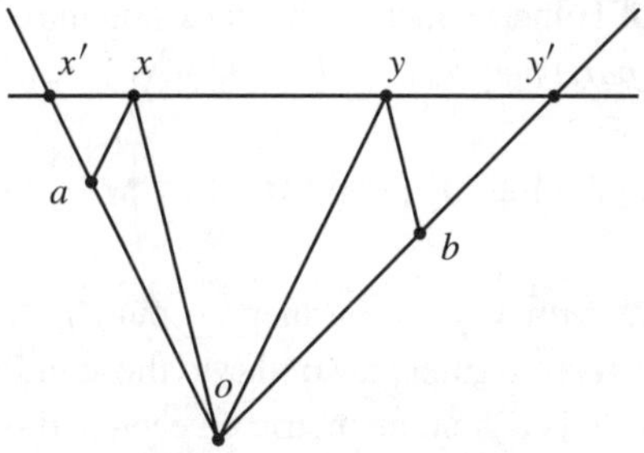

Figure 2.2 Hilbert's metric.

$$y + \sigma(x - y) = x' = \frac{a}{\varphi(a)} = \frac{x - \alpha y}{1 - \alpha}, \tag{2.2}$$

from which we deduce that $\alpha = (\sigma - 1)/\sigma$. Likewise,

$$x + \tau(y - x) = y' = \frac{b}{\varphi(b)} = \frac{y - \beta^{-1}x}{1 - \beta^{-1}} \tag{2.3}$$

yields $\beta = \tau/(\tau - 1)$.

Let $\|\cdot\|$ be any norm on V and $0 < s < t < 1$ be such that $x = x' + s(y' - x')$ and $y = x' + t(y' - x')$.

Then

$$\frac{t}{s} = \frac{\|y - x'\|}{\|x - x'\|} = \frac{\sigma}{1 - \sigma} = \frac{1}{\alpha}$$

and

$$\frac{1 - s}{1 - t} = \frac{\|x - y'\|}{\|y - y'\|} = \frac{\tau}{1 - \tau} = \beta.$$

Thus, we have shown that

$$d_H(x, y) = \log \frac{\beta}{\alpha} = \log\left(\frac{t}{s} \cdot \frac{1 - s}{1 - t}\right) = \kappa(x, y)$$

for all $x \neq y$ in Σ°. To summarize, we have the following result.

Theorem 2.1.2 *If $K \subseteq V$ is a solid cone and $\varphi \in \text{int}(K^*)$, then κ and d_H coincide on $\Sigma^\circ = \{x \in \text{int}(K) : \varphi(x) = 1\}$.*

A useful variant of Hilbert's metric on cones was introduced by Thompson [217]. *Thompson's (part) metric*, $d_T : K \times K \to [0, \infty]$, is defined by

$$d_T(x, y) = \log\big(\max\{M(x/y), M(y/x)\}\big)$$

for all $x \sim_K y$ in K with $y \neq 0$. Further we put $d_T(0, 0) = 0$, and define $d_T(x, y) = \infty$ otherwise. Arguments similar to those in Proposition 2.1.1 can be used to show that d_T is a genuine metric on each part of the cone.

Hilbert's metric and Thompson's metric are particularly useful tools in the study of order-preserving (sub)homogeneous maps on cones. Before discussing this in more detail, we recall several basic notions. A map $f : X \to Y$ from a metric space (X, d_X) into a metric space (Y, d_Y) is called *non-expansive* if

$$d_Y(f(x), f(y)) \leq d_X(x, y) \quad \text{for all } x, y \in X. \tag{2.4}$$

It is called a *contraction* if the inequality in (2.4) is strict for all $x \neq y$ in X. The map $f : X \to Y$ is said to be a *Lipschitz contraction* if there exists $c \in [0, 1)$ such that

$$d_Y(f(x), f(x)) \leq c d_X(x, y) \quad \text{for all } x, y \in X.$$

Furthermore, $f : X \to Y$ is called an *isometry* if equality holds in (2.4) for all $x, y \in X$.

Proposition 2.1.3 *Let $K \subseteq V$ and $K' \subseteq V'$ be closed cones. If $f : K \to K'$ is order-preserving and homogeneous of degree $r > 0$, then*

$$m(x/y)^r \leq m(f(x)/f(y)) \quad \text{and} \quad M(f(x)/f(y)) \leq M(x/y)^r$$

for all $x, y \in K$ with $x \sim_K y$.

Proof Let $x, y \in K$ with $x \sim_K y$. As K is a closed cone, $m(x/y)y \leq x \leq M(x/y)y$, so that $m(x/y)^r f(y) \leq f(x) \leq M(x/y)^r f(y)$. Hence $m(x/y)^r \leq m(f(x)/f(y))$ and $M(f(x)/f(y)) \leq M(x/y)^r$. □

This proposition has the following immediate consequence.

Corollary 2.1.4 *Let $K \subseteq V$ and $K' \subseteq V'$ be closed cones. If $f : K \to K'$ is order-preserving and homogeneous of degree $r > 0$, then*

$$d_H(f(x), f(y)) \leq r d_H(x, y)$$

and

$$d_T(f(x), f(y)) \leq r d_T(x, y) \quad \textit{for all } x, y \in K \textit{ with } x \sim_K y.$$

In particular, every order-preserving homogeneous map $f : K \to K'$ is non-expansive under Hilbert's metric and Thompson's metric. This observation lies at the heart of the work of Birkhoff [25] and Samelson [192], who analyzed linear operators that leave a cone invariant. In fact, Birkhoff defined the *contraction ratio* of a linear operator L on a cone K by

$$\kappa(L) = \inf\{\lambda \geq 0 : d_H(Lx, Ly) \leq \lambda d_H(x, y) \text{ for all } x \sim_K y \text{ in } K\},$$

and proved that

$$\kappa(L) = \tanh\left(\frac{1}{4}\Delta(L)\right),$$

where $\Delta(L) = \sup\{d_H(Lx, Ly) : Lx \sim_K Ly \text{ in } K\}$ is the *projective diameter of L*. Closely related results were obtained by Hopf [90, 91]. The works of Birkhoff and Hopf are discussed in Appendix A in more detail.

Remark that it follows from Corollary 2.1.4 that if $f : K \to K'$ is a bijective linear map, then f is an isometry with respect to Hilbert's metric and Thompson's metric. Arguing as in Proposition 2.1.3, the following result can be shown.

Corollary 2.1.5 *Let $K \subseteq V$ and $K' \subseteq V'$ be closed cones. If $f : K \to K'$ is order-reversing and homogeneous of degree $r < 0$, then*

$$d_H(f(x), f(y)) \leq |r| d_H(x, y)$$

and

$$d_T(f(x), f(y)) \leq |r| d_T(x, y) \quad \textit{for all } x, y \in K \textit{ with } x \sim_K y.$$

Proof Suppose that $x, y \in K$ with $x \sim_K y$. As K is closed, $m(x/y)y \leq x \leq M(x/y)y$, which implies $M(x/y)^r f(y) \leq f(x)$ and $f(x) \leq m(x/y)^r f(y)$. Therefore $m(f(x)/f(y)) \geq M(x/y)^r$ and $M(f(x)/f(y)) \leq m(x/y)^r$. Thus,

$$d_H(f(x), f(y)) \leq \log\left(\frac{m(x/y)^r}{M(x/y)^r}\right) = |r| d_H(x, y)$$

and

$$d_T(f(x), f(y)) \leq \log\left(\max\{M(x/y)^{-r}, M(y/x)^{-r}\}\right) = |r| d_T(x, y)$$

for all $x, y \in K$ with $x \sim_K y$. □

In particular, every order-reversing homogeneous degree -1 map is non-expansive under d_H and d_T.

Hilbert's metric is particularly useful in the analysis of scaled order-preserving homogeneous maps, as the following lemma shows.

Lemma 2.1.6 *Let $K \subseteq V$ be a solid closed cone and let $f : K \to K$ be an order-preserving and homogeneous map such that $f(x) \neq 0$ for all $x \in K \setminus \{0\}$. Suppose that $\varphi : K \to [0, \infty)$ is a homogeneous map with $\varphi(x) \neq 0$ for all $x \in K \setminus \{0\}$. If $\Sigma = \{x \in K : \varphi(x) = 1\}$ and $g : \Sigma \to \Sigma$ is given by*

$$g(x) = \frac{f(x)}{\varphi(f(x))} \quad \textit{for all } x \in \Sigma,$$

then g is non-expansive under Hilbert's metric on $\Sigma \cap P$ for each part P of K. Moreover, if f is strongly order-preserving and homogeneous, then g is a contraction.

Proof Let $x, y \in \Sigma$ with $x \sim_K y$ and write $\alpha = m(x/y)$ and $\beta = M(x/y)$. As K is closed, $\alpha y \leq x$ and $x \leq \beta y$, so that $\alpha f(y) \leq f(x)$ and $f(x) \leq \beta f(y)$. Thus,

$$\alpha \frac{\varphi(f(y))}{\varphi(f(x))} g(y) \leq g(x) \quad \text{and} \quad g(x) \leq \beta \frac{\varphi(f(y))}{\varphi(f(x))} g(y).$$

This implies that

$$d_H(g(x), g(y)) \leq \log(\beta/\alpha) = d_H(x, y).$$

If, in addition, f is strongly order-preserving and $x \neq \lambda y$ for all $\lambda > 0$, then $\alpha y \leq x \leq \beta y$ implies that

$$\alpha f(y) \ll f(x) \ll \beta f(y).$$

Hence there exist $\mu > \alpha$ and $\tau < \beta$ such that $\mu f(y) \leq f(x) \leq \tau f(y)$, so that

$$d_H(g(x), g(y)) \leq \log(\tau/\mu) < \log(\beta/\alpha) = d_H(x, y).$$

□

Typical examples of homogeneous functions $\varphi : K \to [0, \infty)$ in Lemma 2.1.6 are $\varphi(x) = \|x\|$ for some norm $\|\cdot\|$ on V, and $\varphi \in \text{int}(K^*)$.

Thompson's metric is most useful in the analysis of order-preserving subhomogeneous maps on the whole cone.

Lemma 2.1.7 *Let $K \subseteq V$ and $K' \subseteq V'$ be solid closed cones. If $f : K \to K'$ is order-preserving, then f is subhomogeneous if and only if f is non-expansive with respect to Thompson's metric on each part of K. If $f : \text{int}(K) \to \text{int}(K')$ is order-preserving and strictly subhomogeneous, then f is contractive.*

Proof Suppose that f is subhomogeneous and let $x, y \in K$ with $x \sim_K y$. Then $d_T(x, y) = \log \lambda$ for some $\lambda \geq 1$. As K is closed, we know that $y \leq \lambda x$ and $x \leq \lambda y$. Since $f : K \to K$ is order-preserving and subhomogeneous, we get that

$$\lambda^{-1} f(y) \leq f(\lambda^{-1} y) \leq f(x) \quad \text{and} \quad \lambda^{-1} f(x) \leq f(\lambda^{-1} x) \leq f(y). \quad (2.5)$$

This implies that $\max\{M(f(y)/f(x)), M(f(x)/f(y))\} \leq \lambda$ and hence

$$d_T(f(x), f(y)) \leq \log \lambda = d_T(x, y).$$

To prove the opposite implication we suppose that f is non-expansive under d_T on each part of K. Let $x \in K$ and put $y = \lambda^{-1} x$, where $\lambda \geq 1$. Clearly $d_T(x, y) = \log \lambda$ for $x \neq 0$, and $d_T(x, y) = 0 \leq \log \lambda$ for $x = 0$. As f is non-expansive, we have that

$$\log M(f(x)/f(y)) \leq d_T(f(x), f(y)) \leq \log \lambda,$$

so that $f(x) \leq \lambda f(y)$. This implies that $\lambda^{-1} f(x) \leq f(\lambda^{-1} x)$, and hence f is subhomogeneous.

If $f : \text{int}(K) \to \text{int}(K')$ is an order-preserving and strictly subhomogeneous map and $\lambda > 1$, we get, as in (2.5),

$$\lambda^{-1} f(y) \ll f(\lambda^{-1} y) \leq f(x) \quad \text{and} \quad \lambda^{-1} f(x) \ll f(\lambda^{-1} x) \leq f(y)$$

for all $x \neq y$ in $\text{int}(K)$. Therefore there exists $\mu \in [0, \lambda)$ such that

$$\max\{M(f(x)/f(y)), M(f(y)/f(x))\} \leq \mu,$$

and hence $d_T(f(x), f(y)) < d_T(x, y)$ for all $x \neq y$ in $\text{int}(K)$. □

To compute $M(x/y)$ and $m(x/y)$, the following observation is sometimes useful. Given a solid closed cone $K \subseteq V$ and $u \in \text{int}(K)$, define

$$\Sigma_u^* = \{\varphi \in K^* : \varphi(u) = 1\}.$$

Note that Σ_u^* is a compact convex set by Lemma 1.2.4, and that $\varphi(u) > 0$ for all $\varphi \in K^*$, as $u \in \text{int}(K)$. For example, if $K = \mathbb{R}_+^n$ and $u = \mathbb{1}$, then $\Sigma_u^* = \{\varphi \in \mathbb{R}_+^n : \sum_i \varphi_i = 1\}$.

Let $\mathcal{E}_u^*$ be the set of extreme points of Σ_u^*. Recall that a point z in a convex set $S \subseteq V$ is called an *extreme point*, if there exist no $x \neq z$ and $y \neq z$ in S such that $\lambda x + (1-\lambda)y = z$ for some $0 < \lambda < 1$. It is known (see [186, corollary 18.5]) that each compact convex set is the convex hull of its extreme points. This result was proved by Minkowski and later generalized to infinite-dimensional vector spaces by Kreĭn and Milman [116]. Their result, known as the Kreĭn–Milman theorem, states that each compact convex set in a locally convex Hausdorff topological vector space is the closure of the convex hull of its extreme points.

Note that it follows from Lemma 1.2.1 that $x \leq \beta y$ if and only if $\varphi(x) \leq \beta\varphi(y)$ for all $\varphi \in \Sigma_u^*$. Because Σ_u^* is the convex hull of its extreme points, we see that $x \leq \beta y$ is equivalent to $\varphi(x) \leq \beta\varphi(y)$ for all $\varphi \in \mathcal{E}_u^*$. Thus,

$$M(x/y) = \inf\{\beta > 0 : x \leq \beta y\} = \sup_{\varphi \in \mathcal{E}_u^*} \frac{\varphi(x)}{\varphi(y)} \tag{2.6}$$

and

$$m(x/y) = \sup\{\alpha > 0 : \alpha y \leq x\} = \inf_{\varphi \in \mathcal{E}_u^*} \frac{\varphi(x)}{\varphi(y)}. \tag{2.7}$$

Notice that in equations (2.6) and (2.7) any subset S^* of Σ_u^* suffices as long as it contains $\mathcal{E}_u^*$.

In the next section we use this observation to show that Hilbert's metric and Thompson's metric on polyhedral cones are closely related to the *sup-norm*,

$$\|x\|_\infty = \max_i |x_i| \quad \text{for } x \in \mathbb{R}^n.$$

2.2 Polyhedral cones

For $K = \mathbb{R}_+^n$ and $u = \mathbb{1}$, we have that $\Sigma_u^* = \{x \in \mathbb{R}_+^n : \sum_i x_i = 1\}$, which has the standard basis vectors as its extreme points. So, it follows from (2.6) and (2.7) that

$$M(x/y) = \inf\{\beta > 0 : x \leq \beta y\} = \max_i \frac{x_i}{y_i} \tag{2.8}$$

and

$$m(x/y) = \sup\{\alpha > 0 : \alpha y \leq x\} = \min_i \frac{x_i}{y_i} \tag{2.9}$$

for all $x, y \in \text{int}(\mathbb{R}^n_+)$. Let $\mathsf{t} : \mathbb{R}^n \to \mathbb{R}$ be the *top function* defined by $\mathsf{t}(z) = \max_i z_i$. Similarly, define $\mathsf{b} : \mathbb{R}^n \to \mathbb{R}$ to be the *bottom function* given by $\mathsf{b}(x) = \min_i x_i$. Now (2.8) and (2.9) can be rewritten as

$$\log M(x/y) = \max_i(\log x_i - \log y_i) = \mathsf{t}(L(x) - L(y)) \tag{2.10}$$

and

$$\log m(x/y) = \min_i(\log x_i - \log y_i) = \mathsf{b}(L(x) - L(y)), \tag{2.11}$$

where $L : \text{int}(\mathbb{R}^n_+) \to \mathbb{R}^n$ is the coordinatewise log function. Using this observation it is straightforward to prove the following result for Thompson's metric.

Proposition 2.2.1 *The map L is an isometry from $(\text{int}(\mathbb{R}^n_+), d_T)$ onto $(\mathbb{R}^n, \|\cdot\|_\infty)$.*

Proof From Equation (2.10) it follows that

$$d_T(x, y) = \max\{\mathsf{t}(L(x) - L(y)), \mathsf{t}(L(y) - L(x))\} = \|L(x) - L(y)\|_\infty$$

for all $x, y \in \text{int}(\mathbb{R}^n_+)$, as $\|z\|_\infty = \max\{\mathsf{t}(z), \mathsf{t}(-z)\}$ for all $z \in \mathbb{R}^n$. Thus, L is an isometry from $(\text{int}(\mathbb{R}^n_+), d_T)$ onto $(\mathbb{R}^n, \|\cdot\|_\infty)$. □

For polyhedral cones we have the following result.

Lemma 2.2.2 *Suppose that $K \subseteq V$ is a solid polyhedral cone with facet-defining functionals $\psi_1, \ldots, \psi_N$. Let P be a part of K and $I(P) = \{i : \psi_i(x) > 0$ for some $x \in P\}$. Then there exists an isometric embedding from (P, d_T) into $(\mathbb{R}^m, \|\cdot\|_\infty)$, where $m = |I(P)|$.*

Proof By Lemma 1.2.3, we may assume, after relabeling, that

$$P = \{x \in K : \psi_i(x) > 0 \text{ for } 1 \leq i \leq m \text{ and } \psi_i(x) = 0 \text{ otherwise}\}.$$

Define $\Psi : P \to \text{int}(\mathbb{R}^m_+)$ by $\Psi(x) = (\psi_1(x), \ldots, \psi_m(x))$ for $x \in P$. Remark that

$$\begin{aligned} M(x/y; K) &= \inf\{\beta > 0 : \psi_i(x) \leq \beta\psi_i(y) \text{ for } 1 \leq i \leq m\} \\ &= \max_{1 \leq i \leq m} \psi_i(x)/\psi_i(y) \\ &= M(\Psi(x)/\Psi(y); \mathbb{R}^m_+). \end{aligned}$$

This implies that $d_T(\Psi(x), \Psi(y)) = d_T(x, y)$ for all $x, y \in P$, so that Ψ embeds (P, d_T) isometrically into $(\mathrm{int}(\mathbb{R}^m_+), d_T)$. It now follows from Proposition 2.2.1 that (P, d_T) can be isometrically embedded into $(\mathbb{R}^m, \|\cdot\|_\infty)$. $\square$

Notice that in Lemma 2.2.2 we can use any set of functionals $\{\varphi_1, \ldots, \varphi_k\}$ in K^* that determines K, i.e., $K = \{x \in V : \varphi_j(x) \geq 0 \text{ for all } 1 \leq j \leq k\}$, to obtain an isometric embedding.

For Hilbert's metric on polyhedral cones there exists a similar result.

Proposition 2.2.3 *Suppose that $K \subseteq V$ is a solid polyhedral cone with N facets. Let $\varphi : K \to [0, \infty)$ be a homogeneous function with $\varphi(x) > 0$ for all $x \in K \setminus \{0\}$, and let $\Sigma^\circ = \{x \in \mathrm{int}(K) : \varphi(x) = 1\}$. Then (Σ°, d_H) can be isometrically embedded into $(\mathbb{R}^m, \|\cdot\|_\infty)$, where $m = N(N-1)/2$.*

Proof Let $\psi_1, \ldots, \psi_N$ be the facet-defining functionals of K. Remark that for $x, y \in \mathrm{int}(K)$ we have that

$$M(x/y) = \inf\{\beta > 0 : \psi_i(x) \leq \beta\psi_i(y) \text{ for } 1 \leq i \leq N\} = \max_{i \leq i \leq N} \psi_i(x)/\psi_i(y).$$

Similarly,

$$1/m(x/y) = \max_{1 \leq j \leq N} \psi_j(y)/\psi_j(x),$$

and therefore

$$\begin{aligned} d_H(x, y) &= \log\Big(\max_{1 \leq i,j \leq N} \frac{\psi_i(x)\psi_j(y)}{\psi_i(y)\psi_j(x)}\Big) \\ &= \max_{1 \leq i,j \leq N} \Big(\log\frac{\psi_i(x)}{\psi_j(x)} - \log\frac{\psi_i(y)}{\psi_j(y)}\Big) \qquad (2.12) \\ &= \max_{1 \leq i < j \leq N} \Big|\log\frac{\psi_i(x)}{\psi_j(x)} - \log\frac{\psi_i(y)}{\psi_j(y)}\Big| \end{aligned}$$

for all $x, y \in \Sigma^\circ$. Now let $\Psi : \Sigma^\circ \to \mathbb{R}^m$, where $m = N(N-1)/2$, be given by

$$\Psi_{ij}(z) = \log\frac{\psi_i(z)}{\psi_j(z)}$$

for all $z \in \Sigma^\circ$ and $1 \leq i < j \leq N$. From (2.12) we deduce that Ψ embeds (Σ°, d_H) isometrically into $(\mathbb{R}^m, \|\cdot\|_\infty)$, which completes the proof. $\square$

In the special case where K is a simplicial cone (Σ°, d_H) is isometric to a normed space. To show this, define on $\mathbb{R}^{n+1}$ an equivalence relation $\sim$ by $x \sim y$ if $x = y + \lambda\mathbb{1}$ for some $\lambda \in \mathbb{R}$. It is easy to verify that $\mathbb{R}^{n+1}/\sim$ is an n-dimensional vector space, which can be equipped with the *variation norm*,

$$\|x\|_{\mathrm{var}} = \mathsf{t}(x) - \mathsf{b}(x) \quad \text{for } x \in \mathbb{R}^{n+1}/\sim.$$

The following proposition shows that if S_n is an open n-dimensional simplex, then (S_n, κ) is isometric to $(\mathbb{R}^{n+1}/\sim, \|\cdot\|_{\mathrm{var}})$. In the proof we denote the standard n-simplex by $\Delta_n = \{x \in \mathbb{R}^{n+1}_+ : \sum_i x_i = 1\}$ and its relative interior by Δ_n°.

Proposition 2.2.4 *If S_n is an open n-dimensional simplex in $\mathbb{R}^n$, then (S_n, κ) is isometric to $(\mathbb{R}^{n+1}/\sim, \|\cdot\|_{\mathrm{var}})$.*

Proof Let $x^1, \ldots, x^{n+1}$ be the vertices of $\mathrm{cl}(S_n)$ in $\mathbb{R}^n$. We embed S_n into $\mathbb{R}^{n+1}$ by identifying $\mathbb{R}^n$ with the affine hyperplane $\{x \in \mathbb{R}^{n+1} : x_{n+1} = 1\}$. Let $K = \{x \in \mathbb{R}^{n+1} : x = \sum_i \alpha_i x^i \text{ with } \alpha_i \geq 0 \text{ for } 1 \leq i \leq n+1\}$ be the simplicial cone in $\mathbb{R}^{n+1}$ generated by $\mathrm{cl}(S_n)$. Remark that $A : \mathbb{R}^{n+1} \to \mathbb{R}^{n+1}$, given by

$$A_j\left(\textstyle\sum_i \alpha_i x^i\right) = \alpha_j \quad \text{for } 1 \leq j \leq n+1 \text{ and } \textstyle\sum_i \alpha_i x^i \in \mathbb{R}^{n+1},$$

is a bijective linear map that maps K onto $\mathbb{R}^{n+1}_+$. As the inverse of A is a linear map of $\mathbb{R}^{n+1}_+$ onto K, we deduce from Corollary 2.1.4 that A is an isometry from (K, d_H) onto $(\mathbb{R}^{n+1}_+, d_H)$. As the metrics κ and d_H coincide on S_n, and $A(S_n) = \Delta_n^\circ$, we conclude that (S_n, κ) is isometric to (Δ_n°, d_H).

Note that the coordinatewise logarithm, $L : \Delta_n^\circ \to \mathbb{R}^{n+1}/\sim$, is a bijection between Δ_n° and $\mathbb{R}^{n+1}/\sim$. Indeed, if $L(x) = L(y)$ in $\mathbb{R}^{n+1}/\sim$, then $L(x) = L(y) + \lambda \mathbb{1} = L(e^\lambda x)$ in $\mathbb{R}^{n+1}$ for some $\lambda \in \mathbb{R}$, so that $e^\lambda y = x$. As $\sum_i x_i = \sum_i y_i$, we conclude that $x = y$. On the other hand, for each $z \in \mathbb{R}^{n+1}/\sim$ there exists a representative $x \in \mathbb{R}^{n+1}$ such that $\sum_i e^{x_i} = 1$: simply take

$$x = z - \log\left(\sum_i e^{z_i}\right)\mathbb{1}.$$

Clearly $u = (e^{x_1}, \ldots, e^{x_{n+1}}) \in \Delta_n^\circ$ and $L(u) = x$.

It follows from (2.10) and (2.11) that

$$d_H(x, y) = \mathsf{t}(L(x) - L(y)) - \mathsf{b}(L(x) - L(y))$$

for all $x, y \in \Delta_n^\circ$, and hence (Δ_n°, d_H) is isometric to $(\mathbb{R}^{n+1}/\sim, \|\cdot\|_{\mathrm{var}})$. □

Note that we can identify $\mathbb{R}^{n+1}/\sim$ with $\mathbb{R}^n$ by considering representatives $x \in \mathbb{R}^{n+1}$ with $x_{n+1} = 0$. Thus, $(\mathbb{R}^{n+1}/\sim, \|\cdot\|_{\mathrm{var}})$ is isometric to $(\mathbb{R}^n, \|\cdot\|_H)$, where

$$\|x\|_H = \max\{0, \mathsf{t}(x)\} - \min\{0, \mathsf{b}(x)\}$$

for $x \in \mathbb{R}^n$. It can be shown that the unit ball of $\|\cdot\|_H$ is a polyhedron with $n(n+1)$ facets. In particular, if $n = 2$, the unit ball is a hexagon, and for $n = 3$ it is a rhombic dodecahedron.

2.3 Lorentz cones

From Theorem 2.1.2 we know that Hilbert's metric on the interior of the Lorentz cone is closely related to the n-dimensional hyperbolic space. In fact, it is well known (see for instance [37, chapter 1.6]) that if we let

$$B_1^n = \{(x_1, x_2, \ldots, x_{n+1}) \in \mathrm{int}(\Lambda_{n+1}) : x_1 = 1\},$$

then $(B_1^n, \frac{1}{2}d_H)$ corresponds to *Klein's model* of the n-dimensional hyperbolic space. There are several other models of the hyperbolic space, each with its own advantages. To further analyze Hilbert's metric on $\mathrm{int}(\Lambda_{n+1})$ it is particularly beneficial to consider the hyperboloid model of the hyperbolic space, which lives inside $\mathrm{int}(\Lambda_{n+1})$. It will enable us to give explicit formulas for $M(x/y; \Lambda_{n+1})$ and $m(x/y; \Lambda_{n+1})$ for all $x, y \in \mathrm{int}(\Lambda_{n+1})$, and will help us understand Thompson's metric on $\mathrm{int}(\Lambda_{n+1})$.

To work with the hyperboloid model it is convenient to write $x \in \mathbb{R}^{n+1}$ as $x = (s, u) \in \mathbb{R} \times \mathbb{R}^n$; so, $s \in \mathbb{R}$ and $u = (u_1, \ldots, u_n) \in \mathbb{R}^n$. Let $Q : \mathbb{R}^{n+1} \to \mathbb{R}$ be the quadratic form,

$$Q((s, u)) = s^2 - u_1^2 - \cdots - u_n^2 = s^2 - \langle u, u \rangle,$$

for $(s, u) \in \mathbb{R} \times \mathbb{R}^n$. Notice that the Lorentz cone can now be expressed as

$$\Lambda_{n+1} = \{(s, u) \in \mathbb{R} \times \mathbb{R}^n : Q((s, u)) \geq 0 \text{ and } s \geq 0\}.$$

Let $\mathbb{H}^n = \{(s, u) \in \mathbb{R} \times \mathbb{R}^n : Q((s, u)) = 1 \text{ and } s > 0\}$. In other words, $\mathbb{H}^n$ is the upper sheet of the hyperboloid $S = \{x \in \mathbb{R}^{n+1} : Q(x) = 1\}$ inside $\mathrm{int}(\Lambda_{n+1})$; see Figure 2.3. Furthermore, let $B : \mathbb{R}^{n+1} \times \mathbb{R}^{n+1} \to \mathbb{R}$ be the bilinear form,

$$B(x, y) = st - u_1 v_1 - \cdots u_n v_n = st - \langle u, v \rangle$$

for $x = (s, u)$ and $y = (t, v)$ in $\mathbb{R} \times \mathbb{R}^n$.

Each non-zero point $z = (t, v)$ in the tangent space at a point $x = (s, u)$ in $\mathbb{H}^n$ lies outside Λ_{n+1}, when viewed as a vector acting from the origin. Indeed, the gradient at $x = (s, u)$ is given by $(2s, -2u)$, so

$$st - \langle u, v \rangle = 0. \tag{2.13}$$

Using the Cauchy–Schwarz inequality we find that

$$st - \|u\|_2 \|v\|_2 \leq 0, \tag{2.14}$$

where $\|w\|_2 = \left(\sum_i |w_i|^2\right)^{1/2}$. Clearly if $t < 0$, then $z = (t, v) \notin \Lambda_{n+1}$. Furthermore, if $t = 0$, then $z = (0, v)$ is also not in Λ_{n+1}, as $v \neq 0$. So,

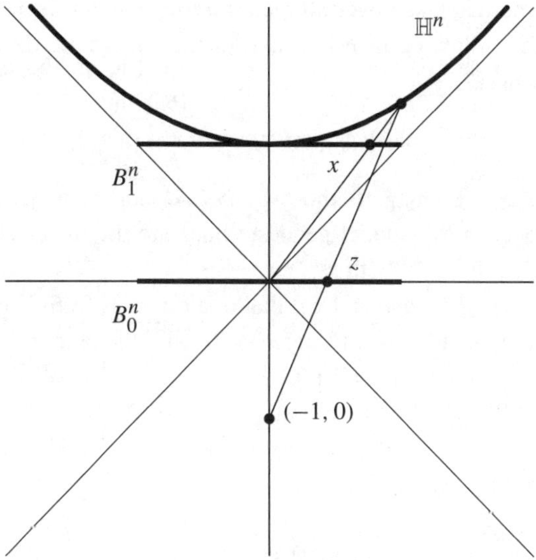

Figure 2.3 Isometries between different models.

assume that $t > 0$. It follows from (2.13), and the fact that $s > 0$, that $u \neq 0$. Now substitute $s = \sqrt{1 + \|u\|_2^2}$ in (2.14) to get

$$\frac{\sqrt{1 + \|u\|_2^2}}{\|u\|_2} t - \|v\|_2 \leq 0,$$

which implies

$$\frac{1}{\|u\|_2^2} t^2 + t^2 \leq \|v\|_2^2.$$

It follows that

$$B(z, z) = t^2 - \|v\|_2^2 \leq -\frac{t^2}{\|u\|_2^2} < 0.$$

One can use $-B$ to define a Riemannian metric d_{hyp} on $\mathbb{H}^n$ by first defining the length of a piecewise C^1 path $\gamma : [a, b] \to \mathbb{H}^n$ by

$$L(\gamma) = \int_a^b \sqrt{-B(\gamma'(t), \gamma'(t))} \mathrm{d}t,$$

and subsequently

$$d_{\text{hyp}}(x, y) = \inf_{\gamma} L(\gamma),$$

where the infimum is taken over all piecewise C^1 paths γ in $\mathbb{H}^n$ from x to y. There exists the following nice formula for the hyperbolic distance between points x and y in $\mathbb{H}^n$:

$$\cosh d_{\text{hyp}}(x, y) = B(x, y).$$

The metric space $(\mathbb{H}^n, d_{\text{hyp}})$ is the *hyperboloid model* of the n-dimensional hyperbolic space. In this model geodesic lines are the intersection of planes through the origin in $\mathbb{R}^{n+1}$ with $\mathbb{H}^n$.

It is known (see [37, chapter 1.6]) that one can pass from Klein's model to the hyperboloid model by identifying $x \in B_1^n$ with the point of intersection of the straight line through the origin in $\mathbb{R}^{n+1}$ and $\mathbb{H}^n$. The resulting map is an isometry from $(B_1^n, \frac{1}{2}d_H)$ onto $(\mathbb{H}^n, d_{\text{hyp}})$.

Recall that $d_H(x, y) = d_H(\lambda x, \mu y)$ for all $\lambda, \mu > 0$ and $x, y \in \text{int}(\Lambda_{n+1})$ by Proposition 2.1.1. So, we see that

$$d_{\text{hyp}}(x, y) = \frac{1}{2} d_H(x, y) \tag{2.15}$$

for all $x, y \in \mathbb{H}^n$. Later in this section we shall prove this equality in a different way, and also show that

$$d_{\text{hyp}}(x, y) = d_T(x, y) \tag{2.16}$$

for all $x, y \in \mathbb{H}^n$.

Another commonly used model of the hyperbolic space is the *Poincaré ball model*. In this model points are elements of the open unit ball B^n in $\mathbb{R}^n$ and the geodesic lines are the diameters and the segments of circles that meet ∂B^n orthogonally. The distance d_P between points in B^n is given by

$$\cosh d_P(u, v) = 1 + \frac{2\|u - v\|_2^2}{(1 - \|u\|_2^2)(1 - \|v\|_2^2)} \quad \text{for } u, v \in B^n.$$

To see how one passes from the Poincaré model to the hyperboloid model it is convenient to identify B^n with $B_0^n = \{(0, u) \in \mathbb{R} \times \mathbb{R}^n : u \in B^n\}$ inside $\mathbb{R}^{n+1}$. The map that associates with $z \in B_0^n$ the point of intersection of the straight line through $(-1, 0) \in \mathbb{R} \times \mathbb{R}^n$ and $\mathbb{H}^n$ is an isometry from (B_0^n, d_P) onto $(\mathbb{H}^n, d_{\text{hyp}})$. It is easy to verify that this map is given by the formula

$$(0, u) \mapsto \left(\frac{1 + \|u\|_2^2}{1 - \|u\|_2^2}, \frac{2u}{1 - \|u\|_2^2}\right).$$

An extensive discussion of the various models of the hyperbolic space can be found in [37].

To further analyze d_H on $\text{int}(\Lambda_{n+1})$ we follow Lim [128] and consider the zeros of the quadratic polynomial

$$p(\lambda) = \lambda^2 - 2B(x, y)\lambda + Q(x)Q(y),$$

which are

$$\alpha(x, y) = B(x, y) - \sqrt{B(x, y)^2 - Q(x)Q(y)}$$

and

$$\beta(x, y) = B(x, y) + \sqrt{B(x, y)^2 - Q(x)Q(y)}.$$

Note that $\alpha(x, y)$ and $\beta(x, y)$ are positively homogeneous in both coordinates. In the course of the proof of the next theorem it will be shown that $B(x, y)^2 - Q(x)Q(y) \geq 0$ for all $x, y \in \mathbb{H}^n$.

Theorem 2.3.1 *For $x, y \in \mathbb{H}^n$ we have that $M(x/y; \Lambda_{n+1}) = \beta(x, y)$ and $m(x/y; \Lambda_{n+1}) = \alpha(x, y)$.*

Proof Clearly the equalities hold if $x = y$. So, suppose that $x = (s, u)$ and $y = (t, v)$ in $\mathbb{H}^n$ are distinct. Remark that $x \leq \lambda y$ if and only if $Q(\lambda y - x) \geq 0$ and $\lambda t - s \geq 0$, which is equivalent to

$$(\lambda t - s)^2 - \langle \lambda v - u, \lambda v - u \rangle = \lambda^2 - 2\lambda B(x, y) + 1 \geq 0$$

and $\lambda \geq s/t$, as $Q(x) = Q(y) = 1$. Thus,

$$M(x/y; \Lambda_{n+1}) = \inf\{\lambda > 0 : p(\lambda) \geq 0 \text{ and } \lambda \geq s/t\}. \tag{2.17}$$

Note that

$$\begin{aligned} 2stB(x, y) &= 2s^2t^2 - 2st\langle u, v \rangle \\ &\geq 2s^2t^2 - 2st\|u\|_2\|v\|_2 \\ &\geq 2s^2t^2 - t^2\langle u, u \rangle - s^2\langle v, v \rangle \\ &= 2s^2t^2 - t^2(s^2 - 1) - s^2(t^2 - 1) \\ &= s^2 + t^2, \end{aligned} \tag{2.18}$$

where equality holds if and only if $u = \mu v$ for some $\mu > 0$ and $t\|u\| = s\|v\|$. In that case $t\mu\|v\| = s\|v\|$, so that $\mu = s/t$ and $x = (s, u) = (s, sv/t)$. This implies

$$1 = Q(x) = s^2 - (s/t)^2\langle v, v \rangle = \frac{s^2}{t^2}(t^2 - \langle v, v \rangle) = \frac{s^2}{t^2}Q(y) = \frac{s^2}{t^2},$$

and hence $s = t$. Thus, equality holds in (2.18) if and only if $x = y$. (Also notice that (2.18) implies that $B(x, y)^2 \geq \left(\frac{s^2+t^2}{2st}\right)^2 \geq 1 = Q(x)Q(y)$ for all $x, y \in \mathbb{H}^n$.)

As $x \neq y$, we conclude that

$$s^2 - 2stB(x, y) + t^2 < 0,$$

which implies that

$$t^2 p(s/t) = t^2\big((s/t)^2 - 2B(x, y)s/t + 1\big) = s^2 - 2stB(x, y) + t^2 < 0,$$

and hence $p(s/t) < 0$. It now follows from (2.17) that $M(x/y; \Lambda_{n+1}) = \beta(x, y)$ for $x, y \in \mathbb{H}^n$.

Similarly,

$$\begin{aligned} m(x/y; \Lambda_{n+1}) &= \sup\{\lambda > 0 : \lambda y \leq x\} \\ &= \sup\{\lambda > 0 : p(\lambda) \geq 0 \text{ and } \lambda \leq s/t\} \\ &= \alpha(x, y). \end{aligned}$$

□

Theorem 2.3.1 has the following consequence.

Corollary 2.3.2 *For* $x, y \in \operatorname{int}(\Lambda_{n+1})$ *we have that*

$$M(x/y) = \frac{\beta(x, y)}{Q(y)} \quad \text{and} \quad m(x/y) = \frac{\alpha(x, y)}{Q(y)}. \tag{2.19}$$

Moreover,

$$d_H(x, y) = \log\big(\beta(x, y)/\alpha(x, y)\big)$$

and

$$d_T(x, y) = \log\Big(\max\big\{\beta(x, y)/Q(x), \beta(x, y)/Q(y)\big\}\Big).$$

Proof It suffices to prove (2.19). Put $\sigma = \sqrt{Q(x)}$ and $\tau = \sqrt{Q(y)}$. Then

$$M(x/y) = \frac{\sigma}{\tau} M(\sigma^{-1}x/\tau^{-1}y) = \frac{\sigma}{\tau}\beta(\sigma^{-1}x, \tau^{-1}y) = \frac{1}{\tau^2}\beta(x, y)$$

by Theorem 2.3.1. Likewise,

$$m(x/y) = \frac{\sigma}{\tau} m(\sigma^{-1}x/\tau^{-1}y) = \frac{\sigma}{\tau}\alpha(\sigma^{-1}x, \tau^{-1}y) = \frac{1}{\tau^2}\alpha(x, y).$$

□

It is now easy to prove (2.16). Simply remark that for $x, y \in \mathbb{H}^n$ we have that $\beta(x, y) = \alpha(x, y)^{-1}$, so that

$$\frac{1}{2} d_H(x, y) = \log \beta(x, y) = d_T(x, y).$$

But

$$\cosh \log \beta(x,y) = \frac{\beta(x,y)+\beta(x,y)^{-1}}{2} = \frac{\beta(x,y)+\alpha(x,y)}{2} = B(x,y),$$

which shows (2.15) and (2.16).

2.4 The cone of positive-semidefinite symmetric matrices

In this section we consider the vector space of $n \times n$ symmetric matrices, Sym_n, and the solid closed cone $\Pi_n(\mathbb{R})$ consisting of all positive-semidefinite matrices in Sym_n. To analyze Hilbert's metric and Thompson's metric on $\Pi_n(\mathbb{R})$, it is useful to remark that the group of invertible $n \times n$ matrices, $\mathrm{GL}_n(\mathbb{R})$, acts transitively on $\mathrm{int}(\Pi_n(\mathbb{R}))$ by conjugation, i.e., for each $A, B \in \mathrm{int}(\Pi_n(\mathbb{R}))$ there exists $Q \in \mathrm{GL}_n(\mathbb{R})$ such that

$$A = Q^T B Q.$$

Furthermore the vector space Sym_n can be equipped with an inner-product $\langle A, B\rangle = \mathrm{tr}(AB)$. Using the spectral decomposition for symmetric matrices, it is easy to see that $\Pi_n(\mathbb{R})$ is self-dual under this inner-product. Indeed, note that if $B \in \Pi_n(\mathbb{R})$, then $B^{1/2} \in \Pi_n(\mathbb{R})$, and hence $\langle A, B\rangle = \mathrm{tr}(AB) = \mathrm{tr}(B^{1/2}AB^{1/2}) \geq 0$ for all $A \in \Pi_n(\mathbb{R})$. Thus, $B \in \Pi_n(\mathbb{R})^*$ in that case. Conversely, if $B \in \mathrm{Sym}_n$, then there exists an orthogonal matrix U such that $B = UDU^T$, where $D = \mathrm{diag}(\lambda_1, \ldots, \lambda_n)$ and $\lambda_1 \leq \lambda_2 \leq \ldots \leq \lambda_n$ are the eigenvalues of B. Suppose that $\mathrm{tr}(AB) = \mathrm{tr}(A(UDU^T)) = \mathrm{tr}((U^TAU)D) \geq 0$ for each $A \in \Pi_n(\mathbb{R})$. By taking $A = UEU^T$ for some appropriate positive diagonal matrix E, we can conclude that all the eigenvalues of D must be nonnegative, and hence $B \in \Pi_n(\mathbb{R})$.

In the following lemma, I denotes the identity matrix in $\mathrm{int}(\Pi_n(\mathbb{R}))$ and $\Sigma_I^* = \{A \in \Pi_n(\mathbb{R}) : \mathrm{tr}(A) = 1\}$.

Lemma 2.4.1 *A matrix $A \in \Sigma_I^*$ is an extreme point if and only if it has exactly one non-zero eigenvalue, which is equal to 1.*

Proof Let $\mathcal{E}_I^*$ denote the set of extreme points of Σ_I^*, and let $\mathcal{D}$ be the set of those $A \in \Sigma_I^*$ having exactly one non-zero eigenvalue. Suppose that A is an extreme point of Σ_I^*. There exists an orthogonal matrix U such that $A = U^TDU$, where $D = \mathrm{diag}(\lambda_1, \ldots, \lambda_n)$, $0 \leq \lambda_1 \leq \ldots \leq \lambda_n$ are the eigenvalues of A and $\sum_i \lambda_i = 1$. Thus,

$$A = U^TDU = \sum_i \lambda_i U^T \mathrm{diag}(e_i) U,$$

where e_i is the i-th unit vector. As A is an extreme point, we conclude that $\lambda_1 = \ldots = \lambda_{n-1} = 0$ and $\lambda_n = 1$, which shows that $\mathcal{E}_I^* \subseteq \mathcal{D}$

To finish the proof we note that the group of orthogonal matrices, $O(n)$, acts transitively on $\mathcal{D}$ by conjugation. Moreover, for a fixed $U \in O(n)$ the map $A \mapsto U^T AU$ is a linear map, which maps Σ_I^* onto itself. Hence it maps extreme points of Σ_I^* to extreme points. As $\mathcal{E}_I^* \subseteq \mathcal{D}$ is non-empty, and $O(n)$ acts transitively on $\mathcal{D}$, we conclude that $\mathcal{E}_I^* = \mathcal{D}$. □

It follows from Lemma 2.4.1 that Σ_I^* has infinitely many extreme points if $n \geq 2$, and hence $\Pi_n(\mathbb{R})$ is not polyhedral. On the other hand, $\Pi_n(\mathbb{R})$ is also not strictly convex if $n \geq 3$, as

$$(1-\lambda)\mathrm{diag}(1, 0, 0, \ldots, 0)+\lambda\mathrm{diag}(0, 1, 0, \ldots, 0) \in \partial\Pi_n(\mathbb{R}) \quad \text{for } 0 \leq \lambda \leq 1.$$

The cone $\Pi_2(\mathbb{R})$ is strictly convex and identical to Λ_3. In fact, the reader can easily verify that the bijective linear map

$$\begin{pmatrix} a & b \\ b & c \end{pmatrix} \in V \mapsto (\frac{a+c}{2}, \frac{a-c}{2}, b) \in \mathbb{R}^3$$

maps $\Pi_2(\mathbb{R})$ onto Λ_3.

We could use Lemma 2.4.1 and (2.6) to compute $M(X/Y; \Pi_n(\mathbb{R}))$ for $X, Y \in \mathrm{int}(\Pi_n(\mathbb{R}))$. There is, however, a more direct way to do this, which goes as follows. For each $X, Y \in \mathrm{int}(\Pi_n(\mathbb{R}))$ we have that $X \leq \beta Y$ if and only if $Y^{-1/2}XY^{-1/2} \leq \beta I$, which is equivalent to $\sigma(Y^{-1}X) = \sigma(Y^{-1/2}XY^{-1/2}) \subseteq (0, \beta]$. Thus,

$$M(X/Y; \Pi_n(\mathbb{R})) = \max\{\lambda : \lambda \in \sigma(Y^{-1}X)\}.$$

Similarly, $\alpha Y \leq X$ if and only if $\alpha I \leq Y^{-1/2}XY^{-1/2}$, which is equivalent to $\sigma(Y^{-1}X) = \sigma(Y^{-1/2}XY^{-1/2}) \subseteq [\alpha, \infty)$. So, we see that

$$m(X/Y; \Pi_n(\mathbb{R})) = \min\{\lambda : \lambda \in \sigma(Y^{-1}X)\}.$$

To summarize, we have the following result.

Proposition 2.4.2 *Suppose that* $X, Y \in \mathrm{int}(\Pi_n(\mathbb{R}))$, *and let* $\lambda_+(X, Y) = \max\{\lambda : \lambda \in \sigma(Y^{-1}X)\}$ *and* $\lambda_-(X, Y) = \min\{\lambda : \lambda \in \sigma(Y^{-1}X)\}$. *Then*

$$d_H(X, Y) = \log\left(\frac{\lambda_+(X, Y)}{\lambda_-(X, Y)}\right)$$

and

$$d_T(X, Y) = \log\left(\max\{\lambda_+(X, Y), 1/\lambda_-(X, Y)\}\right).$$

Similar arguments can be used to show that the same formulas hold for elements in the interior of the cone of positive-semidefinite Hermitian matrices $\Pi_n(\mathbb{C})$.

2.5 Completeness

It follows from the results in Section 2.2 that Hilbert's and Thompson's metrics on parts of polyhedral cones are complete metric spaces, whose topology coincides with the norm topology. It turns out that this is true for general closed cones in V. A key property to prove completeness is normality of the cone. Recall that a cone K in a normed space $(V, \|\cdot\|)$ is *normal* if there exists $\delta > 0$ such that $0 \leq_K x \leq_K y$ implies $\|x\| \leq \delta\|y\|$, and the smallest such $\delta > 0$ is called the *normality constant* of K in $(V, \|\cdot\|)$. Also recall that every closed cone in a finite-dimensional normed space is normal by Lemma 1.2.5.

Most completeness results for Hilbert's and Thompson's metrics discussed in this section carry over to normal cones in infinite-dimensional Banach spaces; see [158]. To prove that Thompson's metric yields a complete topology, it is useful to first establish the following inequality.

Lemma 2.5.1 *Let K be a cone in $(V, \|\cdot\|)$ and let $\delta > 0$ be the normality constant of K. For each $x, y \in K$ with $\|x\| \leq b$ and $\|y\| \leq b$, we have that*

$$\|x - y\| \leq b(1 + 2\delta)(e^{d_T(x,y)} - 1).$$

Proof Put $\lambda = e^{d_T(x,y)} \geq 1$. Clearly the inequality holds if $d_T(x, y) = \infty$. Therefore we may assume that $x \sim_K y$. For each $\varepsilon > 0$, we have that $x \leq (\lambda + \varepsilon)y$ and $y \leq (\lambda + \varepsilon)x$. This implies that $x - y \leq (\lambda + \varepsilon - 1)y$ and $y - x \leq (\lambda + \varepsilon - 1)x$, and hence there exist $u, v \in K$ such that $x - y + u = (\lambda + \varepsilon - 1)y$ and $y - x + v = (\lambda + \varepsilon - 1)x$. We remark that

$$\|u\| \leq \delta\|u + v\| = \delta\|(\lambda + \varepsilon - 1)y + (\lambda + \varepsilon - 1)x\| \leq 2\delta b(\lambda + \varepsilon - 1).$$

This implies that

$$\|x-y\| \leq \|x-y+u\|+\|u\| \leq (\lambda+\varepsilon-1)b+2\delta b(\lambda+\varepsilon-1) = b(1+2\delta)(\lambda+\varepsilon-1),$$

and we are done. □

By using Lemma 2.5.1 we now prove that Thompson's metric induces a complete topology on each part of a closed cone in V.

Proposition 2.5.2 *If $K \subseteq V$ is a closed cone and P is a part of K, then the metric space (P, d_T) is complete.*

Proof Let $(x_k)_k$ be a Cauchy sequence in (P, d_T) and put $\alpha_{k,l} = M(x_k/x_l)$ for $k, l \geq 1$. As $(x_k)_k$ is Cauchy, there exists $N \geq 1$ such that $d_T(x_k, x_l) < 1$ for all $k, l \geq N$. This implies that $\alpha_{k,l} < e$ for all $k, l \geq N$ and therefore $x_k \leq ex_N \leq 3x_N$ for all $k \geq N$. Let $\delta > 0$ be the normality constant of K with respect to a norm $\|\cdot\|$ on V. (Recall that a closed cone in a finite-dimensional normed space is normal by Lemma 1.2.5.) Then $\|x_k\| \leq 3\delta\|x_N\|$ for all $k \geq N$, and hence $(x_x)_k$ is a norm-bounded sequence. Put $c = \max\{3\delta\|x_N\|, \max_{1\leq k\leq N}\|x_k\|\}$.

Next we show that $(x_k)_k$ is a Cauchy sequence in the norm topology. Let $\varepsilon > 0$ and remark that there exists $\mu > 0$ such that $e^\mu \leq 1 + \varepsilon/C$, where $C = c(1+2\delta)$. Since $(x_k)_k$ is Cauchy, there exists $N_\varepsilon \geq 1$ such that $d_T(x_k, x_l) < \mu$ for all $k, l \geq N_\varepsilon$. Therefore, $\alpha_{k,l} \leq e^\mu \leq 1 + \varepsilon/C$ and $\alpha_{l,k} \leq e^\mu \leq 1 + \varepsilon/C$ for all $k, l \geq N_\varepsilon$. This implies that $x_l \leq (1 + \varepsilon/C)x_k$ for all $k, l \geq N_\varepsilon$. By Lemma 2.5.1 we find that

$$\|x_k - x_l\| \leq c(1 + 2\delta)(1 + \varepsilon/C - 1) = \varepsilon \quad \text{for all } k, l \geq N_\varepsilon.$$

Let u be the limit of $(x_k)_k$. To show that $d_T(x_k, u) \to 0$ as $k \to \infty$, we note that $d_T(x_k, x_l) < \varepsilon$ for all k and l sufficiently large, so that $x_k \leq e^\varepsilon x_l$ and $x_l \leq e^\varepsilon x_k$ for all k, l large. As K is closed and $\|x_k - u\| \to 0$, as $k \to \infty$, $x_k \leq e^\varepsilon u$ and $u \leq e^\varepsilon x_k$ for all k large. From these inequalities we conclude that $u \in P$ and $d_T(x_k, u) \leq \varepsilon$ for all k sufficiently large. □

Hilbert's metric also induces a complete topology, if the cone is closed. To prove this, it is convenient to work with a special norm, the so-called order-unit norm. Let P be a part of a solid closed cone K in V and let $u \in P$. Define

$$E_u = \{x \in V : -\alpha u \leq x \leq \alpha u \text{ for some } \alpha > 0\}.$$

It is easy to verify that E_u is a linear subspace of V. On E_u the *order-unit norm* $\|\cdot\|_u$ is given by

$$\|x\|_u = \inf\{\alpha > 0 : -\alpha u \leq x \leq \alpha u\} \quad \text{for } x \in E_u.$$

To see that $\|\cdot\|_u$ is a norm we note that $\|\lambda x\|_u = |\lambda|\|x\|_u$ and $\|x + y\|_u \leq \|x\|_u + \|y\|_u$ for all $\lambda \in \mathbb{R}$ and $x, y \in E_u$. Moreover, $\|x\|_u = 0$ if and only if $x = 0$. Indeed, $\|x\|_u = 0$ implies that $\alpha_k u - x \geq 0$ and $x + \alpha_k u \geq 0$ for some $\alpha_k \to 0$. From this it follows that $x \geq 0$ and $-x \geq 0$, as K is closed, so that $x = 0$. Moreover, $0 \leq x \leq y$ implies $\|x\|_u \leq \|y\|_u$. In particular, we see that if $u \in \text{int}(K)$, then $\|\cdot\|_u$ is a monotone norm on V.

Lemma 2.5.3 *Let $K \subseteq V$ be a solid closed cone and let P be a part of K. If $u \in \text{int}(K)$ and $\Sigma_u = \{x \in K : \|x\|_u = 1\}$, then $(\Sigma_u \cap P, d_H)$ is a complete metric space.*

Proof To prove this lemma we first show that

$$d_T(x, y) \leq d_H(x, y) \leq 2d_T(x, y) \quad \text{for all } x, y \in \Sigma_u \cap P. \tag{2.20}$$

For $x, y \in \Sigma_u \cap P$ we have that $m(x/y)x \leq y \leq M(x/y)x$. This implies that $m(x/y) \leq 1 \leq M(x/y)$, as $\|\cdot\|_u$ is monotone, so that

$$m(x/y)^{-1}M(x/y) \leq \max\{m(x/y)^{-2}, M(x/y)^2\} = \max\{M(y/x)^2, M(x/y)^2\}.$$

Thus, $d_H(x, y) \leq 2d_T(x, y)$ for all $x, y \in \Sigma_u \cap P$. On the other hand,

$$\max\{m(x/y)^{-1}, M(x/y)\} \leq m(x/y)^{-1}M(x/y),$$

so that $d_T(x, y) \leq d_H(x, y)$ for all $x, y \in \Sigma_u \cap P$.

Subsequently, we observe that $\Sigma_u \cap P$ is a closed subset of (P, d_T). Indeed, if x is a limit point of a sequence $(x_k)_k$ in $\Sigma_u \cap P$, then $\|x_k - x\|_u \to 0$ by Lemma 2.5.1, and hence $\|x\|_u = 1$. It follows from Proposition 2.5.2 that (P, d_T) is complete, and hence $(\Sigma_u \cap P, d_T)$ is also complete. The completeness of $(\Sigma_u \cap P, d_H)$ now follows from (2.20). □

Equipped with this lemma it is easy to prove the following completeness result for Hilbert's metric.

Proposition 2.5.4 *Let K be a solid closed in V and $\varphi \in \mathrm{int}(K^*)$. If P is a part of K, with $P \neq \{0\}$, and $\Sigma = \{x \in K : \varphi(x) = 1\}$, then $(\Sigma \cap P, d_H)$ is complete.*

Proof The idea is to show that $(\Sigma_u \cap P, d_H)$ and $(\Sigma \cap P, d_H)$ are isometric for $u \in \mathrm{int}(K)$. To see this we simply note that for each $x \in \Sigma \cap P$ there exists a unique $t_x > 0$ such that $t_x x \in \Sigma_u \cap P$, so that $\rho : \Sigma \cap P \to \Sigma_u \cap P$, given by $\rho(x) = t_x x$ for $x \in \Sigma \cap P$, is an isometry from $(\Sigma \cap P, d_H)$ onto $(\Sigma_u \cap P, d_H)$. From Lemma 2.5.3 we know that $(\Sigma_u \cap P, d_H)$ is complete and hence $(\Sigma \cap P, d_H)$ is also complete. □

Given a norm $\|\cdot\|$ on V, and ρ and Σ as in the proof of Proposition 2.5.4, there exists a constant $M > 0$ such that

$$\|x - y\| \leq M\|\rho(x) - \rho(y)\| \quad \text{for all } x, y \in \Sigma.$$

It therefore follows from Lemma 2.5.1 and inequality (2.20) that there exists a constant $C > 0$ such that

$$\|x - y\| \leq C(e^{d_H(x,y)} - 1) \quad \text{for all } x, y \in \Sigma. \tag{2.21}$$

We can also estimate $d_H(x, y)$ in terms of the norm.

Lemma 2.5.5 *Let K be a solid closed cone in $(V, \|\cdot\|)$. If $x \in \mathrm{int}(K)$ and $r > 0$ are such that the closed ball $B_r(x) = \{z \in V : \|x-z\| \leq r\}$ is contained in* $\mathrm{int}(K)$*, then*

$$d_H(x, y) \leq \log\left(\frac{r + \|x-y\|}{r - \|x-y\|}\right)$$

and

$$d_T(x, y) \leq \log \max\left\{\frac{r + \|x-y\|}{r}, \frac{r}{r - \|x-y\|}\right\} \quad \textit{for all } y \in B_r(x).$$

Proof Let $y \in B_r(x)$ and $x \neq y$. Then there exists $0 < \alpha < 1$ such that $\frac{\alpha}{1-\alpha}\|x-y\| = r$. Indeed, take $\alpha = r/(r+\|x-y\|)$. We note that $x \pm w \in \mathrm{int}(K)$ for $\|w\| \leq r$, and hence $x + \frac{\alpha}{1-\alpha}(x-y) \in K$. This implies that

$$x - \alpha y = (1-\alpha)(x + \frac{\alpha}{1-\alpha}(x-y))$$

is in K, so that $\alpha = \frac{r}{r+\|x-y\|} \leq m(x/y)$. In the same fashion, it can be shown that

$$\frac{r}{r - \|x-y\|} \geq M(x/y).$$

Thus,

$$d_H(x, y) \leq \log\left(\frac{r + \|x-y\|}{r - \|x-y\|}\right)$$

and

$$d_T(x, y) \leq \log \max\left\{\frac{r + \|x-y\|}{r}, \frac{r}{r - \|x-y\|}\right\} \quad \text{for all } y \in B_r(x),$$

which completes the proof. □

By combining the inequalities in Lemmas 2.5.1 and 2.5.5 and (2.21), we see that the topologies induced by Hilbert's metric and Thompson's metric are the same as the norm topology.

Corollary 2.5.6 *If K is a solid closed cone in V, $\varphi \in \mathrm{int}(K^*)$, and $\Sigma^\circ = \{x \in \mathrm{int}(K) : \varphi(x) = 1\}$, then the topologies of (Σ°, d_H) and $(\mathrm{int}(K), d_T)$ coincide with the norm topology.*

2.6 Convexity and geodesics

Hilbert's metric is an example of a projective metric, meaning that straight line segments are geodesics. As we shall see, however, straight lines are in general not the only geodesics. Before we start analyzing the geodesics in Hilbert's and Thompson's metric spaces in more detail, we prove the convexity of the balls.

Lemma 2.6.1 *If $K \subseteq V$ is a solid closed cone, $\varphi \in \mathrm{int}(K^*)$, and $\Sigma = \{x \in K : \varphi(x) = 1\}$, then every closed ball, $B_r(x) = \{y \in \Sigma : d_H(x, y) \leq r\}$, is a convex subset of V.*

Proof Let $y^1, y^2 \in B_r(x)$ and remark that $\alpha_1 x \leq y^1 \leq \beta_1 x$ and $\alpha_2 x \leq y^2 \leq \beta_2 x$, where $0 \leq \beta_1/\alpha_1, \beta_2/\alpha_2 \leq e^r$. For $0 \leq t \leq 1$ we deduce that

$$\alpha_t x = ((1-t)\alpha_1 + t\alpha_2)x \leq (1-t)y^1 + ty^2 \leq ((1-t)\beta_1 + t\beta_2)x = \beta_t x,$$

so that

$$\frac{\beta_t}{\alpha_t} = \Big(\frac{(1-t)\alpha_1}{\alpha_t}\Big)\Big(\frac{\beta_1}{\alpha_1}\Big) + \Big(\frac{t\alpha_2}{\alpha_t}\Big)\Big(\frac{\beta_2}{\alpha_2}\Big) \leq \frac{(1-t)\alpha_1}{\alpha_t}e^r + \frac{t\alpha_2}{\alpha_t}e^r = e^r.$$

Thus, we find that $(1-t)y^1 + ty^2 \in B^r(x)$ for $0 \leq t \leq 1$, and hence $B_r(x)$ is convex. □

A similar argument works for Thompson's metric.

Lemma 2.6.2 *If $K \subseteq V$ is a solid closed cone, then every closed ball, $B_r(x) = \{y \in K : d_T(x, y) \leq r\}$, is a convex subset of V.*

Proof If $y^1, y^2 \in B_r(x)$, then $\alpha_1 x \leq y^1 \leq \beta_1 x$ and $\alpha_2 x \leq y^2 \leq \beta_2 x$, where $\max\{\beta_1, 1/\alpha_1\} \leq e^r$ and $\max\{\beta_2, 1/\alpha_2\} \leq e^r$. Let $0 \leq t \leq 1$ and remark that

$$\alpha_t x = ((1-t)\alpha_1 + t\alpha_2)x \leq (1-t)y^1 + ty^2 \leq ((1-t)\beta_1 + t\beta_2)x = \beta_t x.$$

Clearly $(1-t)\beta_1 + t\beta_2 \leq e^r$ and $(1-t)\alpha_1 + t\alpha_2 \geq e^{-r}$. This implies that $\max\{\beta_t, 1/\alpha_t\} \leq e^r$, and hence $(1-t)y^1+ty^2 \in B_r(x)$ for each $0 \leq t \leq 1$. □

Recall (see [42, 177]) that a *geodesic path* in a metric space (X, d) is a map $\gamma : I \to (X, d)$ such that

$$d(\gamma(s), \gamma(t)) = |s - t| \quad \text{for all } s, t \in I.$$

Here $I \subseteq \mathbb{R}$ is a (possibly unbounded) interval. The image of a geodesic path is called a *geodesic*. A metric space is said to be a *geodesic space* if for each $x, y \in X$ there exists a geodesic path $\gamma : [a, b] \to (X, d)$ joining x and y, i.e, $\gamma(a) = x$ and $\gamma(b) = y$.

Hilbert's metric spaces are geodesic, since every straight-line segment is a geodesic. The proof of this result goes back to Hilbert [86] and uses the cross-ratio metric.

Theorem 2.6.3 (Hilbert) *If X is an open, bounded, convex subset of a finite-dimensional real affine space, then the cross-ratio metric space, (X, κ), is a geodesic space and every straight-line segment is a geodesic.*

Proof Let $u \neq w$ in X and $u', w' \in \partial X$ be such that u is between u' and w, and w is between u and w'. Let $0 < \alpha < \gamma < 1$ be such that

$$u = (1-\alpha)u' + \alpha w' \quad \text{and} \quad w = (1-\gamma)u' + \gamma w'.$$

Note that if $v = (1-\beta)u' + \beta w'$ and $\alpha \leq \beta \leq \gamma$, then

$$[u', u, v, w'] = \frac{1-\alpha}{1-\beta} \cdot \frac{\beta}{\alpha} \quad \text{and} \quad [u', v, w, w'] = \frac{1-\beta}{1-\gamma} \cdot \frac{\gamma}{\beta},$$

so that

$$[u', u, v, w'] \cdot [u', v, w, w'] = [u', u, w, w'].$$

(This is the co-cycle property of the cross-ratio.) Thus,

$$\kappa(u, v) + \kappa(v, w) = \kappa(u, w). \tag{2.22}$$

Now let $x \neq y$ in X. For $0 \leq t \leq 1$ define $z_t = (1-t)x + ty$. It follows from (2.22) that $\kappa(z_s, x) + \kappa(z_s, z_t) = \kappa(z_t, x)$ for all $0 \leq s \leq t \leq 1$. This implies that $\psi : [0, 1] \to [0, \kappa(x, y)]$ with $\psi(t) = \kappa(z_t, x)$ is a strictly increasing map from $[0, 1]$ onto $[0, \kappa(x, y)]$.

To complete the proof we show that $\gamma : [0, \kappa(x, y)] \to (X, \kappa)$ with $\gamma(\mu) = z_{\psi^{-1}(\mu)}$ for $\mu \in [0, \kappa(x, y)]$ is a geodesic path connecting x and y. Clearly $\gamma(0) = x$ and $\gamma(\kappa(x, y)) = y$. From (2.22) it follows that

$$\kappa(\gamma(\nu), \gamma(\mu)) = \kappa(z_{\psi^{-1}(\nu)}, z_{\psi^{-1}(\mu)}) = \kappa(z_{\psi^{-1}(\nu)}, x) - \kappa(z_{\psi^{-1}(\mu)}, x) = \nu - \mu$$

for each $0 \leq \mu \leq \nu \leq \kappa(x, y)$, and we are done. □

Combining Theorems 2.1.2 and 2.6.3 gives the following result.

Corollary 2.6.4 *If $K \subseteq V$ is a solid closed cone, $\varphi \in \mathrm{int}(K^*)$, and $\Sigma^\circ = \{x \in \mathrm{int}(K) : \varphi(x) = 1\}$, then (Σ°, d_H) is a geodesic space and straight-line segments are geodesics.*

In general there can be more than one geodesic connecting distinct points in a Hilbert's metric space. A prime example is the standard positive cone $\mathbb{R}^n_+$. This is no surprise, as Hilbert's metric on a simplex is isometric to a polyhedral normed space in which there are multiple geodesics between most points.

Proposition 2.6.5 *Let* $\Delta_n^\circ = \{x \in \mathrm{int}(\mathbb{R}_+^n : \sum_i x_i = 1\}$. *If* $x \neq y$ *in* Δ_n° *and* $\delta = d_H(x, y)$, *then*

$$\gamma(s) = \frac{x^{(1-s/\delta)} y^{s/\delta}}{\sum_i x_i^{(1-s/\delta)} y_i^{s/\delta}},$$

for $s \in [0, \delta]$ *is a geodesic path connecting* x *and* y *in* (Δ_n°, d_H).

Proof Recall from the proof of Proposition 2.2.4 that the coordinatewise logarithm L is an isometry of (Δ_n°, d_H) onto $(\mathbb{R}^{n+1}/\sim, \|\cdot\|_{\mathrm{var}})$. Every straight line in $\mathbb{R}^{n+1}/\sim$ is a geodesic. So for $u, v \in \mathbb{R}^{n+1}/\sim$ with $\alpha = \|u - v\|_{\mathrm{var}} > 0$, the path $\psi : [0, \alpha] \to (\mathbb{R}^{n+1}/\sim, \|\cdot\|_{\mathrm{var}})$ given by

$$\psi(s) = (1 - s/\alpha)u + (s/\alpha)v \quad \text{for } s \in [0, \alpha]$$

is a geodesic path connecting u and v.

By using the isometry L, and its inverse E, we find for $x \neq y$ in Δ_n° that the path

$$\gamma(s) = E\Big((1 - s/\delta)L(x) + s/\delta L(y) - \log\big(\sum_i x_i^{1-s/\delta} y_i^{s/\delta}\big)\mathbb{1}\Big)$$

$$= \frac{x^{(1-s/\delta)} y^{s/\delta}}{\sum_i x_i^{(1-s/\delta)} y_i^{s/\delta}},$$

where $s \in [0, \delta]$, is a geodesic path connecting x and y. □

A geodesic space (X, d) is said to be *uniquely geodesic* if for each $x, y \in X$ there exists a unique geodesic connecting x and y.

Theorem 2.6.6 *Let* X *be a bounded, open, convex subset of a finite-dimensional real affine space. The cross-ratio metric space* (X, κ) *is uniquely geodesic if and only if* ∂X *does not contain a pair of straight-line segments that span a two-dimensional affine plane.*

Proof Note that by Theorem 2.6.3 it suffices to show that

$$\kappa(x, z) + \kappa(z, y) = \kappa(x, y) \tag{2.23}$$

implies that z lies on the straight-line segment connecting x and y if and only if ∂X does not contain a pair of straight-line segments that span a two-dimensional affine plane. So, suppose that $x, y, z \in X$ are three distinct points such that (2.23) holds. Consider the intersection of X with the affine plane spanned by x, y, and z. Now use the notation as indicated in Figure 2.4, where p may be at infinity.

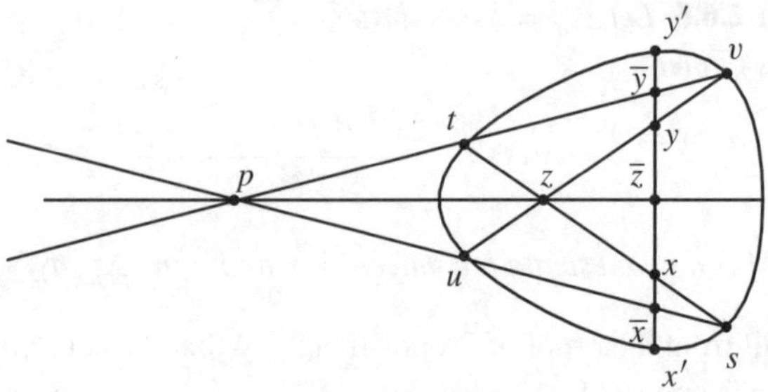

Figure 2.4 Proof of Theorem 2.6.6.

By projective invariance of the cross-ratio,

$$[s, x, z, t] = [\overline{x}, x, \overline{z}, \overline{y}] \quad \text{and} \quad [u, z, y, v] = [\overline{x}, \overline{z}, y, \overline{y}]. \tag{2.24}$$

Note that

$$[\overline{x}, x, \overline{z}, \overline{y}] \geq [x', x, \overline{z}, y']$$

and we have equality if and only if $\overline{x} = x'$ and $\overline{y} = y'$. Likewise,

$$[\overline{x}, \overline{z}, y, \overline{y}] \geq [x', \overline{z}, y, y']$$

and we have equality if and only if $\overline{x} = x'$ and $\overline{y} = y'$. It thus follows from (2.24) that

$$\kappa(x, z) + \kappa(z, y) = \kappa(x, y)$$

if and only if $\overline{x} = x'$ and $\overline{y} = y'$, which implies that the straight-line segment connecting x' and s, and the straight-line segment connecting y' and v, are contained in ∂X. On the other hand, if there exist two straight-line segments in ∂X that span a two-dimensional affine plane, then x, y, and z can be found such that $\overline{x} = x'$ and $\overline{y} = y'$. □

Combining the previous theorem with Theorem 2.1.2 gives the following corollary.

Corollary 2.6.7 *If $K \subseteq V$ is a solid closed cone, $\varphi \in \text{int}(K^*)$, and $\Sigma^\circ = \{x \in \text{int}(K) : \varphi(x) = 1\}$, then (Σ°, d_H) is uniquely geodesic if and only if $\partial\Sigma^\circ$ does not contain a pair of straight-line segments that span a two-dimensional affine plane.*

Another interesting example of a cone that gives rise to a non-uniquely geodesic Hilbert's metric space is the cone of positive-semidefinite matrices, $\Pi_n(\mathbb{R})$, in the vector space of $n \times n$ symmetric matrices.

Proposition 2.6.8 *Let* $\mathrm{Pos}_n(\mathbb{R}) = \{A \in \mathrm{int}(\Pi_n(\mathbb{R})) : \mathrm{tr}(A) = 1\}$ *and for* $A, B \in \mathrm{Pos}_n(\mathbb{R})$ *let*

$$\varphi(t) = A^{1/2}(A^{-1/2}BA^{-1/2})^t A^{1/2} \quad \text{for } 0 \le t \le 1.$$

Then

$$\gamma(t) = \frac{\varphi(t)}{\mathrm{tr}(\varphi(t))} \quad \text{for } 0 \le t \le 1$$

is a geodesic connecting A and B in $(\mathrm{Pos}_n(\mathbb{R}), d_H)$.

Proof We shall first prove the assertion in the special case where $A = I$, and subsequently derive the general case from it. Write $\alpha = m(B/I)$ and $\beta = M(B/I)$. As $\Pi_n(\mathbb{R})$ is closed,

$$\alpha I \le B \le \beta I.$$

From the spectral decomposition theorem it follows that

$$\alpha^s I \le B^s \le \beta^s I \quad \text{for } 0 \le s \le 1,$$

so that

$$d_H(I, B^s) \le \log\left(\frac{\beta}{\alpha}\right)^s = s d_H(I, B). \tag{2.25}$$

Fix $C \in \mathrm{int}(\Pi_n(\mathbb{R}))$ and recall that the linear map $L_C(B) = C^{1/2}BC^{1/2}$ maps $\Pi_n(\mathbb{R})$ onto itself. Thus, L_C is an isometry under d_H by Corollary 2.1.4. Taking $C = B^{-t}$, where $t \in [0, 1]$, we find that

$$d_H(B^t, B) = d_H(I, B^{1-t}) \le (1-t)d_H(I, B). \tag{2.26}$$

As $d_H(I, B^t) + d_H(B^t, B) \ge d_H(I, B)$, we conclude from (2.25) and (2.26) that

$$d_H(I, B^t) = t d_H(I, B)$$

for all $0 \le t \le 1$.

This implies that $\psi : [0, 1] \to [0, d_H(I, B)]$ with $\psi(t) = d_H(I, B^t)$ is strictly increasing and onto. It is now easy to verify that $\varphi_B : [0, d_H(I, B)] \to (\mathrm{Pos}_n(\mathbb{R}), d_H)$ with

$$\varphi_B(s) = \frac{B^{\psi^{-1}(s)}}{\mathrm{tr}(B^{\psi^{-1}(s)})} \tag{2.27}$$

for $0 \le s \le d_H(I, B)$ is a geodesic path connecting I and B.

The general case now follows by remarking that for $A, B \in \mathrm{Pos}_n(\mathbb{R})$ we have that

$$d_H(A, B) = d_H(L_{A^{-1/2}}(A), L_{A^{-1/2}}(B)) = d_H(I, A^{-1/2}BA^{-1/2}).$$

Replacing B by $B_1 = A^{-1/2}BA^{-1/2}$ in (2.27) shows that φ_{B_1} is a geodesic path connecting I and $A^{-1/2}BA^{-1/2}$. As $L_{A^{1/2}}$ is an isometry,

$$\gamma(s) = \frac{L_{A^{1/2}}(\varphi_{B_1}(s))}{\mathrm{tr}(L_{A^{1/2}}(\varphi_{B_1}(s)))} \quad \text{for } 0 \leq s \leq d_H(I, B_1) = d_H(A, B)$$

is a geodesic path connecting A and B. □

The mid-point $A^{1/2}(A^{-1/2}BA^{-1/2})^{1/2}A^{1/2}$ of the geodesic φ is called the *geometric mean* of A and B.

We conclude this section with a short discussion on geodesics in Thompson's metric spaces, which are slightly more subtle.

Theorem 2.6.9 *Let $K \subseteq V$ be a solid closed cone. Suppose that $x, y \in \mathrm{int}(K)$, with $x \neq \lambda y$ for all $\lambda > 0$, and write $\alpha = m(y/x)$ and $\beta = M(y/x)$. Then*

$$\gamma(t) = \frac{\beta^t - \alpha^t}{\beta - \alpha} y + \frac{\beta\alpha^t - \alpha\beta^t}{\beta - \alpha} x,$$

where $0 \leq t \leq 1$ is a geodesic connecting x and y in $(\mathrm{int}(K), d_T)$. Moreover, if $x = \lambda y$, then $\gamma(t) = \lambda^t x$ is a geodesic connecting x and y.

Proof First note that if $x, y \in \mathrm{int}(K)$ are such that $x \neq \lambda y$ for all $\lambda > 0$, then $\alpha < \beta$, so that $\gamma(t)$ is well defined. To show that $\gamma(t) \in \mathrm{int}(K)$ for all $0 < t < 1$, we note that $0 \ll \gamma(t)$ is equivalent to

$$-(\beta\alpha^t - \alpha\beta^t)x \ll (\beta^t - \alpha^t)y.$$

Thus, $\gamma(t) \in \mathrm{int}(K)$ for all $0 < t < 1$ if

$$-\frac{\beta\alpha^t - \alpha\beta^t}{\beta^t - \alpha^t} < \alpha$$

for all $0 < t < 1$. Note that the inequality holds, since $0 < \alpha < \beta$.

Using the definition of α and β we deduce that

$$\gamma(t) \geq \frac{\beta^t - \alpha^t}{\beta - \alpha}(\alpha x) + \frac{\beta\alpha^t - \alpha\beta^t}{\beta - \alpha} x \geq \alpha^t x$$

and

$$\gamma(t) \leq \frac{\beta^t - \alpha^t}{\beta - \alpha}(\beta x) + \frac{\beta\alpha^t - \alpha\beta^t}{\beta - \alpha} x \leq \beta^t x,$$

which implies that

$$d_T(x, \gamma(t)) \leq \log\left(\max\{\beta^t, 1/\alpha^t\}\right) = t d_T(x, y).$$

In the same way we can use the inequalities $\beta^{-1}y \leq x \leq \alpha^{-1}y$ to get that

$$d_T(\gamma(t), y) \leq \log\left(\max\{\beta^{1-t}, 1/\alpha^{1-t}\}\right) = (1-t)d_T(x, y).$$

It now follows from the triangle inequality that

$$d_T(x, \gamma(t)) + d_T(\gamma(t), y) = d_T(x, y)$$

and $d_T(x, \gamma(t)) = t d_T(x, y)$ for all $0 \leq t \leq 1$. The map ψ given by $\psi(t) = d_T(x, \gamma(t))$ is a strictly increasing map from $[0, 1]$ onto $[0, d_T(x, y)]$. It is easy to verify that $\varphi(s) = \gamma(\psi^{-1}(s))$ is a geodesic path connecting x and y, and hence the first assertion holds. We leave the case $x = \lambda y$ to the reader. □

2.7 Topical maps and the sup-norm

Recall that a map $f : \mathbb{R}^n \to \mathbb{R}^n$ is topical if f is order-preserving with respect to $\mathbb{R}^n_+$, and $f(x + \lambda \mathbb{1}) = f(x) + \lambda \mathbb{1}$ for all $x \in \mathbb{R}^n$ and $\lambda \in \mathbb{R}$. Topical maps are closely related to sup-norm non-expansive maps as the following result by Crandall and Tartar [54] shows.

Proposition 2.7.1 *Suppose that $X \subseteq \mathbb{R}^n$ is such that $x + \lambda \mathbb{1} \in X$ for all $\lambda \in \mathbb{R}$ and $x \in X$. A map $f : X \to \mathbb{R}^n$ is topical if and only if*

$$\mathsf{t}(f(x) - f(y)) \leq \mathsf{t}(x - y) \quad \textit{for all } x, y \in X.$$

Proof Suppose that $f : X \to \mathbb{R}^n$ is topical. For $x, y \in X$ we have that $x - y \leq \mathsf{t}(x - y)\mathbb{1}$, so that $f(x) - f(y) \leq \mathsf{t}(x - y)\mathbb{1}$, as f is order-preserving and additively homogeneous. This implies that $\mathsf{t}(f(x) - f(y)) \leq \mathsf{t}(x - y)$.

Now suppose that $\mathsf{t}(f(x) - f(y)) \leq \mathsf{t}(x - y)$ for all $x, y \in X$. If $x \leq y$ in X, then $\mathsf{t}(x - y) \leq 0$, so that $\mathsf{t}(f(x) - f(y)) \leq 0$, and hence $f(x) \leq f(y) + \mathsf{t}(f(x) - f(y))\mathbb{1} \leq f(y)$. To prove that f is additively homogeneous we put $y = x + \lambda \mathbb{1}$. Then $\mathsf{t}(x - y) = -\lambda$ and $\mathsf{t}(y - x) = \lambda$. This implies that $\mathsf{t}(f(x) - f(y)) \leq -\lambda$ and $\mathsf{t}(f(y) - f(x)) \leq \lambda$. Thus, $\lambda \mathbb{1} \leq f(y) - f(x) \leq \lambda \mathbb{1}$, so that $f(y) = f(x) + \lambda \mathbb{1}$. □

Although t is not a norm, we say that $f : X \to \mathbb{R}^n$, with $X \subseteq \mathbb{R}^n$, is *top non-expansive* if

$$\mathsf{t}(f(x) - f(y)) \leq \mathsf{t}(x - y) \quad \text{for all } x, y \in X.$$

The map f is called a *top isometry* if equality holds for all $x, y \in X$. Notice that, as $\|x\|_\infty = \mathsf{t}(x) \vee \mathsf{t}(-x)$ for all $x \in \mathbb{R}^n$, Proposition 2.7.1 implies that every topical map $f : X \to \mathbb{R}^n$ is sup-norm non-expansive.

One can also obtain this result by using the log-exp transform. Recall that if $f : \mathrm{int}(\mathbb{R}^n_+) \to \mathrm{int}(\mathbb{R}^n_+)$ is an order-preserving subhomogeneous map, then the log-exp transform $g : \mathbb{R}^n \to \mathbb{R}^n$ of f, defined by the relation $g \circ L = L \circ f$, is a sub-topical map. This allows us to translate results from the multiplicative homogeneous setting to the additively homogeneous setting and vice versa.

For instance, Lemma 2.1.7 has the following counterpart in the additive setting by Proposition 2.2.1.

Lemma 2.7.2 *If $f : \mathbb{R}^n \to \mathbb{R}^n$ preserves the ordering induced by $\mathbb{R}^n_+$, then f is additively subhomogeneous if and only if f is sup-norm non-expansive.*

2.8 Integral-preserving maps and the ℓ_1-norm

It was also noted by Crandall and Tartar [54] that there exists a relation between order-preserving and integral-preserving maps, and maps that are non-expansive under the ℓ_1-norm. Recall that the *ℓ_1-norm* on $\mathbb{R}^n$ is given by

$$\|x\|_1 = \sum_i |x_i| \quad \text{for } x \in \mathbb{R}^n.$$

In fact, we have the following result.

Proposition 2.8.1 *Let $X \subseteq \mathbb{R}^n$ be such that $x \vee y \in X$ for all $x, y \in X$. If $f : X \to \mathbb{R}^n$ is integral-preserving, then f is order-preserving with respect to $\mathbb{R}^n_+$ if and only if f is non-expansive under the ℓ_1-norm.*

Proof Let $f : X \to \mathbb{R}^n$ be an integral-preserving and order-preserving map. Suppose that $x, y \in X$ and remark that $x \vee y \geq y$. As f is order-preserving and $x \vee y \in X$, it follows that $f(x \vee y) \geq f(y)$, so that $f(x \vee y) - f(y) \geq 0$. Similarly, $x \vee y \geq x$ yields $f(x \vee y) \geq f(x)$. This implies that $f(x \vee y) - f(y) \geq (f(x) - f(y)) \vee 0$. Since f is integral-preserving, it follows that

$$\begin{aligned}\|(f(x) - f(y)) \vee 0\|_1 &= \sum_{i=1}^n (f(x) - f(y)) \vee 0)_i \\ &\leq \sum_{i=1}^n f(x \vee y)_i - f(y)_i \\ &= \sum_{i=1}^n (x \vee y)_i - y_i \\ &= \|(x - y) \vee 0\|_1.\end{aligned}$$

From this inequality we deduce that

$$\begin{aligned}\|f(x) - f(y)\|_1 &= \|(f(x) - f(y)) \vee 0\|_1 + \|(f(y) - f(x)) \vee 0\|_1 \\ &\leq \|(x - y) \vee 0\|_1 + \|(y - x) \vee 0\|_1 \\ &= \|x - y\|_1,\end{aligned}$$

and hence f is non-expansive with respect to the ℓ_1-norm.

To show the opposite implication assume that $f : X \to \mathbb{R}^n$ is ℓ_1-norm non-expansive and integral-preserving. Note that for each $x \in \mathbb{R}^n$,

$$2\|x \vee 0\|_1 = \|x\|_1 + \sum_{i=1}^{n} x_i.$$

Now let $x, y \in X$ be such that $x \leq y$. Clearly,

$$\begin{aligned} 2\|(f(x) - f(y)) \vee 0\|_1 &= \|f(x) - f(y)\|_1 + \sum_{i=1}^{n} f(x)_i - f(y)_i \\ &\leq \|x - y\|_1 + \sum_{i=1}^{n} x_i - y_i = 0, \end{aligned}$$

and hence $f(x) \leq f(y)$, which completes the proof. □

In particular it follows from Proposition 2.8.1 that the sand-shift maps discussed in Section 1.6 are ℓ_1-norm non-expansive.

3

Dynamics of non-expansive maps

In the previous chapter we saw that various classes of order-preserving maps are non-expansive. Non-expansive maps are arguably the most fundamental maps on metric spaces after isometries, and have some extraordinary properties. In this chapter we collect a number of general results for non-expansive maps and their iterative behavior, which are relevant to nonlinear Perron–Frobenius theory. Among other things we discuss ω-limit sets and fixed points of non-expansive maps, the iterative behavior of fixed-point free non-expansive maps, and non-expansive retractions.

3.1 Basic properties of non-expansive maps

We begin by recalling some basic concepts from dynamical systems theory. Let X be a topological space. A point $x \in X$ is called a *periodic point* of $f : X \to X$ if there exists an integer $p \geq 1$ such that $f^p(x) = x$. The minimal such $p \geq 1$ is called the *period* of x under f. If $f(x) = x$, we call x a *fixed point* of f. We denote by Fix_f the set of all fixed points of f. The *orbit* of x under f is given by $\mathcal{O}(x; f) = \{f^k(x) : k = 0, 1, 2, \ldots\}$. If f is clear from the context we simply write $\mathcal{O}(x)$. If x is a periodic point, we say that $\mathcal{O}(x)$ is a *periodic orbit*. If $f : X \to X$ and there exists an integer $p \geq 1$ such that $\lim_{k\to\infty} f^{kp}(x)$ exists for all $x \in X$, then the smallest such p is called the *period of* f.

To understand the long-term behavior of the iterates of a map one has to analyze the structure of its ω-limit sets. For $x \in X$ and a continuous map $f : X \to X$, the ω-*limit set*, denoted $\omega(x; f)$ or simply $\omega(x)$, is defined by

$$\omega(x) = \bigcap_{k\geq 0} \mathrm{cl}\big(\{f^m(x) : m \geq k\}\big),$$

where cl(S) denotes the closure of S in X. It is easy to verify that each ω-limit set is a (possibly empty) closed subset of X and $f(\omega(x)) \subseteq \omega(x)$. We shall mainly work in metric spaces, in which case

$$\omega(x) = \{y \in X : f^{k_i}(x) \to y \text{ for some sequence } (k_i)_i \text{ with } k_i \to \infty\}.$$

The *attractor* of f is defined by

$$\Omega_f = \bigcup_{x \in X} \omega(x).$$

The following basic result from point-set topology will be useful to further analyze the ω-limit sets.

Proposition 3.1.1 *Let X be a topological Hausdorff space and $(A_k)_k$ be a decreasing sequence of non-empty compact subsets of X, with intersection $A_\infty = \bigcap_k A_k$.*

(i) If U is an open subset of X with $A_\infty \subseteq U$, then $A_k \subseteq U$ for all k sufficiently large.
(ii) If $f : X \to X$ is a continuous map, then

$$\bigcap_{k \geq 1} f(A_k) = f(A_\infty).$$

Proof Recall from point-set topology that any decreasing sequence of non-empty compact sets in a topological Hausdorff space is non-empty and compact. By assumption $Y = X \setminus U$ is closed. For $k \geq 1$ let $C_k = A_k \cap Y$. So, $(C_k)_k$ is a decreasing sequence of compact subsets of X. To prove (i) we first remark that if C_m is empty for some m, then $A_m \subseteq U$ and hence $A_k \subseteq U$ for all $k \geq m$, as $(A_k)_k$ is decreasing. On the other hand, if C_k is non-empty for all k, then $C_\infty = \bigcap_k C_k$ is a non-empty subset of $A_\infty \cap Y$, which violates the assumption that $A_\infty \subseteq U$. Thus, $A_k \subseteq U$ for all k sufficiently large.

To prove (ii) let $B_k = f(A_k)$ for $k \geq 1$. So, $(B_k)_k$ is a decreasing sequence of non-empty compact subsets of X, as f is continuous. Let $B_\infty = \bigcap_k B_k$, which is also non-empty and compact. We now need to show that $B_\infty = f(A_\infty)$. Obviously, $f(A_\infty) \subseteq f(A_k) = B_k$ for all k, and hence $f(A_\infty) \subseteq B_\infty$. Suppose, by way of contradiction, that $y \in B_\infty \setminus f(A_\infty)$. Because $f(A_\infty)$ is compact and X is Hausdorff, there exists an open neighborhood W of $f(A_\infty)$ such that $y \in B_\infty \setminus W$. So, $f^{-1}(W)$ is an open neighborhood of A_∞. By part (i) there exists $m \geq 1$ such that $A_k \subseteq f^{-1}(W)$ for all $k \geq m$. But this implies that $B_k = f(A_k) \subseteq W$ for all $k \geq m$, which is impossible as $y \in B_k \setminus W$ for all k. □

From Proposition 3.1.1 it is easy to derive the following property of the ω-limit sets.

Lemma 3.1.2 *Let X be a topological Hausdorff space. If $f : X \to X$ is a continuous map and $\mathcal{O}(x)$ has a compact closure, then $\omega(x)$ is a non-empty compact set and $f(\omega(x)) = \omega(x)$.*

Proof For $k \geq 1$ let $A_k = \mathrm{cl}(\{f^m(x) : m \geq k\})$ and note that $(A_k)_k$ is a decreasing sequence of non-empty compact sets, as $\mathrm{cl}(\mathcal{O}(x))$ is compact. Also remark that $A_\infty = \omega(x)$. It follows from Proposition 3.1.1 that

$$f(\omega(x)) = f(A_\infty) = \cap_k f(A_k) = \cap_k A_{k+1} = \omega(x).$$

□

The following lemma shows that every point with a pre-compact orbit and a finite ω-limit set converges to a periodic point.

Lemma 3.1.3 *Let X be a topological Hausdorff space. If $f : X \to X$ is a continuous map and $x \in X$ is such that $\mathcal{O}(x)$ has a compact closure and $|\omega(x)| = p$, then there exists a periodic point $\xi \in X$ with period p such that $\lim_{k\to\infty} f^{kp}(x) = \xi$ and $\omega(x) = \mathcal{O}(\xi)$.*

Proof As f is continuous and $\mathcal{O}(x)$ has a compact closure, $f(\omega(x)) = \omega(x)$ by Lemma 3.1.2. This implies that each $y \in \omega(x)$ is a periodic point of f, since $\omega(x)$ is finite. As X is Hausdorff, we know that there exist pairwise disjoint open neighborhoods U_y for $y \in \omega(x)$. Furthermore, there exist open neighborhoods $W_y \subseteq U_y$, for $y \in \omega(x)$, such that for each $u \in W_y$ we have that $f^q(u) \in U_y$, where q is the period of y under f.

Now let $\mathrm{cl}(\mathcal{O}(x))$ be the closure of $\mathcal{O}(x)$. We claim that there exists an integer m such that for each $k \geq m$ we have that $f^k(x) \in W_y$ for some $y \in \omega(x)$. To show this let $A_k = \mathrm{cl}(\{f^j(x) : j \geq k\})$ for $k \geq 0$ and recall that $A_\infty = \cap_k A_k = \omega(x)$. It now follows from Proposition 3.1.1 that there exists m such that $A_k \subseteq \cup_{y\in\omega(x)} W_y$ for all $k \geq m$.

Suppose that $f^m(x) \in W_z$ and z has period r under f. By definition of W_z we know that $f^{m+r}(x) \in U_z$. As the open sets U_y are pairwise disjoint for $y \in \omega(x)$, we find that $f^{m+r}(x) \in W_z$. Repeating this argument gives $f^{m+kr}(x) \in W_z$ for all $k \geq 1$.

Now let $V_z \subseteq W_z$ be any open neighborhood of z. By Proposition 3.1.1 there exists $m^* \geq 1$ such that

$$A_k \subseteq \Big(\bigcup_{y\in\omega(x),\, y\neq z} W_y \Big) \cup V_z \quad \text{for all } k \geq m^*.$$

It follows that $f^{m+kr}(x) \in V_z$ for all $k \geq m^*$. So, $\lim_{k\to\infty} f^{m+kr}(x) = z$, and hence $(f^{kr}(x))_k$ converges to $f^{(r-1)m}(z)$, as f is continuous. Thus, if we take $\xi = f^{(r-1)m}(z)$, then $\lim_{k\to\infty} f^{kr}(x) = \xi$ and $\mathcal{O}(\xi) = \omega(x)$. This also implies that $r = p$. □

If, in addition, to the assumptions in Lemma 3.1.2, f is non-expansive, then the restriction of f to $\omega(x)$ is an isometry. This is a consequence of the following result by Freudenthal and Hurewicz [68].

Lemma 3.1.4 *If C is a compact set in a metric space (X, d) and $f : C \to C$ is a surjective non-expansive map, then f is an isometry.*

Proof Let $x, y \in C$ and put $x^1 = f(x)$ and $y^1 = f(y)$. For each $k \geq 0$ select x^{-k} and y^{-k} in C such that $f(x^{-k}) = x^{-k+1}$ and $f(y^{-k}) = y^{-k+1}$. As f is non-expansive,

$$d(x^{-k}, x^{-m}) \geq d(x^{-(k-1)}, x^{-(m-1)}) \tag{3.1}$$

and

$$d(y^{-k}, y^{-m}) \geq d(y^{-(k-1)}, y^{-(m-1)}) \tag{3.2}$$

for all $k, m \geq 0$. Moreover,

$$d(x^{-k}, y^{-k}) \geq d(x^{-(k-1)}, y^{-(k-1)}) \tag{3.3}$$

for all $k \geq 0$. The sequence $(x^{-k})_k$ has a convergent subsequence $(x^{-k_i})_i$, since C is compact. Likewise $(y^{-k_i})_i$ has a convergent subsequence $(y^{-r_j})_j$. Now let $\varepsilon > 0$. As $(x^{-r_j})_j$ and $(y^{-r_j})_j$ are convergent sequences, there exists $N \geq 1$ such that for each $r_p > r_q \geq N$ we have that

$$d(x^{-r_p}, x^{-r_q}) \leq \varepsilon \quad \text{and} \quad d(y^{-r_p}, y^{-r_q}) \leq \varepsilon.$$

Pick $r_p > r_q \geq N$ and put $r = r_p - r_q - 1$. Remark that, by (3.1) and (3.2), $d(x^{-r}, x^1) \leq \varepsilon$ and $d(y^{-r}, y^1) \leq \varepsilon$. Therefore $d(x^{-r}, y^{-r}) \leq 2\varepsilon + d(x^1, y^1)$ and hence it follows from (3.3) that

$$d(x^1, y^1) \leq d(x^0, y^0) \leq 2\varepsilon + d(x^1, y^1).$$

This implies that $d(f(x), f(y)) = d(x, y)$, and we are done. □

A combination of Lemmas 3.1.2 and 3.1.4 immediately yields the following result.

Corollary 3.1.5 *Let (X, d) be a metric space. If $f : X \to X$ is a non-expansive map and $\mathcal{O}(x)$ has a compact closure, then f restricted to $\omega(x)$ is an isometry.*

The following property of ω-limit sets of non-expansive maps is due to Dafermos and Slemrod [55].

Lemma 3.1.6 *Let (X, d) be a metric space. If $f : X \to X$ is a non-expansive map and $x \in X$, then for each $y \in \omega(x)$ we have that $\omega(x) = \omega(y)$.*

Proof To see that $\omega(y) \subseteq \omega(x)$ we let $z \in \omega(y)$. By definition there exist sequences $(k_i)_i$ and $(n_i)_i$ with $k_i, n_i \to \infty$ such that $f^{k_i}(x) \to y$ and $f^{n_i}(y) \to z$. As f is non-expansive, we have that

$$\begin{aligned} d(f^{n_i+k_i}(x), z) &\leq d(f^{n_i+k_i}(x), f^{n_i}(y)) + d(f^{n_i}(y), z) \\ &\leq d(f^{k_i}(x), y) + d(f^{n_i}(y), z) \end{aligned}$$

and hence $f^{n_i+k_i}(x) \to z$, as $i \to \infty$, so that $z \in \omega(x)$. To prove the opposite inclusion we let $z \in \omega(x)$. In that case there exist sequences $(k_i)_i$ and $(n_i)_i$ with $k_i, n_i \to \infty$ such that $f^{k_i}(x) \to y$ and $f^{n_i+k_i}(x) \to z$. As

$$\begin{aligned} d(f^{n_i}(y), z) &\leq d(f^{n_i}(y), f^{n_i+k_i}(x)) + d(f^{n_i+k_i}(x), z) \\ &\leq d(y, f^{k_i}(x)) + d(f^{n_i+k_i}(x), z), \end{aligned}$$

we find that $f^{n_i}(y) \to z$ as $i \to \infty$, and hence $z \in \omega(y)$. □

A metric space is said to be *proper* if every closed ball in X is compact. The following result by Całka [48] shows that if a non-expansive map f on a proper metric space has an orbit that has a bounded subsequence, then every orbit of f is bounded.

Theorem 3.1.7 (Całka) *Let (X, d) be a proper metric space. If $f : X \to X$ is non-expansive and there exists $z \in X$ such that $(f^k(z))_k$ has a bounded subsequence, then $\mathcal{O}(x)$ is bounded for each $x \in X$.*

Proof As f is non-expansive, it suffices to show that $\mathcal{O}(y)$ is bounded for some $y \in X$. For $k \geq 1$ write $z^k = f^k(z)$, so $(z^k)_k$ has a bounded subsequence $(z^{k_i})_i$. There exist $\delta > 0$ such that $(z^{k_i})_i \subseteq B_\delta(z) = \{x \in X : d(x, z) \leq \delta\}$. As (X, d) is proper, it follows that $(z^{k_i})_i$ has a convergent subsequence, with limit $y \in X$. We remark that $y \in \omega(z)$, so that $y \in \omega(y)$ by Lemma 3.1.6. Let $(m_i)_i$ be a sequence of integers such that $m_i \to \infty$ and $\lim_{i\to\infty} f^{m_i}(y) = y$. For each $k \geq 1$ write $y^k = f^k(y)$. Since f is non-expansive, we have for each $r, s \geq 1$ that

$$d(y^{m_i+r}, y^{m_i+s}) \leq d(y^{r+1}, y^{s+1}) \leq d(y^r, y^s)$$

and $\lim_{i\to\infty} d(y^{m_i+r}, y^{m_i+s}) = d(y^r, y^s)$. Therefore

$$d(y^{r+1}, y^{s+1}) = d(y^r, y^s) \quad \text{for all } r, s \geq 1. \tag{3.4}$$

Hence f is an isometry on $\mathcal{O}(y)$. We will show that $\mathcal{O}(y)$ is bounded.

Clearly, if $y^{m_i} = y$ for some i, we immediately obtain the theorem. So, we can assume that y^{m_i} is unequal to y for all i. Let $\varepsilon > 0$. As $y^{m_i} \to y$ as $i \to \infty$, we know that

$$|\{k \geq 1 : d(y^k, y) < \varepsilon\}| = \infty. \tag{3.5}$$

Put $Y = B_\varepsilon(y) \cap \mathcal{O}(y)$, where $B_\varepsilon(y)$ denotes the closed ball with radius ε around y. There exist $x^1, \ldots, x^q \in X$ such that $Y \subseteq \cup_{i=1}^q B_{\varepsilon/4}(x^i)$ and $B_{\varepsilon/4}(x^i) \cap \mathcal{O}(y) \neq \emptyset$ for all $1 \leq i \leq q$, as (X, d) is proper. For each i, pick $y^{n_i} \in Y$ such that $y^{n_i} \in B_{\varepsilon/4}(x^i)$. Clearly

$$Y \subseteq \bigcup_{i=1}^{q} B_{\varepsilon/2}(y^{n_i}). \tag{3.6}$$

By (3.5) there exists $1 \leq p \leq q$ such that $|B_{\varepsilon/2}(y^{n_p}) \cap \mathcal{O}(y)| = \infty$. Again, as $y^{m_i} \to y$ as $i \to \infty$, we know that

$$|B_{\varepsilon/2}(y) \cap \mathcal{O}(y)| = \infty. \tag{3.7}$$

Therefore there exists $n_0 \geq \max\{1, n_1, \ldots, n_q\}$ and $d(y^{n_0}, y) \leq \varepsilon/2$.

Put $B = \cup_{i=1}^{n_0} B_\varepsilon(y^i)$. To prove that $\mathcal{O}(y)$ is bounded, we show that $\mathcal{O}(y) \subseteq B$. Let $m \geq 1$ be fixed. If $m \leq n_0$, then clearly $y^m \in B$. So assume that $m > n_0$. By (3.7) there exists $m_0 \geq m$ such that $d(y^{m_0}, y) \leq \varepsilon/2$. As $d(y^{n_0}, y) \leq \varepsilon/2$, we have that $d(y^{m_0}, y^{n_0}) \leq \varepsilon$, and hence (3.4) gives

$$d(y^{m_0 - j}, y^{n_0 - j}) = d(y^{m_0}, y^{n_0}) \leq \varepsilon \quad \text{for } 0 \leq j \leq n_0.$$

From this we deduce that $y^j \in B$ for $m_0 - n_0 \leq j \leq m_0$ and $y^{m_0 - n_0} \in B_\varepsilon(y) \cap \mathcal{O}(y) = Y$. By (3.6) there exists $1 \leq p \leq q$ such that $d(y^{m_0 - n_0}, y^{n_p}) \leq \varepsilon/2$, so that

$$d(y^{m_0 - n_0 - j}, y^{n_p - j}) \leq \varepsilon/2 \quad \text{for } 0 \leq j \leq \min\{n_p, m_0 - n_0\}, \tag{3.8}$$

by (3.4). There are two cases: $n_p \geq m_0 - n_0$ and $n_p < m_0 - n_0$. If $n_p \geq m_0 - n_0$, then $y^j \in B$ for $0 \leq j \leq m_0$. On the other hand, if $n_p < m_0 - n_0$, then we set $m_1 = m_0 - n_0 - n_p$. By definition, $n_0 \geq 1$, so that $0 < m_1 < m_0$. Remark that $y^j \in B$ for $m_1 \leq j \leq m_0$ and $d(y^{m_1}, y) \leq \varepsilon/2$ by (3.8). Thus, we can replace m_0 by m_1 and repeat the argument until we get that $y^j \in B$ for each $0 \leq j \leq m_0$. □

Edelstein [60] has observed that Całka's Theorem 3.1.7 may fail for general metric spaces. Before discussing Edelstein's elegant example, we make a few general observations. Suppose that X is a reflexive Banach space and $T : X \to X$ is a continuous affine linear map; so,

$$T(x) = L(x) + b \quad \text{for } x \in X,$$

where $b \in X$ is fixed and $L : X \to X$ is a bounded linear map. If $\|L\| = 1$, or more generally $\|L^k\| \le M < \infty$ for all $k \ge 1$, then every orbit of T is bounded if and only if T has a fixed point. Indeed, it is easy to see that if T has a fixed point, then every orbit of T is bounded. On the other hand, if T has a bounded orbit $\mathcal{O}(z)$, then we can consider the closure of the convex hull of $\mathcal{O}(z)$, which we shall denote by C. Remark that C is a bounded closed convex subset of X, which is compact in the weak topology on X, as X is reflexive. Also note that if $x = \sum_{i=1}^m \lambda_i T^{k_i}(z)$, where $\sum_{i=1}^m \lambda_i = 1$ and $\lambda_i \ge 0$ for all i, then

$$T(x) = \sum_{i=1}^{m} \lambda_i L(T^{k_i}(z)) + b = \sum_{i=1}^{m} \lambda_i (L(T^{k_i}(z)) + b) = \sum_{i=1}^{m} \lambda_i T^{k_i+1}(z) \in C.$$

This implies that $T(C) \subseteq C$. Furthermore, T is continuous in the weak topology on C. So, we can apply the Schauder–Tychonoff theorem [58] to conclude that T has a fixed point in C.

Edelstein's example is an affine linear map T on the complex sequence space $\ell_2 = \{(x_1, x_2, x_3, \ldots) \in \mathbb{C}^{\mathbb{N}} : \sum_i |x_i|^2 < \infty\}$ of the form

$$T(x)_k = a_k(x_k - 1) + 1 \quad \text{for all } k \ge 1 \text{ and } x \in \ell_2,$$

where $|a_k| = 1$ and $a_k \neq 1$ for infinitely many k. Furthermore, we assume that $\sum_k |1 - a_k|^2 < \infty$, so that T maps ℓ_2 into itself. It is easy to see that T is an isometry, which has no fixed point. It follows from the previous remarks that every orbit of T is unbounded. In Edelstein's example the idea is to choose the coefficients $a_k \in \mathbb{C}$ such that $(T^k(x))_k$ has a subsequence which converges to 0. For $k \ge 1$ let $a_k = e^{2\pi \mathrm{i}/k!}$. In that case

$$\sum_k |1 - a_k|^2 = \sum_k |1 - e^{2\pi \mathrm{i}/k!}|^2 = 4 \sum_k \sin^2(\pi/k!) < \infty.$$

Also notice that

$$T^m(x)_k = e^{2\pi \mathrm{i} m/k!}(x_k - 1) + 1$$

for all $k, m \ge 1$. Thus, if we consider the subsequence $(T^{m!}(0))_m$ we find that

$$\|T^{m!}(0)\|^2 = \sum_{j \ge 1} |-e^{2\pi \mathrm{i} m!/(m+j)!} + 1|^2 = 4 \sum_{j \ge 1} \sin^2\left(\frac{\pi}{(m+1)\cdots(m+j)}\right) \to 0$$

as $m \to \infty$.

Całka's Theorem 3.1.7 has the following consequence.

Corollary 3.1.8 *If $f : X \to X$ is a non-expansive map on a proper metric space, then the following assertions are equivalent:*

(i) $\Omega_f = \emptyset$.
(ii) $\lim_{k\to\infty} d(f^k(x), x) = \infty$ *for all* $x \in X$.
(iii) *For each compact set* $C \subseteq X$ *and* $x \in C$*, there exists* $m \in \mathbb{N}$ *such that* $f^k(x) \notin C$ *for all* $k \geq m$.

Proof Obviously the second and third assertions are equivalent. Suppose that $\Omega_f \neq \emptyset$. Then there exists $z \in X$ such that $(f^k(z))_k$ has a convergent subsequence. By Theorem 3.1.7 the orbit of each $x \in X$ is bounded. Thus, for each $x \in X$ there exists $\delta_x > 0$ such that $f^k(x)$ is in the closed ball with radius δ_x around x for all $k \geq 1$, which shows that (iii) implies (i). To prove the opposite implication, we remark that if there exists $z \in X$ and $C \subseteq X$ compact with $z \in C$ and $f^{k_i}(z) \in C$ for some subsequence (k_i) with $k_i \to \infty$, then $(f^{k_i}(z))_i$ has a convergent subsequence. Therefore $\omega(z)$ is non-empty, and we are done. □

Limit sets of non-expansive maps on proper metric spaces have a transitive abelian group of isometries. Recall that a metric space (X, d) has a *transitive group of isometries* if there exists a group Γ of isometries $g : X \to X$ that map X onto itself such that Γ acts transitively on X, that is to say, for each $x, y \in X$ there exists $g \in \Gamma$ with $g(x) = y$.

Theorem 3.1.9 *If* $f : (X, d) \to (X, d)$ *is a non-expansive map on a proper metric space and* $\Omega_f \neq \emptyset$*, then* $\omega(x)$ *has a transitive abelian group of isometries for each* $x \in X$.

Before proving theorem 3.1.9 we give, for the reader's convenience, a variant of the classical Arzelà–Ascoli theorem; cf. [42], [154].

Lemma 3.1.10 *If* (X, d) *is a proper metric space and* $f : X \to X$ *is a non-expansive map with* $\Omega_f \neq \emptyset$*, then every subsequence of* $(f^k)_k$ *has a convergent subsequence that converges uniformly on compact subsets of* X.

Proof First note that any proper metric space is separable. Indeed, if we fix $y_0 \in X$, then for each integer $n \geq 1$ the closed ball $B_n(y_0)$ with radius n and center y_0 is compact. Thus, the open covering of $B_n(y_0)$ with open balls with radius $1/n$ around each point in $B_n(y_0)$ has a finite subcover. Let Y_n be centers of the open balls in the subcover. Clearly $Y = \cup_n Y_n$ is a countable dense subset of (X, d)

Write $Y = \{y^k : k = 1, 2, 3, \ldots\}$. As $f : X \to X$ is non-expansive and $\Omega_f \neq \emptyset$, there exists $z \in X$ such that $\omega(z) \neq \emptyset$. Hence $(f^k(z))_k$ has a bounded subsequence, so that $(f^k(x))_k$ is bounded for all $x \in X$ by Theorem 3.1.7. As X is proper, $(f^k(y^1))_k$ has a convergent subsequence, which we denote by $(f^{k_{i,1}}(y^1))_i$. Similarly $(f^{k_{i,1}}(y^2))_i$ has a convergent subsequence $(f^{k_{i,2}}(y^2))_i$.

By repeating this argument we obtain for each $m \geq 1$ a subsequence $(f^{k_{i,m}})_i$ of $(f^k)_k$ such that $(f^{k_{i,m}}(y^l))_i$ converges for all $l \leq m$. Now consider the diagonal sequence $(f^{k_{i,i}})_i$. This subsequence has the property that $(f^{k_{i,i}}(y^m))_i$ converges for each $m \geq 1$. Given $x \in X$ and $\varepsilon > 0$, there exists $y \in Y$ such that $d(x, y) < \varepsilon$. As $f : X \to X$ is non-expansive, we get that

$$\begin{aligned} d(f^{k_{i,i}}(x), f^{k_{j,j}}(x)) &\leq d(f^{k_{i,i}}(x), f^{k_{i,i}}(y)) + d(f^{k_{i,i}}(y), f^{k_{j,j}}(y)) \\ &\quad + d(f^{k_{j,j}}(y), f^{k_{j,j}}(x)) \\ &\leq 2\varepsilon + d(f^{k_{i,i}}(y), f^{k_{j,j}}(y)). \end{aligned}$$

Clearly the right-hand side converges to 2ε as $i, j \to \infty$, and hence $(f^{k_{i,i}}(x))_i$ is a Cauchy sequence in X. This implies that $(f^{k_{i,i}}(x))_i$ converges, since X is proper.

To complete the proof we need to verify that the convergence is uniform on compact subsets of X. Let $C \subseteq (X, d)$ be compact and let $\{y^1, \ldots, y^r\} \subseteq Y$ be such that $C \subseteq \cup_{i=1}^r B_\varepsilon(y^i)$. Obviously, for each $x \in C$,

$$d(f^{k_{i,i}}(x), f^{k_{j,j}}(x)) \leq 2\varepsilon + d(f^{k_{i,i}}(y^s), f^{k_{j,j}}(y^s))$$

for some $1 \leq s \leq r$. Now let $I \geq 1$ be such that, for all $i, j \geq I$,

$$\max_{1 \leq s \leq r} d(f^{k_{i,i}}(y^s), f^{k_{j,j}}(y^s)) < \varepsilon.$$

This implies that $d(f^{k_{i,i}}(x), f^{k_{j,j}}(x)) \leq 3\varepsilon$ for all $i, j \geq I$ and $x \in C$. Therefore $(f^{k_{i,i}})_i$ converges uniformly on C. □

We now prove Theorem 3.1.9.

Theorem 3.1.9 Let $y, z \in \omega(x)$. By Lemma 3.1.6 there exists a subsequence $(k_i)_i$ such that $f^{k_i}(y) \to z$ as $i \to \infty$. By the previous lemma we may assume that $(f^{k_i})_i$ converges uniformly on compact subsets of X to g_{yz}. Clearly $g_{yz}(y) = z$ and $g_{yz}(\omega(x)) \subseteq \omega(x)$. Moreover g_{yz} maps $\omega(x)$ onto itself. Indeed, if $w \in \omega(x)$, then for each k_i there exists $v^i \in \omega(x)$ such that $f^{k_i}(v^i) = w$, as $f(\omega(x)) = \omega(x)$ by Lemma 3.1.2. Now let $v \in \omega(x)$ be the limit point of a convergent subsequence, $(v^{i_j})_j$, of $(v^i)_i$ in $\omega(x)$. (Recall that $\omega(x)$ is compact.) Then

$$\begin{aligned} d(g_{yz}(v), w) &= \lim_{j\to\infty} d(g_{yz}(v), f^{k_{i_j}}(v^{i_j})) \\ &\leq \lim_{j\to\infty} d(g_{yz}(v), f^{k_{i_j}}(v)) + d(f^{k_{i_j}}(v), f^{k_{i_j}}(v^{i_j})) \\ &\leq \lim_{j\to\infty} d(g_{yz}(v), f^{k_{i_j}}(v)) + d(v, v^{i_j}) \\ &= 0. \end{aligned}$$

Thus, $g_{yz}(v)=w$ and hence g_{yz} maps $\omega(x)$ onto itself. As g_{yz} is non-expansive on $\omega(x)$, we deduce from Lemma 3.1.4 that g_{yz} is an isometry on $\omega(x)$.

Now put $\Gamma = \{g_{yz} : \omega(x) \to \omega(x) \mid y, z \in \omega(x)\}$. Clearly Γ acts transitively on $\omega(x)$ and the elements of Γ commute, as the iterates of f commute. To verify that Γ is a group we remark that for each $g_{yz} \in \Gamma$ there exists a constant $d(g_{yz}) \geq 0$ such that $d(g_{yz}(u), u) = d(g_{yz})$ for all $u \in \omega(x)$. Indeed,

$$d(g_{yz}(u), u) = d(g_{uv}(g_{yz}(u)), g_{uv}(u)) = d(g_{yz}(v), v)$$

for all $u, v \in \omega(x)$. This implies that $g_{yy}(z) = z$ for each $y, z \in \omega(x)$, and therefore Γ contains a unit. We also find that $g_{yz}^{-1} = g_{zy}$ for all $y, z \in \omega(x)$, and hence Γ is a group. □

3.2 Fixed-point theorems for non-expansive maps

There are many fixed-point theorems for non-expansive maps, some of which will be useful to us. Let us begin by recalling the classical Banach contraction theorem [13].

Theorem 3.2.1 (Banach) *Let (X, d) be a complete metric space. If $f : X \to X$ is a Lipschitz contraction, then there exists a unique fixed point $x^* \in X$ of f such that* $\lim_{k\to\infty} f^k(x) = x^*$ *for all $x \in X$.*

Proof Let $c \in [0, 1)$ be the contraction factor of f. For each $1 \leq k < m$ we have that

$$\begin{aligned} d(f^k(x), f^m(x)) &\leq \sum_{j=k}^{m-1} d(f^j(x), f^{j+1}(x)) \\ &\leq \sum_{j=k}^{m-1} c^k d(x, f(x)) \\ &\leq \frac{c^k}{1-c} d(x, f(x)). \end{aligned}$$

Therefore $(f^k(x))_k$ is a Cauchy sequence, which converges to a fixed point $x^* \in X$, as X is complete. To finish the proof, we remark that f has at most one fixed point. Indeed, if $x^*, y^* \in X$ are two fixed points of f, then $d(x^*, y^*) = d(f(x^*), f(y^*)) \leq cd(x^*, y^*)$, where $0 \leq c < 1$, so that $x^* = y^*$. □

For contractions on compact metric spaces there exists the following fixed-point result.

Theorem 3.2.2 *If $f : (X, d) \to (X, d)$ is a contraction on a compact metric space, then there exists a unique $x^* \in X$ such that $f(x^*) = x^*$ and $\lim_{k\to\infty} f^k(x) = x^*$ for all $x \in X$.*

Proof Let $x \in X$ and remark that $\omega(x)$ is a non-empty compact subset of X. We claim that each $x^* \in \omega(x)$ is a fixed point. Let $x^* \in \omega(x)$ and suppose that $f^{k_i}(x) \to x^*$, as $i \to \infty$, and $k_{i+1} > k_i$ for all i. Then

$$\begin{aligned} d(x^*, f(x^*)) &\leq \lim_{i\to\infty} d(f^{k_i}(x), f^{k_i}(f(x))) \\ &\leq \lim_{i\to\infty} d(f^{k_i-1}(f(x)), f^{k_i-1}(f^2(x))) \\ &\leq d(f(x^*), f^2(x^*)). \end{aligned}$$

As f is non-expansive, we find that $d(x^*, f(x^*)) = 0$, and hence $x^* \in \omega(x)$ is a fixed point of f. Since f is a contraction, the fixed point must be unique. □

Note that Theorem 3.2.2 may fail if (X, d) is not compact (e.g., take $X = [1, \infty)$ and $f(x) = x + 1/x$).

A fixed point x^* in (X, d) of a continuous map $f : X \to X$ is said to be *locally attracting* if there exists an open neighborhood U of x such that

$$\lim_{k\to\infty} f^k(u) = x^*$$

for all $u \in U$. The following proposition shows that locally attracting fixed points of non-expansive maps on geodesic metric spaces are globally attracting.

Proposition 3.2.3 *If $f : X \to X$ is a non-expansive map on a geodesic metric space (X, d) and $x^* \in X$ is locally attracting fixed point of f, then*

$$\lim_{k\to\infty} f^k(x) = x^*$$

for all $x \in X$.

Proof By assumption there exists $\varepsilon > 0$ such that for each $u \in B_\varepsilon(x^*) = \{x \in X : d(x, x^*) \leq \varepsilon\}$ we have that $\lim_{k\to\infty} f^k(u) = x^*$. Now let $x \in X$ and denote $\delta = d(x, x^*)$. As (X, d) is geodesic, there exists a geodesic path $\gamma : [0, \delta] \to (X, d)$ connecting x^* and x. So, $\gamma(0) = x^*$, $\gamma(\delta) = x$, and $d(\gamma(s), \gamma(t)) = |s - t|$ for all $0 \leq s, t \leq \delta$.

If $\delta \leq \varepsilon$, we are done. Now if $\delta > \varepsilon$, then $\gamma(\varepsilon) \in B_\varepsilon(x^*)$, and hence there exists an integer $k_1 \geq 1$ such that

$$d(f^{k_1}(\gamma(\varepsilon)), x^*) \leq d(\gamma(\varepsilon), x^*)/2 = \varepsilon/2.$$

Since f is non-expansive and $d(x, x^*) = d(x, \gamma(\varepsilon)) + d(\gamma(\varepsilon), x^*)$, we get that

$$\begin{aligned} d(f^{k_1}(x), x^*) &\leq d(f^{k_1}(x), f^{k_1}(\gamma(\varepsilon))) + d(f^{k_1}(\gamma(\varepsilon)), x^*) \\ &\leq d(x, \gamma(\varepsilon)) + d(\gamma(\varepsilon), x^*)/2 \\ &= d(x, x^*) - \varepsilon/2. \end{aligned}$$

Now if $d(f^{k_1}(x), x^*) \leq \varepsilon$, then $f^{k_1}(x) \in B_\varepsilon(x^*)$, so that

$$\lim_{k \to \infty} d(f^k(x), x^*) = 0.$$

Otherwise we replace x by $f^{k_1}(x)$ and repeat the argument to find an integer $k_2 > k_1$ such that

$$d(f^{k_2}(x), x^*) \leq d(x, x^*) - 2\varepsilon/2.$$

Iterating this argument yields, after finitely many steps, an integer $k_m \geq 1$ such that

$$d(f^{k_m}(x), x^*) \leq d(x, x^*) - m\varepsilon/2 \leq \varepsilon.$$

Thus, $f^{k_m}(x) \in B_\varepsilon(x^*)$, and we are done. □

For many metric spaces (X, d) it is true that if $f : X \to X$ is non-expansive and f has a bounded orbit, then f has a fixed point, which in general need not be unique. In particular, this is true for finite-dimensional normed spaces and Hilbert's and Thompson's metric spaces. To see this we use Całka's Theorem 3.1.7.

Proposition 3.2.4 *Let X be a closed convex subset of a finite-dimensional normed space $(V, \|\cdot\|)$. If $f : X \to X$ is a non-expansive map with $\Omega_f \neq \emptyset$, then f has a fixed point.*

Proof Let $z \in X$ be such that $\omega(z) \neq \emptyset$. Then $(f^k(z))_k$ has a bounded subsequence and hence $\mathcal{O}(z)$ is bounded by Theorem 3.1.7. It follows from Lemma 3.1.2 that $\omega(z)$ is non-empty compact set and $f(\omega(z)) = \omega(z)$. Let $\delta = \operatorname{diam}(\omega(z)) < \infty$. Remark that if $\delta = 0$, then f has a fixed point, and we are done. If $\delta > 0$, define $B = \bigcap_{y \in \omega(z)} B_\delta(y)$, where $B_\delta(y)$ is the closed ball with radius δ and center y. Let $C = B \cap X$ and note that C is a compact convex set containing $\omega(z)$.

If $x \in C$, then $f(x) \in X$, because $f(X) \subseteq X$. Furthermore $\|x - y\| \leq \delta$ for all $y \in B$, so $\|f(x) - f(y)\| \leq \delta$ for all $y \in \omega(z)$. Using the fact that $f(\omega(z)) = \omega(z)$, we find that $\|f(x) - w\| \leq \delta$ for all $w \in \omega(z)$. Thus, $f(x) \in C$, and hence $f(C) \subseteq C$. It now follows from the Brouwer fixed-point theorem that f has a fixed point. □

The example by Edelstein [60], which was discussed after the proof of Theorem 3.1.7, shows that Proposition 3.2.4 fails for infinite-dimensional normed spaces.

The balls in Hilbert's and Thompson's metric spaces are convex subsets of V by Lemmas 2.6.1 and 2.6.2. We can therefore apply exactly the same argument as in Proposition 3.2.4 to obtain the following result.

Corollary 3.2.5 *Let $K \subseteq V$ be a solid closed cone, $\varphi \in \mathrm{int}(K^*)$, and $\Sigma^\circ = \{x \in \mathrm{int}(K) : \varphi(x) = 1\}$.*

(i) If $f : \Sigma^\circ \to \Sigma^\circ$ is non-expansive with respect to d_H and $\Omega_f \neq \emptyset$, then f has a fixed point.
(ii) If $f : \mathrm{int}(K) \to \mathrm{int}(K)$ is non-expansive with respect to d_T and $\Omega_f \neq \emptyset$, then f has a fixed point.

3.3 Horofunctions and horoballs

If $f : X \to X$ is a non-expansive map on a proper metric space (X, d) and $\Omega_f = \emptyset$, then every orbit of f is unbounded, by Corollary 3.1.8. In that case one might wonder how the orbits wander off to infinity. Such situations arise naturally in nonlinear Perron–Frobenius theory. Consider, for example, a continuous order-preserving homogeneous map $f : K \to K$, where K is a solid closed cone in V. If $f(\mathrm{int}(K)) \subseteq \mathrm{int}(K)$ and f has no eigenvector in $\mathrm{int}(K)$, then the scaled map $g : \Sigma^\circ \to \Sigma^\circ$ given by

$$g(x) = \frac{f(x)}{\varphi(f(x))} \quad \text{for } x \in \Sigma^\circ,$$

where $\Sigma^\circ = \{x \in \mathrm{int}(K) : \varphi(x) = 1\}$ and $\varphi \in \mathrm{int}(K^*)$, is non-expansive under Hilbert's metric and has no fixed point in Σ°. In that case $\Omega_g = \emptyset$ by Corollary 3.2.5, and the limit points of the orbit of each $x \in \Sigma^\circ$ under g lie in $\partial\Sigma^\circ$. It is interesting to understand the structure of the set of limit points of the orbits of g for $x \in \Sigma^\circ$. To analyze this problem it is useful to recall several fundamental concepts from metric geometry; see [12, 42, 177].

Let (X, d) be a proper metric space and denote by $C(X)$ the space of continuous real-valued functions on X equipped with the topology of uniform convergence on compact sets. Fix a point $y \in X$ and define for each $x \in X$ a function $b_y(x) : X \to \mathbb{R}$ by

$$b_y(x)(z) = d(z, x) - d(x, y) \quad \text{for } z \in X. \tag{3.9}$$

The point $y \in X$ is called the *base point* of $b_y(x)$.

Lemma 3.3.1 *The map $b_y(x) : (X, d) \to \mathbb{R}$ in (3.9) has the following properties for each $x, x', y, y', z, z' \in X$:*

(i) $|b_y(x)(z) - b_y(x)(z')| \leq d(z, z')$,
(ii) $|b_y(x)(z) - b_y(x')(z)| \leq 2d(x, x')$, *and*
(iii) $b_{y'}(x)(z) = b_y(x)(z) - b_y(x)(y')$.

Proof These properties are easy consequences of the triangle inequality. □

Note that the map $b_y : X \to C(X)$ is continuous by Lemma 3.3.1(ii). A sequence $(x_k)_k \subseteq X$ is said to *converge at infinity* if $d(x_k, y) \to \infty$ and $b_y(x_k)$ converges in $C(X)$ for some $y \in X$. Note that this definition is independent of the base point $y \in X$, because $d(x_k, y') \geq |d(x_k, y) - d(y, y')| \to \infty$ and $b_{y'}(x_k) = b_y(x_k) - b_y(x_k)(y')$ by Lemma 3.3.1(iii). Two sequences $(x_k)_k$ and $(x'_k)_k$ in X that converge at infinity are called *equivalent* if

$$\lim_{k\to\infty} b_y(x_k) = \lim_{k\to\infty} b_y(x'_k)$$

for some (and hence for all) $y \in X$. The set of equivalence classes of the set of sequences in X that converges at infinity is denoted by $X(\infty)$ and is called the *horofunction boundary*. Thus, for each $y \in X$ and $\sigma \in X(\infty)$ there is a well-defined function $b_y(\sigma) \in C(X)$, which is called the *horofunction* at σ based at y. The sub-level sets of $b_y(\sigma)$,

$$H_y(\sigma, r) = \{z \in X : b_y(\sigma)(z) \leq r\},$$

are called (closed) *horoballs* and the level sets of $b_y(\sigma)$ are called *horospheres* centered at σ.

Proposition 3.3.2 *Every unbounded sequence $(x_k)_k$ in a proper metric space (X, d) has a subsequence that converges at infinity.*

Proof As $(x_k)_k$ is unbounded, it has a subsequence $(x_{k_i})_i$ with $d(x_{k_i}, y) \to \infty$, as $i \to \infty$. Now let r_k be an increasing sequence of nonnegative numbers such that $r_k \to \infty$ and denote by $B_{r_k}(y)$ the closed ball around $y \in X$ with radius r_k. Remark that $B_{r_k}(y)$ is compact, as (X, d) is proper. It follows from Lemma 3.3.1(i) that the set $\{b_y(x_{k_i}) : B_{r_1}(y) \to \mathbb{R} \mid i \geq 1\}$ is equicontinuous and uniformly bounded, since $|b_y(x_{k_i})(z)| \leq d(y, z)$ for each i. Thus, we can apply the Arzelà–Ascoli theorem [58] to find a subsequence $(m_i^1)_i$ of $(k_i)_i$ such that $(b_y(x_{m_i^1}))_i$ converges uniformly on $B_{r_1}(y)$. By repeating this argument, we obtain a sequence $((x_{m_i^k}))_k$ of successive subsequences such that $(b_y(x_{m_i^k}))_i$ converges uniformly on $B_{r_k}(y)$. Now remark that the diagonal $(x_{m_i^i})_i$ has the property that $(b_y(x_{m_i^i}))_i$ converges uniformly on each compact subset of X, because $r_k \to \infty$. Thus, $(x_{m_i^i})_i$ converges at infinity. □

In our subsequent analysis the shape of the horoballs plays an important role. In particular the following lemma will be useful.

Lemma 3.3.3 *If X is a convex subset of an affine space and (X, d) is a proper metric space in which the balls are convex sets, then each horoball in (X, d) is convex.*

Proof Let $H_y(\sigma, r) = \{z \in X : b_y(\sigma)(z) \leq r\}$ be a horoball in (X, d) centerd at $\sigma = (x_k)_k$. Let $u, v \in H_y(\sigma, r)$ and $0 < \lambda < 1$. Put $w = \lambda u + (1 - \lambda)v$. Remark that for each $\varepsilon > 0$ there exists $N \geq 1$ such that

$$b_y(x_k)(u) = d(u, x_k) - d(x_k, y) \leq r + \varepsilon$$

and

$$b_y(x_k)(v) = d(v, x_k) - d(x_k, y) \leq r + \varepsilon$$

for all $k \geq N$. This implies that u and v are in the ball around x_k with radius $r + \varepsilon + d(x_k, y)$ for all $k \geq N$. As the balls in (X, d) are convex, w is also in the ball with radius $r + \varepsilon + d(x_k, y)$ around x_k for all $k \geq N$. This implies that $b_y(x_k)(w) = d(w, x_k) - d(x_k, y) \leq r + \varepsilon$ for all $k \geq N$. Thus,

$$b_y(\sigma)(w) = \lim_{k \to \infty} d(w, x_k) - d(x_k, y) \leq r + \varepsilon$$

for all $\varepsilon > 0$, and therefore $w \in H_y(\sigma, r)$. □

We shall now see how these notions can be used to provide information about the asymptotic behavior of the iterates of fixed-point free non-expansive maps. Let (X, d) be a proper metric space and let $f : X \to X$ be a non-expansive map with Ω_f empty. Take $x \in X$ and put $a_k = d(f^k(x), x)$ for each $k \geq 1$. We note that $(a_k)_k$ is sub-additive, as

$$a_{k+m} \leq d(f^{m+k}(x), f^m(x)) + d(f^m(x), x) \leq a_k + a_m.$$

Fekete's sub-additive lemma [64]

$$\lim_{k \to \infty} \frac{a_k}{k} = \inf_k \frac{a_k}{k}$$

exists, and is independent of $x \in X$, since

$$\begin{aligned} d(f^k(y), y) &\leq d(f^k(y), f^k(x)) + d(f^k(x), x) + d(x, y) \\ &\leq d(f^k(x), x) + 2d(x, y). \end{aligned}$$

As Ω_f is empty, it follows from Corollary 3.1.8 that $\lim_{k \to \infty} a_k = \infty$. Unbounded sequences in $\mathbb{R}$ possess the following almost obvious properties.

Lemma 3.3.4 *If $(b_k)_k \subseteq \mathbb{R}$ is such that $\sup_k b_k = \infty$, then there exists a subsequence $(b_{k_i})_i$ such that for each i we have that $b_k < b_{k_i}$ for all $k < k_i$. Moreover, if $B = \limsup_{k\to\infty} b_k/k < \infty$, then for each $\varepsilon > 0$ there exists a subsequence $(b_{k_i})_i$ such that $b_{k_i} - b_{k_i-k} \geq (B-\varepsilon)k$ for all $1 \leq k \leq k_i$.*

Proof The first assertion is obvious. To prove the second one, we remark that $(b_k - (B-\varepsilon)k)_k$ is an unbounded sequence, and hence it has a subsequence such that

$$b_{k_i} - (B-\varepsilon)k_i - b_{k_i-k} + (B-\varepsilon)(k_i - k) \geq 0 \quad \text{for all } 1 \leq k \leq k_i.$$

This implies that $b_{k_i} - b_{k_i-k} \geq (B-\varepsilon)k$ for all $1 \leq k \leq k_i$. □

The following theorem is due to Karlsson [99].

Theorem 3.3.5 *Let $f : X \to X$ be a non-expansive map on a proper metric space (X, d) and denote $A = \lim_{k\to\infty} d(f^k(x), x)/k$. If $\Omega_f = \emptyset$, then for each $x \in X$ there exists a subsequence σ of $(f^k(x))_k$ such that σ converges at infinity,*

$$b_x(\sigma)(f^k(x)) \leq -Ak \quad \text{for all } k \geq 1,$$

and

$$\lim_{k\to\infty} \frac{b_x(\sigma)(f^k(x))}{k} = -A.$$

Proof Let $x \in X$ and $a_k = d(f^k(x), x)$ for all $k \geq 1$. Let $\varepsilon > 0$. As Ω_f is empty, $(a_k)_k$ is unbounded, by Corollary 3.1.8. By Lemma 3.3.4 there exists a subsequence such that $a_{k_i} - a_{k_i-k} \geq (A-\varepsilon)k$ for all $0 \leq k \leq k_i$. It follows from Proposition 3.3.2 that we may assume, after taking a further subsequence, that $\sigma = (f^{k_i}(x))_i$ converges at infinity, as (X, d) is proper. Moreover, the following inequalities hold:

$$\begin{aligned}
-d(f^k(x), x) &\leq b_x(\sigma)(f^k(x)) \\
&= \lim_{i\to\infty} d(f^k(x), f^{k_i}(x)) - d(f^{k_i}(x), x) \\
&\leq \liminf_{i\to\infty} a_{k_i-k} - a_{k_i} \\
&\leq -(A-\varepsilon)k.
\end{aligned}$$

Thus, $b_x(\sigma)(f^k(x)) \leq -(A-\varepsilon)k$ for $k \geq 1$, as $\varepsilon > 0$ was arbitrary, $\lim_{k\to\infty} b_x(\sigma)(f^k(x))/k = -A$. □

The point of Theorem 3.3.5 is that every unbounded orbit of a non-expansive map on a proper metric space lies inside a horoball. As we shall see in the next section, this constrains the location of its limit points in the boundary at infinity.

3.4 A Denjoy–Wolff type theorem

In Klein's model of the hyperbolic plane, i.e., the open disc $D \subseteq \mathbb{R}^2$ with the cross-ratio metric, the horoballs are tangent to ∂D; see Figure 3.1. Therefore it follows from Theorem 3.3.5 that if $f : (D, \kappa) \to (D, \kappa)$ is a non-expansive map and $\Omega_f = \emptyset$, then each orbit of f converges to a single point in ∂D. In fact, it will be shown later in this section that there exists a unique $\zeta \in \partial D$ such that

$$\lim_{k\to\infty} f^k(x) = \zeta \quad \text{for all } x \in D.$$

Results of this type hold more generally for metric spaces whose geometry resembles that of a hyperbolic space. More precisely, Beardon [19] proved that the following two axioms suffice.

Definition 3.4.1 A metric space (X, d) is said to satisfy *Axiom I* if it is an open dense subset of a sequentially compact Hausdorff topological space $\overline{X}$ and the relative topology from $\overline{X}$ coincides with the metric topology of X. Moreover, for each convergent sequence $(x_k)_k$ in X with limit $x \in \partial X = \overline{X} \setminus X$, we have that $d(x_k, y) \to \infty$, as $k \to \infty$ for all $y \in X$.

Note that the second condition holds either for all $y \in X$ or for no $y \in X$, by the triangle inequality.

Definition 3.4.2 A metric space (X, d) satisfying Axiom I is said to satisfy *Axiom II* if for every two convergent sequences $(x_k)_k$ and $(z_k)_k$ in X with distinct limits x and z in ∂X, respectively, we have that

$$d(x_k, z_k) - \max\{d(x_k, y), d(z_k, y)\} \to \infty \quad \text{as } k \to \infty, \tag{3.10}$$

for each $y \in X$.

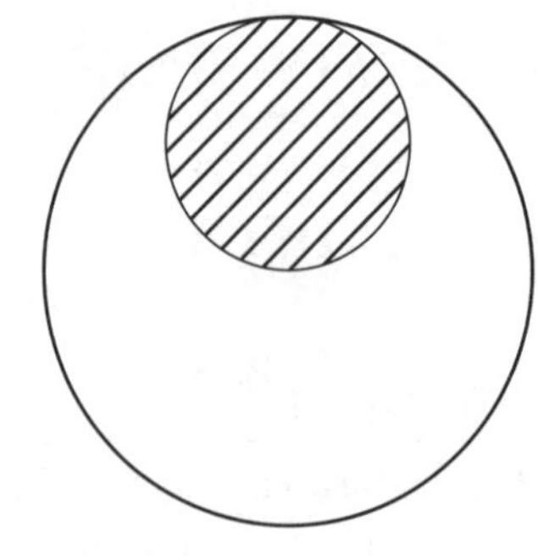

Figure 3.1 Horoballs in Klein's model.

It is again an easy consequence of the triangle inequality that (3.10) either holds for all $y \in X$ or for no $y \in X$.

Relevant examples of metric spaces in nonlinear Perron–Frobenius theory that satisfy Axiom I are Hilbert's metric spaces (X, κ), which are open dense subsets of the norm closure $\overline{X}$ according to Theorem 2.1.2 and Corollary 2.5.6. Indeed,

$$\kappa(x_k, y) = \log[x_k', x_k, y, y_k'] \to \infty \quad \text{as } k \to \infty,$$

since $|x_k' - x_k| \to 0$ when (x_k) converges to $x \in \partial X$. (Here x_k' and y_k' are the end-points of the chord through x_k and y.) In general, however, (X, κ) need not satisfy Axiom II, as the following example shows.

Example 3.4.3 Let $X = (0, 1) \times (0, 1)$ be an open square in $\mathbb{R}^2$. If $x_k = (1/3, 1/k)$ and $z_k = (2/3, 1/k)$ for all $k \geq 1$, then

$$\kappa(x_k, z_k) = \log\left(\frac{2/3}{1/3} \cdot \frac{2/3}{1/3}\right) = \log 4$$

for all $k \geq 1$, so that (3.10) cannot hold for any $y \in X$.

It will be shown in Section 8.3 that (X, κ) satisfies Axiom II if X is strictly convex. The following lemma lists some elementary properties of metric spaces that satisfy Axioms I and II.

Lemma 3.4.4 *If (X, d) is a proper metric space that satisfies Axioms I and II, then the following assertions hold:*

(i) If $(x_k)_k$ and $(z_k)_k$ are convergent sequences in X with distinct limits x and z in ∂X, respectively, then $d(x_k, z_k) \to \infty$ as $k \to \infty$.
(ii) If $(x_k)_k$ and $(z_k)_k$ are sequences in X such that $x_k \to x \in \partial X$ and $d(x_k, z_k) \leq M$ for all $k \geq 1$, then $z_k \to x$ as $k \to \infty$.

Proof The first assertion is a direct consequence of Axiom II. Now suppose that (ii) does not hold. As $\overline{X}$ is sequentially compact, $(z_k)_k$ has a convergent subsequence with limit $z \neq x$ in $\overline{X}$. If $z \in X$, then

$$\lim_{k\to\infty} d(x_k, z) \leq \lim_{k\to\infty} d(x_k, z_k) + d(z_k, z) \leq M,$$

which contradicts Axiom I. On the other hand, if $z \in \partial X$, then it contradicts the first assertion. □

The hyperbolic nature of metric spaces that satisfy Axioms I and II implies that each horoball in X meets ∂X in exactly one point.

Lemma 3.4.5 *If (X, d) is a proper metric space that satisfies Axioms I and II and H is a horoball in X, then there exists $\zeta \in \partial X$ such that each limit point $z \in \partial X$ of a sequence $(z_k)_k \subseteq H$ satisfies $z = \zeta$.*

Proof Let $H = \{x \in X : b_y(\sigma)(x) \leq r\}$ be a horoball with radius r, centered at $\sigma = (x_k)_k$, and base-point y. As $\overline{X}$ is sequentially compact, we may assume that $x_k \to \zeta \in \partial X$ as $k \to \infty$. Suppose that $(z_k)_k$ is a convergent sequence in H with limit $z \in \partial X$. Fix $\varepsilon > 0$ and recall that

$$b_y(\sigma)(x) = \lim_{k\to\infty} d(x, x_k) - d(x_k, y).$$

Therefore for each $m \geq 1$ there exists $k_m \geq 1$ such that

$$d(z_m, x_k) - d(x_k, y) \leq r + \varepsilon$$

for all $k \geq k_m$ and $k_m > k_{m-1}$. Consider the subsequence $(x_{k_m})_m$ of $(x_k)_k$. Clearly

$$d(z_m, x_{k_m}) - d(x_{k_m}, y) \leq r + \varepsilon$$

for all $m \geq 1$, and hence

$$d(z_m, x_{k_m}) - \max\{d(z_m, y), d(x_{k_m}, y)\} \leq r + \varepsilon$$

for all $m \geq 1$. It now follows from Axiom II that $z = \zeta$. □

The following result due to Beardon [19] can be derived from Lemmas 3.4.4 and 3.4.5.

Theorem 3.4.6 *Let (X, d) be a proper metric space that satisfies Axioms I and II. If $f : X \to X$ is a non-expansive map and $\Omega_f = \emptyset$, then there exists $\zeta \in \partial X$ such that*

$$\lim_{k\to\infty} f^k(x) = \zeta \quad \textit{for all } x \in X$$

and the convergence is uniform on compact subsets of X.

Proof Let $x, z \in X$ and remark that there exist horoballs H_x and H_z in X such that $\mathcal{O}(x) \subseteq H_x$ and $\mathcal{O}(z) \subseteq H_z$ by Theorem 3.3.5. As (X, d) is a proper metric space satisfying Axioms I and II and $\Omega_f = \emptyset$, it follows from Lemma 3.4.5 that $\omega(x) = \{\zeta_x\} \in \partial X$ and $\omega(z) = \{\zeta_z\} \in \partial X$. As $\overline{X}$ is sequentially compact this implies that $\lim_{k\to\infty} f^k(x) = \zeta_x$ and $\lim_{k\to\infty} f^k(z) = \zeta_z$. Note that $d(f^k(x), f^k(z)) \leq d(x, z)$ for all $k \geq 1$. It therefore follows from Lemma 3.4.4(ii) that $\zeta_x = \zeta_z$.

To see that the convergence is uniform on compact sets we need to show that if $U \subseteq \overline{X}$ is an open neighborhood of $\zeta \in \partial X$ and B is a closed ball in

(X, d), then there exists $K \geq 1$ such that $f^k(x) \in U$ for all $x \in B$ and $k \geq K$. Suppose by way of contradiction that such K does not exist. Then there exists a sequence $(x_k)_k \in B$ such that $f^k(x_k) \notin U$ for all $k \geq 1$. As $\overline{X} \setminus U$ is sequentially compact, $(x_k)_k$ has a convergent subsequence with limit $\xi \neq \zeta$. But $d(f^k(x), f^k(x_k)) \leq d(x, x_1)$ for all $x \in B$ and $k \geq 1$, as f is non-expansive. This contradicts Lemma 3.4.4(ii). □

Theorem 3.4.6 can be viewed as a generalization of the classical Denjoy–Wolff theorem [56, 226, 227] concerning fixed-point free holomorphic maps on the open unit disc in $\mathbb{C}$, which asserts the following: If $f : D \to D$ is a fixed-point free holomorphic map on the open unit disc $D \subseteq \mathbb{C}$, then there exists $\zeta \in \partial D$ such that

$$\lim_{k \to \infty} f^k(x) = \zeta \quad \text{for all } x \in D,$$

and the convergence is uniform on compact subsets of D. To see how the Denjoy–Wolff theorem fits the framework one first notes that a holomorphic map on the open unit disc is non-expansive under the Poincaré metric by the Schwartz–Pick lemma. Secondly, the open unit disc with the Poincaré metric is a proper metric space that satisfies Axioms I and II; see [19]. The idea to generalize the Denjoy–Wolff theorem to metric spaces is due to Beardon [18–20] and was further developed by Karlsson [99], Karlsson and Noskov [101], Lins [130, 131], and Nussbaum [168]. In Sections 8.3 and 8.4 we will discuss Denjoy–Wolff type theorems for non-expansive maps on Hilbert's metric spaces and their application to the asymptotic behavior of the iterates of order-preserving homogeneous maps that have no eigenvector in the interior of the cone.

3.5 Non-expansive retractions

To analyze the iterates of non-expansive maps $f : X \to X$ it is useful to understand the geometry of the attractor, and the way f acts on it. Intuitively one may expect that Ω_f is the range of a non-expansive *retraction* on X, i.e., a non-expansive map $r : X \to X$ such that $r(X) = \Omega_f$ and $r^2(x) = r(x)$ for all $x \in X$. The purpose of this section is to prove the existence of such a retraction under appropriate conditions on the metric space and show that f acts as an isometry on Ω_f. In particular, we shall see that this result holds if X is a closed subset of a finite-dimensional normed space, from which we can deduce various geometric properties of Ω_f.

Let us first prove the existence of a non-expansive retraction onto Ω_f.

Theorem 3.5.1 *If (X,d) is a proper metric space and $f: X \to X$ is a non-expansive map with $\Omega_f \neq \emptyset$, then the following assertions hold:*

(i) There exists a subsequence $(m_i)_i$ such that the map $r: X \to X$ defined by

$$r(x) = \lim_{i\to\infty} f^{m_i}(x)$$

for $x \in X$ is well defined, and $(f^{m_i})_i$ converges uniformly to r on compact subsets of X.

(ii) The map $r: X \to X$ is a non-expansive retraction onto Ω_f.

Proof By Lemma 3.1.10 we know that $(f^k)_k$ has a convergent subsequence $(f^{k_i})_i$, which converges uniformly on compact sets. By taking a further subsequence we may assume that $k_{i+1} - k_i \to \infty$ as $i \to \infty$. If we apply Lemma 3.1.10 again we may also assume that $(f^{k_{i+1}-k_i})_i$ converges uniformly on compact sets in X. Now define

$$r(x) = \lim_{i\to\infty} f^{k_{i+1}-k_i}(x) \tag{3.11}$$

for all $x \in X$. This proves the first assertion.

It follows from the definition of $r: X \to X$ that r is non-expansive. Let us now show that r is a retraction that maps X onto Ω_f. Clearly $r(X) \subseteq \Omega_f$. Let $y \in \Omega_f$. From Lemma 3.1.6 we know that $y \in \omega(y)$ and hence $(f^k(y))_k$ has a bounded subsequence. It therefore follows from Theorem 3.1.7 that $\mathcal{O}(x)$ is bounded for each $x \in X$, and, as (X,d) is a proper metric space, the closure of $\mathcal{O}(x)$ is compact for each $x \in X$. Thus, we deduce from Lemma 3.1.2 that for each $k \geq 1$ there exists $y^k \in \omega(y)$ such that $f^k(y^k) = y$. Let g denote the pointwise limit of $(f^{k_i})_i$ and let $(y^{k_{i_j}})_j$ be a convergent subsequence of $(y^{k_i})_i$, with limit $z \in \omega(y)$. Then

$$d(f^{k_{i_j}}(y^{k_{i_j}}), g(z)) \to 0,$$

as $j \to \infty$, since $(f^{k_{i_j}})_j$ converges uniformly on compact subsets to g. This implies that $g(z) = y$. Now consider the equality

$$f^{k_{i+1}}(z) = f^{k_{i+1}-k_i}(f^{k_i}(z))$$

for all $i \geq 1$. Clearly the left-hand side converges to $g(z)$ and the right-hand side to $r(g(z))$, as $(f^{k_{i+1}-k_i})_i$ converges uniformly to r on $\omega(y)$. Thus, $r(y) = y$ and hence r is a retraction onto Ω_f. □

Next we show that f is an isometry on Ω_f.

Corollary 3.5.2 *If* (X, d) *is a proper metric space and* $f : X \to X$ *is a non-expansive map with* $\Omega_f \neq \emptyset$, *then the restriction of* f *to* Ω_f *is an isometry and*

$$\lim_{k\to\infty} d(f^k(x), \Omega_f) = 0 \quad \textit{for all } x \in X.$$

Proof Let r be the non-expansive retraction defined in (3.11). For each $x, y \in \Omega_f$ we have that

$$d(x, y) = d(r(x), r(y)) \leq d(f(x), f(y)) \leq d(x, y),$$

so that f is an isometry on Ω_f. Let $x \in X$ and remark that

$$d(f^k(x), \Omega_f) \leq d(f^k(x), f^{k-(k_{i+1}-k_i)}(r(x))) \leq d(f^{k_{i+1}-k_i}(x), r(x)),$$

for all $k \geq k_{i+1} - k_i$. This implies that $d(f^k(x), \Omega_f) \to 0$, as $k \to \infty$. □

Ranges of non-expansive retractions often possess some special geometric properties, especially when X is a closed subset of normed space. For instance, if X is a closed convex subset of a vector space V equipped with a strictly convex norm, then Ω_f is convex for each non-expansive map on X. Recall that a norm $\|\cdot\|$ is *strictly convex* if the unit sphere does not contain any straight-line segments, or, equivalently, $\|x + y\| = \|x\| + \|y\|$ and $y \neq 0$ implies $x = \|x\|/\|y\|y$.

Lemma 3.5.3 *If* $X \subseteq V$ *is closed and convex, and* $f : X \to X$ *is non-expansive with respect to a strictly convex norm, then* Ω_f *is convex.*

Proof Let $r : X \to X$ be the non-expansive retraction defined by (3.5.1) in the proof of Theorem 3.5.1. Take $x, y \in \Omega_f$ and $0 \leq \lambda \leq 1$. Note that $z = \lambda x + (1-\lambda)y$ is the unique point that satisfies $\|x - z\| = (1-\lambda)\|x - y\|$ and $\|y - z\| = \lambda\|x - y\|$, as $\|\cdot\|$ is strictly convex. This implies that $\|x - r(z)\| \leq (1 - \lambda)\|x - y\|$ and $\|y - r(z)\| \leq \lambda\|x - y\|$, as r is non-expansive. Thus, $\|x - r(z)\| + \|r(z) - y\| = \|x - y\|$ and hence $r(z) = z$. □

If, in addition, $f : X \to X$ is order-preserving, then the retraction $r : X \to X$ in (3.11) is also order-preserving. If the underlying norm is *strictly monotone*, meaning that $\|x\| < \|y\|$ for all $0 \leq x < y$, then we have the following result.

Lemma 3.5.4 *Let* $X \subseteq V$ *be a lattice and* $f : X \to X$ *be order-preserving with respect to a minihedral cone* $K \subseteq V$. *If* f *is non-expansive under a strictly monotone norm, then* Fix_f *and* Ω_f *are both lattices.*

Proof Let $x, y \in \text{Fix}_f$. As f is order-preserving, $f(\inf(x, y)) \leq f(x)$ and $f(\inf(x, y)) \leq f(y)$, so that $f(\inf(x, y)) \leq \inf(f(x), f(y))$. Suppose that $f(\inf(x, y)) < \inf(f(x), f(y))$. By strict monotonicity we get that

$$\begin{aligned}\|y - \inf(x, y)\| &\geq \|f(y) - f(\inf(x, y))\| \\ &> \|f(y) - \inf(f(x), f(y))\| \\ &= \|y - \inf(x, y)\|,\end{aligned}$$

which is a contradiction. Thus, $f(\inf(x, y)) = \inf(f(x), f(y)) = \inf(x, y)$. In a similar way it can be shown that $f(\sup(x, y)) = \sup(f(x), f(y)) = \sup(x, y)$ for all $x, y \in \text{Fix}_f$, and hence Fix_f is a lattice.

By applying the first statement to the order-preserving non-expansive retraction defined in Theorem 3.5.1 we deduce that Ω_f is a lattice. □

A prime example of a strictly monotone norm for the standard positive cone is the ℓ_1-norm. So, we can deduce from Lemma 3.5.4 that Fix_f and Ω_f are lattices if $f : \mathbb{R}^n_+ \to \mathbb{R}^n_+$ is an order-preserving map which is non-expansive under the ℓ_1-norm. This observation plays an important role in Chapter 9, where we examine the iterative behavior of such maps.

4

Sup-norm non-expansive maps

In Chapter 2 it was shown that order-preserving (sub)homogeneous maps are non-expansive under Hilbert's metric or Thompson's metric. We also saw that if the underlying cone is polyhedral, Hilbert's and Thompson's metric spaces can be isometrically embedded into $(\mathbb{R}^n, \|\cdot\|_\infty)$, as long as n is sufficiently large. Therefore sup-norm non-expansive maps play a special role in nonlinear Perron–Frobenius theory. The main purpose of this chapter is to prove that the iterates of sup-norm non-expansive maps have some striking properties. Among other facts it will be shown that every bounded orbit of a sup-norm non-expansive map converges to a periodic orbit, and, moreover, there exists an a-priori upper bound for the periods of its periodic points in terms of the dimension of the underlying space. Consequently, every bounded orbit of a sub-topical map converges to a periodic orbit. In the final section it will be shown that the period of each periodic point of a sub-topical map on $\mathbb{R}^n$ does not exceed $\binom{n}{\lfloor n/2 \rfloor}$, which is a sharp upper bound.

4.1 The size of the ω-limit sets

In this section an upper bound for the size of ω-limit sets of sup-norm non-expansive maps is given in terms of the dimension of the underlying space. To begin we recall that $\omega(x)$ has a transitive abelian group of sup-norm isometries by Theorem 3.1.9, if $\mathcal{O}(x)$ is bounded. Sets that have an abelian group of isometries enjoy the following basic property.

Lemma 4.1.1 *If (X, d) is a metric space with a transitive abelian group Γ of isometries, then for each $g \in \Gamma$ there exists a constant $d(g) \geq 0$ such that $d(x, g(x)) = d(g)$ for all $x \in X$. Moreover, $|X| = |\Gamma|$.*

Proof Let $g \in \Gamma$ and note that, as Γ is abelian, $d(x, g(x)) = d(f(x), g(f(x)))$ for all $f \in \Gamma$. This implies that $d(x, g(x))$ is constant for all $x \in X$, since Γ acts transitively on X. To see that $|X| = |\Gamma|$, we fix $x \in X$ and consider $\varphi_x : \Gamma \to X$ given by $\varphi_x(g) = g(x)$ for $g \in \Gamma$. As Γ acts transitively, φ_x is onto. The map φ_x is also injective. Indeed, $g(x) = f(x)$ implies that $f^{-1}(g(x)) = x$, so that $f(y) = g(y)$ for all $y \in X$ by the first assertion of the lemma. We conclude that $|X| = |\Gamma|$. □

A sequence $x^1, x^2, \ldots, x^m$ in a metric space (X, d) is called an *additive chain* if

$$d(x^1, x^m) = \sum_{k=1}^{m-1} d(x^k, x^{k+1}).$$

An additive chain has *length m* if it consists of m distinct elements. Using Lemma 4.1.1 the following property of additive chains can be derived.

Proposition 4.1.2 *Let (X, d) be a metric space with a transitive abelian group Γ of isometries. If $x^1, x^2, \ldots, x^m$ is an additive chain of length m in (X, d) and $f_k \in \Gamma$ is such that $f_k(x^k) = x^{k+1}$ for $1 \leq k < m$, then for every permutation π on $\{1, 2, \ldots, m\}$ and every $x \in X$ the sequence*

$$x,\ f_{\pi(1)}(x),\ f_{\pi(2)} \circ f_{\pi(1)}(x),\ \ldots,\ f_{\pi(m-1)} \circ \cdots \circ f_{\pi(1)}(x)$$

is an additive chain of length m.

Proof Let $f = f_{m-1} \circ \cdots \circ f_1$ and remark that $f = f_{\pi(m-1)} \circ \cdots \circ f_{\pi(1)}$, as Γ is abelian. For simplicity write $z^k = f_{\pi(k-1)} \circ \cdots \circ f_{\pi(1)}(x)$ for $2 \leq k \leq m$ and put $z^1 = x$. From Lemma 4.1.1 it follows that

$$d(z^1, z^m) = d(x^1, x^m) = \sum_{k=1}^{m-1} d(f_k) = \sum_{k=1}^{m-1} d(f_{\pi(k)}) = \sum_{k=1}^{m-1} d(z^k, z^{k+1}),$$

and hence $z^1, z^2, \ldots, z^m$ is an additive chain of length m. □

We now consider the case where the metric is induced by the sup-norm. As the sup-norm is a polyhedral norm, the assumption that x, y, z forms an additive chain need not imply that y lies on the straight-line segment connecting x and z. In fact, in a sup-norm metric space there are usually infinitely many geodesics connecting two distinct points. To determine whether a sequence is an additive chain with respect to the sup-norm it is convenient to introduce for each $1 \leq i \leq n$ a partial ordering $\leq_i$ on $\mathbb{R}^n$ by $x \leq_i y$ if $\|x - y\|_\infty = y_i - x_i$. A sequence $x^1, x^2, \ldots, x^m$ in $(\mathbb{R}^n, \|\cdot\|_\infty)$ is called an *i-chain* if either $x^1 \leq_i x^2 \leq_i \cdots \leq_i x^m$ or $x^m \leq_i \cdots \leq_i x^2 \leq_i x^1$. It is easy to verify that $x^1, x^2, \ldots, x^m$ is an additive chain in $(\mathbb{R}^n, \|\cdot\|_\infty)$ if and only if it is an i-chain for some i. Moreover, if $x^1, x^2, \ldots, x^m$ is an additive chain in $(\mathbb{R}^n, \|\cdot\|_\infty)$ and $\|x^1 - x^m\|_\infty = |x_i^1 - x_i^m|$, then it is an i-chain.

Another useful concept is the notion of an extreme pair. A pair $\{x, y\}$ in a metric space (X, d) called an *extreme pair* in X if there exists no $z \in X$ such that z, x, y or x, y, z form an additive chain of length 3. Using the extreme pairs the following upper bound for the size of compact sets in $(\mathbb{R}^n, \|\cdot\|_\infty)$ that have a transitive abelian group of isometries can be proved.

Theorem 4.1.3 *If X is a compact set in $(\mathbb{R}^n, \|\cdot\|_\infty)$ and X has a transitive abelian group of isometries, then*

$$|X| \leq \max_{1 \leq k \leq n} 2^k \binom{n}{k}.$$

Before proving this theorem we isolate a combinatorial lemma concerning anti-chains in a certain finite partially ordered set. Recall that a subset $\mathcal{A}$ of a partially ordered set $(S, \preceq)$ is called an *anti-chain* if there exist no $x, y \in \mathcal{A}$ with $x \preceq y$ and $x \neq y$. Let $\preceq$ be the partial ordering on $\{0, 1, 2\}^n$ given by

$$a \preceq b \text{ if } a_i = b_i \text{ for all } i \text{ for which } b_i \in \{0, 1\}.$$

It is easy to verify that $\preceq$ is reflexive, anti-symmetric, and transitive. Using a standard method from extremal set theory to derive a so-called LYM inequality [28, section 3], we prove the following upper bound for the cardinality of anti-chains in the partially ordered set $(\{0, 1, 2\}^n, \preceq)$. (LYM stands for Lubell [136], Yamamoto [229], and Meshalkin [146], who independently discovered this method.)

Lemma 4.1.4 *If $\mathcal{A}$ is an anti-chain in $(\{0, 1, 2\}^n, \preceq)$, then*

$$|\mathcal{A}| \leq \max_k 2^k \binom{n}{k}.$$

Proof We call a chain $a^0 \preceq a^1 \preceq \ldots \preceq a^n$ a *maximal chain* in$(\{0, 1, 2\}^n, \preceq)$ if it consists of $n + 1$ distinct elements. There are $n!2^n$ maximal chains. Indeed, to obtain a maximal chain one has to select a point a^0 such that a^0 has no coordinates equal to 2. There are 2^n possibilities. Subsequently we change, one by one, the coordinates of a^0 to 2. There are $n!$ ways to do this. Thus, in total we have $n!2^n$ maximal chains.

Given an element a in $\mathcal{A}$ with k coordinates equal to 2, we now determine how many maximal chains $a^0 \preceq a^1 \preceq \ldots \preceq a^n$ contain a. Clearly there are $k!2^k$ chains $a^0 \preceq \ldots \preceq a^{k-1} \preceq a$, with $k + 1$ distinct elements, and each of these can be extended in $(n - k)!$ ways to a maximal chain. Thus a is contained in $k!(n - k)!2^k$ maximal chains.

To finish the argument we let m_k denote the number of elements in $\mathcal{A}$ with k coordinates equal to 2. Obviously $|\mathcal{A}| = \sum_{k=0}^{n} m_k$. As $\mathcal{A}$ is an anti-chain, each maximal chain contains at most one element of $\mathcal{A}$. Therefore

$$\sum_{k=0}^{n} m_k k!(n-k)!2^k \leq n!2^n,$$

which yields the (LYM) inequality

$$\sum_{k=0}^{n} m_k 2^{k-n} \binom{n}{k}^{-1} \leq 1.$$

From this it follows that

$$|\mathcal{A}| = \sum_{k=0}^{n} m_k \leq \max_k \, 2^{n-k} \binom{n}{k} = \max_k \, 2^k \binom{n}{k}.$$

□

Remark that the upper bound in Lemma 4.1.4 is sharp, as the collection of points in $\{0, 1, 2\}^n$ with exactly $n - k$ coordinates equal to 2 is an anti-chain containing $2^k \binom{n}{k}$ elements. It can be shown, using Stirling's formula, that there exist constants $0 < \delta_1 \leq \delta_2$ such that

$$\delta_1 3^n / \sqrt{n} \leq \max_{1 \leq k \leq n} \, 2^k \binom{n}{k} \leq \delta_2 3^n / \sqrt{n} \quad \text{for all } n \in \mathbb{N}.$$

Let us now prove Theorem 4.1.3.

Proof of Theorem 4.1.3 Suppose that $|X| > \max_k 2^k \binom{n}{k}$. Then there exists $\varepsilon > 0$ and $X_\varepsilon \subseteq X$ such that $\|x - y\|_\infty \geq \varepsilon$ for all $x \neq y$ in X_ε and $|X_\varepsilon| > \max_k 2^k \binom{n}{k}$. We call $z^1, z^2, \ldots, z^m$ an ε-additive chain if it is an additive chain and $\|z^k - z^{k+1}\|_\infty \geq \varepsilon$ for all $1 \leq k < m$. We say that $\{x, y\}$ is an ε-extreme pair in X if $\|x - y\|_\infty \geq \varepsilon$ and there exists no $z \in X$ such that z, x, y or x, y, z is an ε-additive chain.

Define $c \colon X_\varepsilon \to \{0, 1, 2\}^n$ by

$$c(x)_i = \begin{cases} 0 & \text{if there exists } y \in X \text{ with } \{x, y\} \text{ an } \varepsilon\text{-extreme pair in } X \\ & \text{and } \|x - y\|_\infty = y_i - x_i, \\ 1 & \text{if there exists } y \in X \text{ with } \{x, y\} \text{ an } \varepsilon\text{-extreme pair in } X \\ & \text{and } \|x - y\|_\infty = x_i - y_i, \\ 2 & \text{otherwise,} \end{cases}$$

for all $1 \leq i \leq n$ and $x \in X_\varepsilon$. We claim that c is well defined. Indeed, if $c(x)_i = 0$ and $c(x)_i = 1$, then there exist $y, z \in X$ such that $\{x, y\}$ and $\{x, z\}$ are ε-extreme pairs in X with $\|x - y\|_\infty = y_i - x_i$ and $\|x - z\|_\infty = x_i - z_i$.

This implies that $z \leq_i x \leq_i y$, so that z, x, y is an ε-additive chain. But this contradicts the fact that $\{x, y\}$ is an ε-extreme pair in X.

Next we show that $c(y) \not\preceq c(x)$ for all $x \neq y$ in X_ε. Let $x \neq y$ in X_ε and let $\mathcal{F}$ be the collection of ε-additive chains $z^1, z^2, \ldots, z^m$ in X with $z^1 = x$ and $z^2 = y$. As X is a compact set in $(\mathbb{R}^n, \|\cdot\|_\infty)$, it is bounded and hence there exists an ε-additive chain, $z^1, z^2, \ldots, z^r$, in $\mathcal{F}$ with maximal length r. We claim that $\{z^1, z^r\}$ is an ε-extreme pair in X. Indeed, clearly there exists no $u \in X$ such that z^1, z^r, u is an ε-additive chain, as r is maximal. On the other hand, if there exists $u \in X$ with u, z^1, z^r an ε-additive chain, then $z^1, z^r, f(z^r)$, where $f \in \Gamma$ is such that $f(u) = z^1$, is also an ε-additive chain by Proposition 4.1.2 and Lemma 4.1.1. Thus, $\{z^1, z^r\}$ is an ε-extreme pair in X.

There are two cases: $\|z^1 - z^r\|_\infty = z_i^r - z_i^1$ and $\|z^1 - z^r\|_\infty = z_i^1 - z_i^r$. In the first case $c(x)_i = 0$. We now argue, by contradiction, that $c(y)_i \neq 0$ in that case. If $c(y)_i = 0$, then there exists $u \in X$ such that $\{y, u\}$ is an ε-extreme pair in X and $\|y - u\|_\infty = u_i - y_i$. As $\|z^1 - z^r\|_\infty = z_i^r - z_i^1$, we know that $\|x - y\|_\infty = y_i - x_i$, so that $x \leq_i y \leq_i u$. Thus, x, y, u is an ε-additive chain of length 3, which is a contradiction. In the second case we have that $c(x)_i = 1$ and it can be shown, in the same fashion, that $c(y)_i \neq 1$ in that case. Therefore $c(y) \not\preceq c(x)$, and hence $c(X_\varepsilon)$ is an anti-chain in $(\{0, 1, 2\}^n, \preceq)$ whose size exceeds $\max_k 2^k \binom{n}{k}$, which contradicts Lemma 4.1.4. □

From Theorem 4.1.3 it is easy to deduce the following upper bound of ω-limit sets of bounded orbits of sup-norm non-expansive maps.

Theorem 4.1.5 *If $f: X \to X$, with $X \subseteq \mathbb{R}^n$, is a sup-norm non-expansive map and $x \in X$ has a bounded orbit, then $|\omega(x)| \leq \max_k 2^k \binom{n}{k}$.*

Proof As $\mathcal{O}(x)$ is bounded, its closure is compact. Therefore it follows from Theorem 3.1.9 that $\omega(x)$ has a transitive abelian group of sup-norm isometries, so that $|\omega(x)| \leq \max_k 2^k \binom{n}{k}$ by Theorem 4.1.3. □

4.2 Periods of periodic points

A combination of Theorem 4.1.5 and Lemma 3.1.3 immediately gives the following result for the iterative behavior of sup-norm non-expansive maps.

Theorem 4.2.1 *If $X \subseteq \mathbb{R}^n$ is closed and $f: X \to X$ is a sup-norm non-expansive map, then either all orbits of f are unbounded, or for each $x \in X$ there exist an integer $p \geq 1$ and a periodic point $\xi \in X$ of f with period p such that*

$$\lim_{k\to\infty} f^{kp}(x) = \xi$$

and $p \leq \max_k 2^k \binom{n}{k}$.

Theorem 4.2.1 shows that the period of each periodic point of a sup-norm non-expansive map $f\colon X \to X$, with $X \subseteq \mathbb{R}^n$, does not exceed $\max_k 2^k \binom{n}{k}$. It is believed, however, that this upper bound is not optimal. In fact, Nussbaum [160] has made the following conjecture.

Conjecture 4.2.2 (Nussbaum) *The optimal upper bound for the periods of periodic points of sup-norm non-expansive maps $f\colon X \to X$, with $X \subseteq \mathbb{R}^n$, is 2^n.*

At present the conjecture is proved for $n = 1$, 2, and 3 only. The case $n = 3$ is already nontrivial and was established by Lyons and Nussbaum [137]. The upper bound in Theorem 4.2.1 is due to Lemmens and Scheutzow [123] and is currently the strongest estimate. Other upper bounds were obtained in [27, 140, 160, 204]. Evidence supporting Conjecture 4.2.2 is provided by the following result of Lyons and Nussbaum.

Theorem 4.2.3 *A sup-norm non-expansive map $f\colon X \to X$, where $X \subseteq \mathbb{R}^n$, cannot have a periodic point whose period p is prime and satisfies $p > 2^n$.*

Proof Let $\xi \in X$ be a periodic point with period p of a sup-norm non-expansive map $f\colon X \to X$, where $X \subseteq \mathbb{R}^n$. Suppose that p is prime. Let g denote the restriction of f to $\mathcal{O}(\xi)$. Clearly, $\mathcal{O}(\xi)$ consists of p points and $\Gamma = \{g^k : 0 \leq k < p\}$ is a transitive cyclic group of sup-norm isometries on $\mathcal{O}(\xi)$.

For $1 \leq i \leq n$, let $l_i = \min_{z\in\mathcal{O}(\xi)} z_i$ and $u_i = \max_{z\in\mathcal{O}(\xi)} z_i$. Put $l = (l_1, \ldots, l_n)$ and $u = (u_1, \ldots, u_n)$. Let Q be the parallelepiped with vertex set $\{(x_1, \ldots, x_n) \in \mathbb{R}^n : x_i = l_i \text{ or } x_i = u_i\}$. (It is not hard to show that $\mathcal{O}(\xi)$ is contained in the boundary of Q, but we shall not need that here.)

The proof uses the idea of a maximal tight subset. A subset T of $\mathcal{O}(\xi)$ is called *tight* if the (sup-norm) diameter of T is strictly smaller than the diameter of $\mathcal{O}(\xi)$. It is said to be a *maximal tight* subset of $\mathcal{O}(\xi)$ if it is tight and not properly contained in another tight subset of $\mathcal{O}(\xi)$. Now let T be a maximal tight subset of $\mathcal{O}(\xi)$. We claim that $T, g(T), \ldots, g^{p-1}(T)$ are all distinct. Indeed, if $g^k(T) = g^m(T)$ for some $0 \leq k < m < p$, then $g^{m-k}(T) = T$. But as p is prime, there exists $q \geq 1$ such that $q(m-k) \equiv 1 \bmod p$, so that $g(T) = g^{q(m-k)}(T) = T$. This implies that $T = \cup_k g^k(T) = \mathcal{O}(\xi)$, which is a contradiction, as T is tight. We also remark that $T, g(T), \ldots, g^{p-1}(T)$ are all maximal tight subsets of $\mathcal{O}(\xi)$, as g is an isometry of $\mathcal{O}(\xi)$ onto itself. To complete the proof it suffices to show that the number of maximal tight subsets of $\mathcal{O}(\xi)$ does not exceed 2^n. To see this we associate with each maximal tight subset S of $\mathcal{O}(\xi)$ a vertex v^S of the parallelepiped Q as follows: $v_i^S = l_i$ if there exists $s \in S$ with $s_i = l_i$ and $v_i^S = u_i$ otherwise. We claim that $v^S \neq v^T$ if S and T are distinct maximal tight subsets of $\mathcal{O}(\xi)$. Indeed, if $S \neq T$, then there exist $s \in S$ and $t \in T$ such that $\|s - t\|_\infty = \mathrm{diam}(\mathcal{O}(\xi))$, as S and T are distinct maximal tight subsets. We know that $|s_i - t_i| = \mathrm{diam}(\mathcal{O}(\xi))$ for some i. This implies that either $s_i = l_i$ and $t_i = u_i$ or $s_i = u_i$ and $t_i = l_i$. In the first case $v_i^S = l_i$, but $v_i^T \neq l_i$, as T is a tight subset. In the second case, $v_i^T = l_i$ and $v_i^S \neq l_i$, as S is a tight subset. As Q contains at most 2^n vertices, we are done. □

It is easy to see that one cannot expect to do better than 2^n in Conjecture 4.2.2. Simply consider the set of vertices of the n-dimensional cube, $H_n = \{(x_1, \ldots, x_n) \in \mathbb{R}^n : x_i = \pm 1 \text{ for } 1 \leq i \leq n\}$. As $\|x - y\|_\infty = 2$ for all $x \neq y$ in H_n, any cyclic permutation σ of the elements of H_n of order 2^n is a sup-norm isometry, and each point of H_n has period 2^n under σ.

It is known [223, chapter III] that every sup-norm non-expansive map $g : X \to \mathbb{R}^n$, with $X \subseteq \mathbb{R}^m$, admits a sup-norm non-expansive extension to the whole of $\mathbb{R}^m$. This is a special case of the so-called Aronszajn–Panitchpakdi theorem [10]. In fact, for sup-norm non-expansive maps there exists the following explicit construction.

Lemma 4.2.4 *If $g : X \to \mathbb{R}^n$, with $X \subseteq \mathbb{R}^m$, is a sup-norm non-expansive map, then $G : \mathbb{R}^m \to \mathbb{R}^n$ given by*

$$G_i(z) = \sup_{x \in X} \Big(g_i(x) - \|x - z\|_\infty \Big) \quad \text{for } z \in \mathbb{R}^m$$

is a sup-norm non-expansive extension of g.

Proof To see that $G_i(z) < \infty$ for all i and $z \in \mathbb{R}^m$ we take $x^* \in X$ fixed. Remark that

$$\begin{aligned} g_i(x) - \|x - z\|_\infty &\leq g_i(x) - \|x - x^*\|_\infty + \|x^* - z\|_\infty \\ &\leq g_i(x) - \|g(x) - g(x^*)\|_\infty + \|x^* - z\|_\infty \\ &\leq g_i(x^*) + \|x^* - z\|_\infty, \end{aligned}$$

for each $x \in X$, so that $G_i(z) < \infty$.

Clearly for each $z \in X$ we have that $G(z) \geq g(z)$. On the other hand, if $z \in X$ and $1 \leq i \leq n$, then for each $\varepsilon > 0$ there exists $x \in X$ such that

$$G_i(z) - \varepsilon \leq g_i(x) - \|x - z\|_\infty \leq g_i(x) - \|g(x) - g(z)\|_\infty \leq g_i(z).$$

Thus, $G(z) = g(z)$ for all $z \in X$.

To show that G is sup-norm non-expansive, let $\varepsilon > 0$ and $y, z \in \mathbb{R}^m$. Assume, without loss of generality, that $\|G(y) - G(z)\|_\infty = G_i(y) - G_i(z)$. There exists $x \in X$ such that

$$\begin{aligned}\|G(y) - G(z)\|_\infty &= G_i(y) - G_i(z) \\ &\leq g_i(x) - \|x - y\|_\infty + \varepsilon - (g_i(x) - \|x - z\|_\infty) \\ &\leq \|y - z\|_\infty + \varepsilon,\end{aligned}$$

which completes the proof. □

As

$$\|x - z\|_\infty = \max\{\mathsf{t}(x - z), \mathsf{t}(z - x)\} = -\min\{\mathsf{b}(z - x), \mathsf{b}(x - z)\},$$

we see that if X is a finite set, the sup-norm non-expansive map G in Lemma 4.2.4 is of the form

$$G_i(z) = \max_x \min_j \big(r_i(x, j) \pm z_j\big),$$

where $r_i(x, j) \in \mathbb{R}$. Special examples are so-called Boolean maps. A *Boolean map* is a map $h\colon \mathbb{R}^n \to \mathbb{R}^n$ where each coordinate function is a finite min or max combination of expressions of the type x_i or $1 - x_i$, e.g., the map $h\colon \mathbb{R}^3 \to \mathbb{R}^3$ given by

$$f\begin{pmatrix} x_1 \\ x_2 \\ x_3 \end{pmatrix} = \begin{pmatrix} (x_2 \wedge x_3) \vee (1 - x_1) \\ x_1 \vee (1 - x_3) \\ ((1 - x_2) \wedge (1 - x_3)) \vee x_1 \end{pmatrix} \quad \text{for } x \in \mathbb{R}^3.$$

They are called Boolean maps because the restriction of each coordinate function $h_i(x)$ to $\{0, 1\}^n$ is a Boolean sentence with variables $x_1, \ldots, x_n$, where min is interpreted as AND, max as OR, and $(1 - x_i)$ as NOT x_i. It is clear that a Boolean map leaves $\{0, 1\}^n$ invariant and that each of its periodic points in $\{0, 1\}^n$ has period at most 2^n. Boolean maps also leave $\{0, 1/2, 1\}^n$ invariant. The value $1/2$ may be interpreted as a fuzzy value when one is not sure. It is unknown if each of its periodic points in $\{0, 1/2, 1\}^n$ has period at most 2^n, as is predicted by Conjecture 4.2.2.

Theorem 4.2.1 has consequences for the dynamics of other non-expansive maps as well.

Corollary 4.2.5 *Let (X, d) be a complete metric space that can be isometrically embedded into $(\mathbb{R}^n, \|\cdot\|_\infty)$. If $f\colon X \to X$ is non-expansive, then every bounded orbit of f converges to a periodic orbit whose period does not exceed* $\max_k 2^k \binom{n}{k}$.

Proof Let φ be the isometric embedding of (X, d) into $(\mathbb{R}^n, \|\cdot\|)$ and put $Y = \varphi(X)$. We remark that Y is a closed subset of $(\mathbb{R}^n, \|\cdot\|_\infty)$, as (X, d) is complete. Define $g\colon Y \to Y$ by $g = \varphi \circ f \circ \varphi^{-1}$. As f is non-expansive, g is sup-norm non-expansive. It thus follows from Theorem 4.2.1 that every bounded orbit of g converges to a periodic orbit whose period does not exceed $\max_k 2^k \binom{n}{k}$. The same is true for f, since $g = \varphi \circ f \circ \varphi^{-1}$. □

In Chapter 2 we encountered various complete metric spaces that can be isometrically embedded into $(\mathbb{R}^n, \|\cdot\|_\infty)$. In particular we saw that $(\mathrm{int}(\mathbb{R}^n_+), d_T)$ is isometric to $(\mathbb{R}^n, \|\cdot\|_\infty)$. We also showed that if $K \subseteq V$ is a solid polyhedral cone with N facets and $\Sigma^\circ = \{x \in \mathrm{int}(K)\colon \varphi(x) = 1\}$ where $\varphi \in \mathrm{int}(K^*)$, then (Σ°, d_H) is a complete metric space that can be isometrically embedded into $(\mathbb{R}^m, \|\cdot\|_\infty)$ for $m = N(N-1)/2$ by Proposition 2.2.3. Other interesting examples of metric spaces that can be isometrically embedded into $(\mathbb{R}^m, \|\cdot\|_\infty)$ are finite-dimensional normed spaces with a *polyhedral norm*; that is to say, its unit ball is a polyhedron.

Lemma 4.2.6 *If $\|\cdot\|$ is a polyhedral norm on $\mathbb{R}^n$ with N facets, then $(\mathbb{R}^n, \|\cdot\|)$ can be isometrically embedded into $(\mathbb{R}^{N/2}, \|\cdot\|_\infty)$.*

Proof Let B denote the unit ball of the polyhedral norm $\|\cdot\|$. As B has N facets, there exist N linear functionals $\varphi_1, \dots, \varphi_N$ on $\mathbb{R}^n$ such that

$$B = \bigcap_{i=1}^{N} \{x \in \mathbb{R}^n\colon \varphi_i(x) \leq 1\}.$$

Since B is symmetric, we know that N is even and we may assume that $\varphi_i = -\varphi_{i+N/2}$ for $1 \leq i \leq N/2$. Now define $\Phi\colon \mathbb{R}^n \to \mathbb{R}^{N/2}$ by $\Phi(x)_i = \varphi_i(x)$ for $1 \leq i \leq N/2$. We claim that Φ is an isometry from $(\mathbb{R}^n, \|\cdot\|)$ into $(\mathbb{R}^{N/2}, \|\cdot\|_\infty)$. As Φ is linear, it suffices to show that $\|\Phi(x)\|_\infty = 1$ for all $x \in \mathbb{R}^n$ with $\|x\| = 1$. Note that $\|x\| = 1$ implies that there exists $1 \leq i \leq N/2$ such that $\varphi_i(x) = \pm 1$ and $|\varphi_j(x)| \leq 1$ for all $j \neq i$. Therefore $\|\Phi(x)\|_\infty = |\varphi_i(x)| = 1$, and we are done. □

An important example of a polyhedral norm on $\mathbb{R}^n$ is the ℓ_1-norm, whose unit ball is the cross-polytope, which has 2^n facets. So, we can combine Corollary 4.2.5 and Lemma 4.2.6 to derive an upper bound (in terms of n) on the periods of the periodic points of ℓ_1-norm non-expansive maps $f\colon X \to X$, where X can be any subset of $\mathbb{R}^n$. Although this upper bound significantly improves old results by Misiurewicz [149], it is believed to be far from sharp; see [121]. In Chapter 9, however, we shall see that much stronger results are known for ℓ_1-norm non-expansive maps $f\colon X \to X$, with $f(0) = 0$, if one assumes that $X = \mathbb{R}^n_+$ or $X = \mathbb{R}^n$.

4.3 Iterates of topical maps

Special examples of sup-norm non-expansive maps are topical maps. Topical maps are not only non-expansive under the sup-norm, but also under the top function by Proposition 2.7.1. To find a suitable upper bound for the periods of periodic points of topical maps $f : \mathbb{R}^n \to \mathbb{R}^n$, we remark that if x is a periodic point of f, then $\mathcal{O}(x)$ is an anti-chain in $(\mathbb{R}^n, \leq)$, where $\leq$ is induced by the standard positive cone. Moreover, the restriction of f to $\mathcal{O}(x)$ is a top isometry, since f is top non-expansive. This implies that $\mathcal{O}(x)$ has a transitive cyclic group of top isometries. Conversely, if X is a finite set in $\mathbb{R}^n$ with a transitive cyclic group of top isometries, then there exists a topical map $g : \mathbb{R}^n \to \mathbb{R}^n$ that has X as one of its periodic orbits. Indeed, we have the following proposition, which is similar to Lemma 4.2.4.

Proposition 4.3.1 *If $X \subseteq \mathbb{R}^n$ and $h: X \to \mathbb{R}^n$ is a top non-expansive map, then the topical map $H : \mathbb{R}^n \to \mathbb{R}^n$, defined by*

$$H_i(z) = \sup_{x \in X} \big(h_i(x) - \mathsf{t}(x - z)\big) \quad \textit{for } z \in \mathbb{R}^n,$$

extends h.

Proof The proof is identical to the proof of Lemma 4.2.4 and is left to the reader. □

Thus, to find a suitable estimate for the possible periods of periodic points of topical maps on $\mathbb{R}^n$, it suffices to find a sequence $x^0, x^1, \ldots, x^{p-1}$ in $\mathbb{R}^n$ such that the map $x^k \mapsto x^{k+1 \bmod p}$ is a top isometry, or, equivalently,

$$\mathsf{t}(x^{k+m} - x^k) = \mathsf{t}(x^m - x^0) \quad \text{for all } k, m \geq 0.$$

(Here the indices are counted modulo p.) We note that if X is an anti-chain in the partially ordered set $(\{0, 1\}^n, \leq)$, where $\leq$ is induced by $\mathbb{R}^n_+$, then $\mathsf{t}(x - y) = 1$ for all $x \neq y$ in X. Hence any cyclic permutation of the elements of X is a top isometry in that case. An obvious example of an anti-chain of size $\binom{n}{m}$ is the set $\{x \in \{0, 1\}^n : \sum_i x_i = m\}$. It is a classic result in combinatorics that the maximal size of an anti-chain in $(\{0, 1\}^n, \leq)$ is $\binom{n}{\lfloor n/2 \rfloor}$. This result is known as Sperner's theorem [209] and is usually formulated as follows (cf. [28, section 3]).

Theorem 4.3.2 (Sperner) *If $\mathcal{A}$ is a collection of subsets of $\{1, \ldots, n\}$ such that there are no $A, B \in \mathcal{A}$ with $A \subseteq B$ and $A \neq B$, then $|\mathcal{A}| \leq \binom{n}{\lfloor n/2 \rfloor}$.*

Proof The proof of this theorem is classical and uses the LYM technique. We call a chain $A_0 \subseteq A_1 \subseteq \ldots \subseteq A_n$ a maximal chain in the partially ordered set $(\{1,\ldots,n\},\subseteq)$ if $|A_i| = i$ for all $0 \le i \le n$. We note that there are $n!$ maximal chains. If we consider $A \in \mathcal{A}$ and $|A| = k$, then A is contained in $k!(n-k)!$ maximal chains. Indeed, there are $k!$ chains $A_0 \subseteq \ldots \subseteq A_{k-1} \subseteq A$ with $k+1$ distinct elements, and each of these can be extended in $(n-k)!$ ways to a maximal chain. Now let m_k denote the number of elements of $\mathcal{A}$ with size k. Clearly $|\mathcal{A}| = \sum_k m_k$. As $\mathcal{A}$ is an anti-chain in $(\{1,\ldots,n\},\subseteq)$, each maximal chain contains at most one element of $\mathcal{A}$. Therefore

$$\sum_k m_k k!(n-k)! \le n!, \quad \text{so that} \quad \sum_k m_k \binom{n}{k}^{-1} \le 1.$$

From this (LYM) inequality it follows that

$$|\mathcal{A}| = \sum_k m_k \le \max_k \binom{n}{k} = \binom{n}{\lfloor n/2 \rfloor}.$$

□

By using the anti-chain $\{x \in \{0,1\}^n : \sum_i x_i = \binom{n}{\lfloor n/2 \rfloor}\}$, we see that there exists a topical map on $\mathbb{R}^n$ which has a periodic point with period $\binom{n}{\lfloor n/2 \rfloor}$. For instance, for $n = 4$ the map

$$f\begin{pmatrix} x_1 \\ x_2 \\ x_3 \\ x_4 \end{pmatrix} = \begin{pmatrix} (x_1 \wedge x_2) \vee (x_1 \wedge x_3) \vee (x_3 \wedge x_4) \\ (x_1 \wedge x_4) \vee (x_2 \wedge x_3) \vee (x_3 \wedge x_4) \\ (x_1 \wedge x_2) \vee (x_1 \wedge x_4) \vee (x_2 \wedge x_4) \\ (x_1 \wedge x_3) \vee (x_2 \wedge x_3) \vee (x_2 \wedge x_4) \end{pmatrix}$$

has $(1,1,0,0)$ as a period 6 point. Analogous to Conjecture 4.2.2 it seems reasonable to hypothesize that $\binom{n}{\lfloor n/2 \rfloor}$ is the optimal upper bound for the periods of periodic points of topical maps on $\mathbb{R}^n$, and this turns out to be true.

To prove this, we use arguments similar to the ones used in Section 4.1, but instead of the sup-norm we use the top function. To begin, we need the notion of a top-additive chain. A sequence $x^1, x^2, \ldots, x^m$ in $\mathbb{R}^n$ is called a *top-additive chain* if

$$\mathsf{t}(x^1 - x^m) = \sum_{k=1}^{m-1} \mathsf{t}(x^k - x^{k+1}).$$

We say that the top-additive chain has *length* m if it consists of m distinct points. Note that if $x^1, x^2, \ldots, x^m$ is a top-additive chain, then $x^m, \ldots, x^2, x^1$ need not be a top-additive chain, as $\mathsf{t}(x) \neq \mathsf{t}(-x)$ in general.

Lemma 4.3.3 *A sequence $x^1, x^2, \ldots, x^m$ in $\mathbb{R}^n$ is a top-additive chain with $\mathsf{t}(x^1 - x^m) = x_i^1 - x_i^m$ if and only if $\mathsf{t}(x^k - x^{k+1}) = x_i^k - x_i^{k+1}$ for all $1 \le k < m$.*

Proof First we remark that if $x^1, x^2, \dots, x^m$ is a top-additive chain and $\mathsf{t}(x^1 - x^m) = x_i^1 - x_i^m$, then

$$\mathsf{t}(x^1 - x^m) = \sum_{k=1}^{m-1} \mathsf{t}(x^k - x^{k+1}) \geq \sum_{k=1}^{m-1} x_i^k - x_i^{k+1} = x_i^1 - x_i^m = \mathsf{t}(x^1 - x^m),$$

so that $\mathsf{t}(x^k - x^{k+1}) = x_i^k - x_i^{k+1}$ for $1 \leq k < m$. On the other hand, if $\mathsf{t}(x^k - x^{k+1}) = x_i^k - x_i^{k+1}$ for $1 \leq k < m$, then

$$x_i^1 - x_i^m \leq \mathsf{t}(x^1 - x^m) \leq \sum_{k=1}^{m-1} \mathsf{t}(x^k - x^{k+1}) = \sum_{k=1}^{m-1} x_i^k - x_i^{k+1} = x_i^1 - x_i^m.$$

□

The following basic property can be proved by an argument analogous to the proof of Lemma 4.1.1.

Lemma 4.3.4 *If $X \subseteq \mathbb{R}^n$ has a transitive abelian group Γ of top isometries, then for each $g \in \Gamma$ there exists a constant $d(g)$ such that $\mathsf{t}(x - g(x)) = d(g)$ for all $x \in X$.*

By arguing as in Proposition 4.1.2 we deduce the following result.

Proposition 4.3.5 *Let $X \subseteq \mathbb{R}^n$ and suppose that X has a transitive abelian group Γ of top isometries. If x, y, z is a top-additive chain in X and $g \in \Gamma$ is such that $g(x) = y$, then $y, z, g(z)$ is also a top-additive chain in X.*

Proof Let $f \in \Gamma$ be such that $f(y) = z$. As Γ is abelian, it follows from Lemma 4.3.4 that $\mathsf{t}(y - g(z)) = \mathsf{t}(y - f(g(y))) = \mathsf{t}(x - f(g(x))) = \mathsf{t}(x - y) + \mathsf{t}(y - z) = \mathsf{t}(z - g(z)) + \mathsf{t}(y - z)$. Thus, $y, z, g(z)$ is a top-additive chain in X. □

We also need the analogous notion of an extreme pair for the top function. An ordered pair $(x, y) \in X \times X$, with $x \neq y$, is called a *top-extreme pair in* X if there exists no $z \in X$ such that z, x, y or x, y, z is a top-additive chain of length 3 in X. Remark that (x, y) being a top-extreme pair need not imply that (y, x) is a top-extreme pair. This asymmetry is useful and allows us to consider ordered pairs instead of unordered pairs. We also need the following proposition.

Proposition 4.3.6 *Let X be an anti-chain in $(\mathbb{R}^n, \leq)$, where $\leq$ is induced by $\mathbb{R}^n_+$. If X has a transitive abelian group Γ of top isometries, then $(x, y) \in X \times X$, with $x \neq y$, is a top-extreme pair if and only if there exists no $z \in X$ such that x, y, z is a top-additive chain of length 3 in X.*

Proof One implication follows directly from the definition. To prove the other one we suppose that $(x, y) \in X \times X$, with $x \neq y$, is not a top-extreme pair in X. Then there exists $z \in X$ such that either z, x, y or x, y, z is a top-additive chain of length 3 in X. If x, y, z is a top-additive chain of length 3 we are done. On the other hand, if z, x, y is top-additive chain of length 3, then $x, y, g(y)$, where $g \in \Gamma$ is such that $g(z) = x$, is also a top-additive chain by Proposition 4.3.5. To see that it has length 3 we observe that $t(u - v) > 0$ for all $u \neq v$, as X is an anti-chain in $(\mathbb{R}^n, \leq)$. Thus, $\mathsf{t}(y - g(y)) = \mathsf{t}(z - x) > 0$, as $z \neq x$, and $\mathsf{t}(x - g(y)) = \mathsf{t}(z - y) > 0$, as $z \neq y$. Therefore $x, y, g(y)$ has length 3. □

By using these preliminary observations we now prove the following result.

Theorem 4.3.7 *If X is a finite anti-chain in $(\mathbb{R}^n, \leq)$, where $\leq$ is induced by $\mathbb{R}^n_+$, and X has a transitive abelian group Γ of top isometries, then*

$$|X| \leq \binom{n}{\lfloor n/2 \rfloor}.$$

Proof The proof is quite similar to the proof of Theorem 4.1.3. Define a coding $c\colon X \to \{0, 1\}^n$ by

$$c(x)_i = \begin{cases} 1 & \text{if there exists } y \in X \text{ such that } (x, y) \text{ is a top-} \\ & \text{extreme pair in } X \text{ and } \mathsf{t}(x - y) = x_i - y_i, \\ 0 & \text{otherwise,} \end{cases} \tag{4.1}$$

for $1 \leq i \leq n$ and $x \in X$. By Sperner's Theorem 4.3.2, it suffices to show that $c(x) \not\leq c(y)$ for all $x \neq y$ in X. So let $x \neq y$ in X and consider a top-additive chain $z^1, z^2, \ldots, z^r$ in X, with $z^1 = x$ and $z^2 = y$, of maximal length r. We claim that (z^1, z^r) is a top-extreme pair in X. Indeed, if we suppose, for the sake of contradiction, that (z^1, z^r) is not a top-extreme pair, then there exists $u \in X$ such that z^1, z^r, u is a top-additive chain of length 3 in X by Proposition 4.3.6. This implies that $z^1, \ldots, z^r, u$ is a top-additive chain. To obtain the contradiction we show that it has length $r + 1$. If $u = z^k$, then $1 < k < r$ and

$$\begin{aligned} \mathsf{t}(z^1 - u) &= \mathsf{t}(z^1 - z^k) + \mathsf{t}(z^k - z^r) + \mathsf{t}(z^r - u) \\ &= \mathsf{t}(z^1 - u) + \mathsf{t}(u - z^r) + \mathsf{t}(z^r - u), \end{aligned}$$

so that $\mathsf{t}(z^r - u) + \mathsf{t}(u - z^r) = 0$. This contradicts the fact that $\mathsf{t}(z^r - u) > 0$ and $\mathsf{t}(u - z^r) > 0$, as $z^r \neq u$ and X is an anti-chain in $(\mathbb{R}^n, \leq)$. Thus, (z^1, z^r) is a top-extreme pair in X.

We know $\mathsf{t}(z^1 - z^r) = z^1_j - z^r_j$ for some j, so that $c(x)_j = 1$. We claim that $c(y)_j \neq 1$. If $c(y)_j = 1$, then there exists $v \in X$ such that (y, v) is top-extreme pair in X and $\mathsf{t}(y - v) = y_j - v_j$. As $\mathsf{t}(z^1 - z^r) = z^1_j - z^r_j$, it follows from Lemma 4.3.3 that $\mathsf{t}(x - y) = x_j - y_j$. Applying Lemma 4.3.3 once more gives that x, y, v is a top-additive chain in X. This additive chain has length 3, as $x = v$ implies $y_j - x_j = y_j - v_j = \mathsf{t}(y - v) = \mathsf{t}(y - x) > 0$, which contradicts the fact that $x_j - y_j = \mathsf{t}(x - y) > 0$. Now note that x, y, v cannot be a top-additive chain of length 3, since (y, v) is a top-extreme pair in X. Hence $c(y)_j \neq 1$ and therefore $c(y)_j = 0$. By interchanging the roles of x and y, there also exists $1 \leq i \leq n$ such that $c(y)_i = 1$ and $c(x)_i = 0$. This implies that $c(x) \not\leq c(y)$, and we are done. □

As each periodic orbit of a topical map $f : \mathbb{R}^n \to \mathbb{R}^n$ has a transitive cyclic group of top isometries, Theorem 4.3.7 has the following consequence.

Corollary 4.3.8 *The optimal upper bound for the periods of periodic points of topical maps* $f : \mathbb{R}^n \to \mathbb{R}^n$ *is* $\binom{n}{\lfloor n/2 \rfloor}$.

It turns out that this result can be generalized to sub-topical maps. Recall that a map $g : \mathbb{R}^n \to \mathbb{R}^n$ is sub-topical if it is additively subhomogeneous and order-preserving with respect to $\mathbb{R}^n_+$. Any sub-topical map $g : \mathbb{R}^n \to \mathbb{R}^n$ can be turned into a topical map $G : \mathbb{R}^{n+1} \to \mathbb{R}^{n+1}$ by adding one extra coordinate in the following way. Define

$$G\begin{pmatrix} x \\ x_{n+1} \end{pmatrix} = \begin{pmatrix} x_{n+1}\mathbb{1} + g(x - x_{n+1}\mathbb{1}) \\ x_{n+1} \end{pmatrix} \quad \text{for} \quad \begin{pmatrix} x \\ x_{n+1} \end{pmatrix} \in \mathbb{R}^{n+1}. \tag{4.2}$$

The map G has the following properties.

Lemma 4.3.9 *If* $g : \mathbb{R}^n \to \mathbb{R}^n$ *is sub-topical and* $G : \mathbb{R}^{n+1} \to \mathbb{R}^{n+1}$ *is given by (4.2), then*

(i) G is topical, and
(ii) if $x \in \mathbb{R}^n$ *is a periodic point of g with period p, then* $(x, 0)$ *is a periodic point with period p of G.*

Proof It is clear from the definition that G is additively homogeneous. To see that G is order-preserving, assume that $(x, x_{n+1}) \leq (y, y_{n+1})$. Note that

$$\begin{aligned}(y_{n+1} - x_{n+1})\mathbb{1} + g(y - y_{n+1}\mathbb{1}) - g(x - x_{n+1}\mathbb{1}) &\geq g(y - x_{n+1}\mathbb{1}) \\ &\quad - g(x - x_{n+1}\mathbb{1}) \geq 0,\end{aligned}$$

as g is additively subhomogeneous and order-preserving. So, G is order-preserving.

Obviously, $G^k(x, 0) = (g^k(x), 0)$ for all $k \geq 1$ and therefore $(x, 0)$ is periodic point of G with period p, if x is a periodic point of g with period p. □

Using the map G it is easy to prove the following generalization of Corollary 4.3.8.

Theorem 4.3.10 *The optimal upper bound for the periods of periodic points of sub-topical maps* $g: \mathbb{R}^n \to \mathbb{R}^n$ *is* $\binom{n}{\lfloor n/2 \rfloor}$.

Proof Let $\xi \in \mathbb{R}^n$ be a periodic point of a sub-topical map $g: \mathbb{R}^n \to \mathbb{R}^n$, with period p. Let $G: \mathbb{R}^{n+1} \to \mathbb{R}^{n+1}$ be given by (4.2). It follows from Lemma 4.3.9 that G is a topical map, which has $(\xi, 0)$ as a periodic point with period p. Let $X \subseteq \mathbb{R}^{n+1}$ denote the periodic orbit of $(\xi, 0)$ under G. As G is order-preserving and non-expansive with respect to the top function, X is an anti-chain in $(\mathbb{R}^{n+1}, \leq)$ and X has a transitive abelian group of top isometries.

Now define $c: X \to \{0, 1\}^{n+1}$ as in (4.1). We know from the proof of Theorem 4.3.7 that $c(X)$ is an anti-chain in $(\{0, 1\}^{n+1}, \leq)$ of size $|X|$. We also know that $c(x)_{n+1} = 0$ for all $x \in X$, as $x_{n+1} = 0$ for all $x \in X$. By restricting each element of $c(X)$ to the first n coordinates, we obtain an anti-chain in $(\{0, 1\}^n, \leq)$ of size $|c(X)| = |X| = p$. Thus, it follows from Sperner's Theorem 4.3.2 that $p \leq \binom{n}{\lfloor n/2 \rfloor}$. □

5

Eigenvectors and eigenvalues of nonlinear cone maps

In this chapter we will lay the foundations for the spectral theory of continuous order-preserving homogeneous maps $f: K \to K$, where K is a solid closed cone in a finite-dimensional vector space V. Two sensible definitions for the cone spectral radius, $r_K(f)$, of f are given, and it is shown that they coincide. The main result, which generalizes the Kreĭn–Rutman Theorem 1.1.6, tells us that there exists an eigenvector $x \in K$, with $x \neq 0$, corresponding to $r_K(f)$; so,

$$f(x) = r_K(f)x.$$

It turns out that the spectral radius $r_K(f)$ satisfies a "minimax" variational formula analogous to the Collatz–Wielandt formula for the cone spectral radius of nonnegative matrices. We will also discuss continuity properties of $r_K(f)$, and analyze the cone spectrum,

$$\sigma_K(f) = \{\lambda \geq 0 \colon f(x) = \lambda x \text{ for some } x \in K \setminus \{0\}\}.$$

Before we begin developing the theory we need to clarify a practical matter concerning continuous extensions of order-preserving maps.

5.1 Extensions of order-preserving maps

It frequently happens that we are given an order-preserving subhomogeneous map that is defined on the interior of a closed cone. Natural examples arise when we have a sub-topical map $g: \mathbb{R}^n \to \mathbb{R}^n$. In that case the log-exp transform of g is an order-preserving subhomogeneous map on the interior of $\mathbb{R}^n_+$. In such situations it is often useful to know if the map admits a continuous order-preserving subhomogeneous extension to the boundary of the cone. In this section we discuss this problem. It turns out that polyhedral cones play a key role in the solution.

Let K be a solid closed cone in V and $x \in \partial K$. We say that $(x_k)_k$ in $\text{int}(K)$ *converges to x from above* if $x_k \to x$, as $k \to \infty$, and $x \ll_K x_k$ for all $k \geq 1$. Such sequences always exist if K is solid; a simple example is the sequence $x_k = x + u/k$, where $u \in \text{int}(K)$.

Lemma 5.1.1 *Let $K \subseteq V$ be a solid closed cone and $x \in \partial K$. If $(x_k)_k$ converges to x from above, then there exists a subsequence $(x_{k_i})_i$ such that $x_m \ll_K x_{k_i}$ for all $m \geq k_{i+1}$.*

Proof The subsequence $(x_{k_i})_i$ is defined inductively. Put $x_{k_1} = x_1$ and suppose that we have found $k_1 < k_2 < \ldots < k_i$ that satisfy the assertion. As $x_{k_i} - x \in \text{int}(K)$, there exists a neighborhood U_i of $x_{k_i} - x$ with $U_i \subseteq \text{int}(K)$. Now let V_i be an open symmetric neighborhood of 0 such that $x_{k_i} - x - v \in U_i$ for all $v \in V_i$. Since $(x_k)_k$ converges to x, there exists $k_{i+1} > k_i$ such that $x_m - x \in V_i$ for all $m \geq k_{i+1}$. This implies that $x_{k_i} - x_m = x_{k_i} - x - (x_m - x) \in U_i$ for all $m \geq k_{i+1}$, and we are done. □

By using this lemma the following extension result from [41] is proved.

Theorem 5.1.2 *If $K \subseteq V$ and $K' \subseteq V'$ are closed cones, where K is solid, and $f: \text{int}(K) \to K'$ is an order-preserving continuous map, then the following assertions are true:*

(i) If $(x_k)_k$ converges to x from above, then there exists $y_x \in K'$ such that

$$\lim_{k \to \infty} f(x_k) = y_x. \tag{5.1}$$

(ii) The map $F: K \to K'$ defined by $F(x) = y_x$ is a well-defined order-preserving extension of f to ∂K.

(iii) If, in addition, f is (sub)homogeneous, then $F: K \to K'$ is also (sub)homogeneous.

Proof By Lemma 5.1.1 there exists a subsequence $(x_{k_i})_i$ such that $x_m \ll_K x_{k_i}$ for all $m \geq k_{i+1}$. In particular, we get that $0 \leq_K x_{k_{i+1}} \ll_K x_{k_i}$ for all $i \geq 1$, and hence $0 \leq_{K'} f(x_{k_{i+1}}) \leq_{K'} f(x_{k_i})$, as f is order-preserving. This implies that there exists $y_x \in K'$ such that $\lim_{i \to \infty} f(x_{k_i}) = y_x$, since K' is a normal cone.

Let $\varepsilon > 0$ and $i \geq 1$ be such that $\|f(x_{k_i}) - y_x\| < \varepsilon$. Suppose that $m \geq k_{i+1}$; so, $x_m \ll_K x_{k_i}$. As $(x_{k_i})_i$ converges to x from above, there exists $j \geq i+1$, such that $x_{k_j} \leq_K x_m$. Otherwise, $x_m \leq_K x_{k_j}$ for all $j \geq i+1$, so that $x_m \leq_K x$, as K is closed. This contradicts the fact that $0 \ll_K x_m - x$. Thus, for each $j \geq 1$

sufficiently large, we have that $x_{k_j} \leq_K x_m \leq_K x_{k_i}$. As f is order-preserving, $f(x_{k_j}) \leq_{K'} f(x_m) \leq_{K'} f(x_{k_i})$. This implies that

$$\|f(x_m) - f(x_{k_j})\| \leq M\|f(x_{k_i}) - f(x_{k_j})\|,$$

since K' is normal. By letting $j \to \infty$, we get that

$$\|f(x_m) - y_x\| \leq M\|f(x_{k_i}) - y_x\| < M\varepsilon.$$

Thus, $(f(x_k))_k$ converges to y_x, which completes the proof of the first assertion.

To verify that F is well defined, we let $(x'_k)_k$ be another sequence that converges to x from above. Define $(z_k)_k$ by $z_{2k} = x_k$ and $z_{2k-1} = x'_k$ for all $k \geq 1$. Clearly $(z_k)_k$ also converges to x from above. Moreover $\lim_{k\to\infty} f(x_k) = \lim_{k\to\infty} f(z_k) = \lim_{k\to\infty} f(x'_k)$ by the first assertion. Thus, F is a well-defined extension of f. To show that F is order-preserving, we remark that if $x \leq_K y$, then $(x + u/k)_k$ converges to x from above and $(y + u/k)_k$ converges to y from above for $u \in \text{int}(K)$. For each $k \geq 1$ we have that $x + u/k \leq_K y + u/k$, so that $f(x + u/k) \leq_{K'} f(y + u/k)$. Thus, $\lim_{k\to\infty} f(x + u/k) \leq_{K'} \lim_{k\to\infty} f(y + u/k)$ and hence $F(x) \leq_{K''} F(y)$.

Suppose that f is subhomogeneous and let $x \in K$ and $0 < \lambda < 1$. If $(x_k)_k$ converges to x from above, then $(\lambda x_k)_k$ also converges to λx from above. Therefore $\lambda F(x) = \lim_{k\to\infty} \lambda f(x_k) \leq_{K'} \lim_{k\to\infty} f(\lambda x_k) = F(\lambda x)$. A similar argument can be given if f is homogeneous. □

In general the extension $F\colon K \to K'$ given in Theorem 5.1.2 need not be continuous, as can be seen from the following example. Let Sym_n denote the real vector space of $n \times n$ symmetric matrices. Denote by $\Pi_n(\mathbb{R})$ the set of all positive-semidefinite matrices in Sym_n and put $K = \Pi_n(\mathbb{R}) \times \Pi_n(\mathbb{R})$. Note that K is a closed cone in $\text{Sym}_n \times \text{Sym}_n$ and recall that $\Pi_n(\mathbb{R})$ is a solid closed cone in Sym_n. Consider the map $f\colon \text{int}(K) \to \Pi_n(\mathbb{R})$ given by

$$f(A, B) = (A^{-1} + B^{-1})^{-1} \quad \text{for all } (A, B) \in \text{int}(K).$$

It follows from Lemma 1.4.2 that $g(A) = A^{-1}$ is order-reversing with respect to $\Pi_n(\mathbb{R})$. Therefore f is order-preserving. It is easy to see that f is homogeneous. Thus, f is non-expansive with respect to Thompson's metric, and hence continuous on $\text{int}(K)$. However, we shall see that even for $n = 2$ the extension of f given in Theorem 5.1.2 is not continuous. Take

$$A = B = \begin{pmatrix} 1 & 0 \\ 0 & 0 \end{pmatrix} \in \partial\Pi_n(\mathbb{R}),$$

and for $k \in \mathbb{N}$ let $A_k = A + k^{-1}I$ and $B_k = B + k^{-1}I$. Thus, $((A_k, B_k))_k$ converges to (A, B) from above. By using the equality $f(X, Y) = X(X+Y)^{-1}Y = Y(X+Y)^{-1}X$ for $X, Y \in \operatorname{int}(\Pi_n(\mathbb{R}))$, a simple calculation shows that

$$F(A, B) = \lim_{k\to\infty} f(A_k, B_k) = \begin{pmatrix} 1/2 & 0 \\ 0 & 0 \end{pmatrix}.$$

On the other hand, if we consider

$$A'_k = \begin{pmatrix} 1 & 1/k \\ 1/k & m/k^2 \end{pmatrix} \quad \text{and} \quad B'_k = \begin{pmatrix} 1 & 2/k \\ 2/k & 4m/k^2 \end{pmatrix},$$

where $k, m \in \mathbb{N}$ and $m > 1$, then $((A'_k, B'_k))_k$ also converges to (A, B), but

$$\lim_{k\to\infty} f(A'_k, B'_k) \neq \begin{pmatrix} 1/2 & 0 \\ 0 & 0 \end{pmatrix}.$$

Indeed, a straightforward computation shows that

$$\lim_{k\to\infty} f(A'_k, B'_k) = \begin{pmatrix} a & 0 \\ 0 & 0 \end{pmatrix},$$

where $a = (5m-5)/(10m-9) < 1/2$. Consequently F is not continuous.

The cone $\Pi_n(\mathbb{R})$ is not polyhedral by Lemma 2.4.1. It turns out that the extension $F\colon K \to K'$ defined in Theorem 5.1.2 is always continuous, if K is polyhedral. This is due to the following geometric property, which, as we shall prove, characterizes polyhedral cones.

Definition 5.1.3 A closed cone $K \subseteq V$ is said to satisfy *condition* G *at* $x \in K$ if, for every sequence $(x_k)_k \subseteq K$ with $\lim_{k\to\infty} x_k = x$ and each $0 < \lambda < 1$, there exists $m \in \mathbb{N}$ such that $\lambda x \leq x_k$ for all $k \geq m$. We say that K satisfies *condition* G if condition G is satisfied at every point in K.

Note that if K is a solid closed cone, condition G is satisfied at every point $x \in \operatorname{int}(K)$. However, for K to satisfy condition G at each point in ∂K, it needs to be polyhedral, as the following result from [41] shows.

Lemma 5.1.4 *A solid closed cone $K \subseteq V$ is polyhedral if and only if it satisfies condition* G*.*

Proof If $K \subseteq V$ is a polyhedral cone, there exist finitely many linear functionals $\psi_1, \ldots, \psi_N$ such that $K = \{x \in V : \psi_i(x) \geq 0 \text{ for } i = 1, \ldots, N\}$. Let $(x_k)_k$ be a convergent sequence in K with limit x, and let $0 < \lambda < 1$. As $x_k \to x$, we have that $\psi_i(x_k) \to \psi_i(x)$ for $1 \leq i \leq m$. If $\psi_i(x) > 0$, then $\psi_i(x_k) - \psi_i(\lambda x) \to (1-\lambda)\psi_i(x) > 0$. In that case there exists $m_i \geq 1$ such that $\psi_i(x_k) \geq \psi_i(\lambda x)$ for all $k \geq m_i$. If $\psi_i(x) = 0$, then clearly $0 = \psi_i(\lambda x) \leq \psi_i(x_k)$ for all $k \geq 1$ and we put $m_i = 1$. Now let

$m = \max\{m_i : i = 1, \dots, N\}$. Then $\psi_i(\lambda x) \le \psi_i(x_k)$ for all $k \ge m$. This implies that $x_k - \lambda x \in K$ for all $k \ge m$.

To show the opposite implication, we take $\varphi \in \mathrm{int}(K^*)$ and let $\Sigma = \{x \in K : \varphi(x) = 1\}$. Recall from Lemma 1.2.4 that such a positive functional φ exists and Σ is a compact and convex set. Clearly K is polyhedral if Σ is a polyhedron. Thus, it suffices to prove that Σ has finitely many extreme points.

Let $x \in \Sigma$ and $0 < \lambda < 1$. It follows from condition G that there exists a neighborhood U_x of x such that $\lambda x \le u$ for all $u \in U_x \cap K$. Let $y \ne x$ in $U_x \cap \Sigma$ and for $0 < \alpha \le \lambda$, put $z = y/(1-\alpha) - \alpha x/(1-\alpha)$. Remark that $z \in \Sigma$, as $\varphi(z) = 1$ and $0 < \alpha \le \lambda$. Thus, $y = (1-\alpha)z + \alpha x$ is not an extreme point of Σ. This implies that U_x contains at most one extreme point of Σ. As Σ is compact, $\cup_{x\in\Sigma} U_x$ has a finite subcover of Σ, and hence Σ has finitely many extreme points. □

By exploiting property G of polyhedral cones, we now prove the following result.

Theorem 5.1.5 *Let $K \subseteq V$ and $K' \subseteq V'$ be closed cones and suppose that K is a solid polyhedral cone. If $f : \mathrm{int}(K) \to K'$ is order-preserving and (sub)homogeneous, then $F : K \to K'$ given in Theorem 5.1.2 is a continuous order-preserving (sub)homogeneous extension of f.*

Proof By Theorem 5.1.2 it remains to be shown that F is continuous. As f is non-expansive under Thompson's metric on $\mathrm{int}(K)$, F is continuous on $\mathrm{int}(K)$. Now let $(x_k)_k \subseteq K$ be a sequence that converges to $x \in \partial K$. Take an increasing sequence $(\lambda_k)_k \subseteq (0, 1)$ with $\lambda_k \to 1$. For each $k \ge 1$ there exists $m_k \ge 1$ such that $\lambda_k x \le_K x_m$ for all $m \ge m_k$. On the other hand, there exists $u \in \mathrm{int}(K)$ and a decreasing sequence $(\delta_k)_k$ with $\delta_k \to 0$ such that

$$x - \delta_k u \le_K x_m \le_K x + \delta_k u \quad \text{for all } m \ge m_k,$$

and hence $\lambda_k F(x) \le_{K'} F(\lambda_k x) \le_{K'} F(x_m) \le_{K'} F(x + \delta_k u)$ for all $m \ge m_k$. This implies that

$$0 \le_{K'} F(x_m) - \lambda_k F(x) \le_{K'} f(x + \delta_k u) - \lambda_k F(x) \quad \text{for all } m \ge m_k.$$

and hence $\lim_{k\to\infty} F(x_k) = F(x)$. □

5.2 The cone spectrum

The analysis of the eigenvalues and eigenvectors of order-preserving homogeneous maps on closed cones in finite-dimensional vector spaces has numerous similarities with the classical Perron–Frobenius theory, but also some marked contrasts. For instance, there exist continuous order-preserving homogeneous maps $f: K \to K$ that have infinitely many distinct eigenvalues. This may come as a surprise, especially since K is only assumed to be finite-dimensional. The purpose of this section is to discuss this issue in detail. Given an order-preserving homogeneous map $f: K \to K$ we say that $\lambda \geq 0$ is an *eigenvalue* of f if there exists $x \in K$ with $x \neq 0$ such that $f(x) = \lambda x$, and denote the *cone spectrum* by

$$\sigma_K(f) = \{\lambda \geq 0\colon \lambda \text{ is an eigenvalue of } f\}.$$

The following result is elementary, but useful.

Lemma 5.2.1 *Suppose that $f: K \to K$ is an order-preserving homogeneous map on a closed cone $K \subseteq V$ and $x, y \in K \setminus \{0\}$ are such that y dominates x. If $\lambda x \leq_K f(x)$ and $f(y) \leq_K \mu y$, then $\lambda \leq \mu$.*

Proof The case $\lambda = 0$ is trivial. So, assume that $\lambda > 0$. Note that as y dominates x, there exists $\beta > 0$ such that $x \leq_K \beta y$. This implies that $\lambda^k x \leq_K f^k(x) \leq_K \beta f^k(y) \leq_K \beta \mu^k y$, so that

$$\left(\frac{\mu}{\lambda}\right)^k y - \frac{1}{\beta} x \in K$$

for all $k \geq 1$. Letting $k \to \infty$ we see that if $\mu < \lambda$, then $-x/\beta \in K$, which is impossible. Thus, $\lambda \leq \mu$, which completes the proof. □

As a consequence we derive the following result.

Corollary 5.2.2 *If $f: K \to K$ is an order-preserving homogeneous map on a closed cone $K \subseteq V$, and f has eigenvectors x and y in K with $x \sim_K y$, then the eigenvalues of x and y are equal.*

Proof Suppose that $f(x) = \lambda x$ and $f(y) = \mu y$. As $x \sim_K y$, we know that x dominates y, and y dominates x. So, it follows from Lemma 5.2.1 that $\lambda = \mu$. □

As $\{0\}$ is a part of any cone K, we see that the number of distinct eigenvalues of an order-preserving homogeneous map $f: K \to K$ is bounded by $m - 1$, where m is the number of parts of K. In case $K = \mathbb{R}^n_+$ it is easy to construct an example that shows that the upper bound is sharp. Indeed, given $I \subseteq \{1, \ldots, n\}$, define $\mathbb{1}^I \in \mathbb{R}^n_+$ by $\mathbb{1}^I_i = 1$ if $i \in I$, and $\mathbb{1}^I_i = 0$ otherwise. Furthermore, let $\lambda_I > 0$ be such that $\lambda_I < \lambda_J$ if $I \subseteq J$ and $I \neq J$. Now consider $f: \mathbb{R}^n_+ \to \mathbb{R}^n_+$ given by

$$f(x) = \bigvee_{\emptyset \neq I \subseteq \{1,\ldots,n\}} \lambda_I (\min_{i \in I} x_i) \mathbb{1}^I \qquad \text{for } x \in \mathbb{R}^n_+.$$

Clearly f is a continuous order-preserving homogeneous map on $\mathbb{R}^n_+$. Moreover, for $J \subseteq \{1, \ldots, n\}$ with $J \neq \emptyset$ we have that

$$f(\mathbb{1}^J) = \bigvee_{\emptyset \neq I \subseteq J} \lambda_I \mathbb{1}^I = \lambda_J \mathbb{1}^J,$$

as $\lambda_I < \lambda_J$ for $I \subseteq J$ with $I \neq J$. Thus, we can arrange f to have $2^n - 1$ distinct eigenvalues.

It turns out that the upper bound, $m - 1$, is sharp for every polyhedral cone. Before we prove this we introduce some notation. Let $K \subseteq V$ be a solid polyhedral cone with N facets and facet-defining functionals $\psi_1, \ldots, \psi_N$. For $r < 0$ and $I \subseteq \{1, \ldots, N\}$ non-empty, define $M_r(I)\colon \mathrm{int}(K) \to [0, \infty)$ by

$$M_r(I)(x) = \Big(\sum_{i \in I} \psi_i(x)^r\Big)^{1/r} \qquad \text{for } x \in \mathrm{int}(K).$$

Note that $M_r(I)$ has a continuous order-preserving homogeneous extension to ∂K by Theorem 5.1.2. Moreover, if $x \in \partial K$ and there exists $i \in I$ with $\psi_i(x) = 0$, then $M_r(I)(x) = 0$, since $r < 0$. For $r = -\infty$ we let $M_{-\infty}(I)\colon K \to [0, \infty)$ be given by

$$M_{-\infty}(I)(x) = \min_{i \in I} \psi_i(x) \qquad \text{for } x \in K.$$

Theorem 5.2.3 *Let $K \subseteq V$ be a solid polyhedral cone with m faces. If $f\colon K \to K$ is a homogeneous order-preserving map, then $|\sigma_K(f)| \leq m - 1$. Moreover, there exists a continuous homogeneous order-preserving map on K with $m - 1$ distinct eigenvalues.*

Proof It follows from Lemma 1.2.2 that if K is a polyhedral cone with m faces, then K has m parts. Omitting the trivial part $\{0\}$ and using Corollary 5.2.2 we find that $m - 1$ is an upper bound for the size of $\sigma_K(f)$.

To construct an example with $m - 1$ distinct eigenvalues on a polyhedral cone K with m faces, we let $\psi_1, \ldots, \psi_N$ denote the facet-defining functionals of K. Let $\mathcal{P}(K)$ denote the collection of parts of K, so $|\mathcal{P}(K)| = m$. Recall that each part of K is the relative interior of a face of K; see Lemma 1.2.2. Moreover by Lemma 1.2.3 we know that for each $P \in \mathcal{P}(K)$ we have that

$$P = \{x \in K : \psi_i(x) > 0 \text{ if and only if } i \in I(P)\},$$

where $I(P) = \{i : \psi_i(x) > 0 \text{ for some } x \in P\}$

For each $P \in \mathcal{P}(K)$ with $P \neq \{0\}$ select $z^P \in P$. The idea is to construct a continuous order-preserving homogeneous map which has the points

z^P as eigenvectors with distinct eigenvalues. Let $r \in [-\infty, 0)$. We will use a continuous order-preserving homogeneous map $f_r : K \to K$ of the form

$$f_r(x) = \sum_{P \in \mathcal{P}(K), P \neq \{0\}} \lambda_P M_r(I(P))(x) u^P \qquad \text{for } x \in K, \tag{5.2}$$

where $\lambda_P > 0$ and $u^P \in P$ are chosen appropriately.

Recall that $\mathcal{P}(K)$ has a natural partial ordering $\trianglelefteq$ given by $P \trianglelefteq Q$ if for each $y \in Q$ and $x \in P$ there exists $\beta > 0$ such that $x \leq_K \beta y$. Furthermore recall from Lemma 1.2.3 that $P \trianglelefteq Q$ if and only if $I(P) \subseteq I(Q)$. It follows that for each $x \in Q$ and $P \in \mathcal{P}(K)$ with $P \neq \{0\}$ we have that $M_r(I(P))(x) > 0$ if $P \trianglelefteq Q$, and $M_r(I(P))(x) = 0$ otherwise.

We will define $\lambda_P > 0$ and $u^P \in P$ inductively using the height of P with respect to $\trianglelefteq$. For $P \in \mathcal{P}(K)$ with $\dim P = 1$, so that P has height 1, take $u^P = z^P$ and chose $\lambda_P > 0$ such that the positive numbers $\mu_P = \lambda_P M_r(I(P))(u^P)$ are all distinct for $P \in \mathcal{P}(K)$ with height 1. So, $f(z^P) = \mu_P z^P$ for those parts P.

Now suppose that we have already selected $u^P \in P$ and $\lambda_P > 0$ for all $P \trianglelefteq Q$ with $\{0\} \neq P \neq Q$. Consider

$$f_r(z^Q) = \sum_{P \in \mathcal{P}(K), P \neq \{0\}} \lambda_P M_r(I(P))(z^Q) u^P = \sum_{P \trianglelefteq Q, P \neq \{0\}} \lambda_P M_r(I(P))(z^Q) u^P$$

and write

$$w^Q = \sum_{P \trianglelefteq Q, \{0\} \neq P \neq Q} \lambda_P M_r(I(P))(z^Q) u^P.$$

Note that for each $\mu_Q > 0$ sufficiently large we have that $\mu_Q z^Q - w^Q \in Q$. Take $\lambda_Q > 0$ and $u^Q \in Q$ such that

$$\lambda_Q M_r(I(Q))(z^Q) u^Q = \mu_Q z^Q - w^Q.$$

Recall that $M_r(I(Q))(z^Q) > 0$; so, once we have fixed $\mu_Q > 0$ and $\lambda_Q > 0$ the vector $u^Q \in Q$ is uniquely determined.

It follows from the construction that $f_r(z^Q) = w^Q + \mu_Q z^Q - w^Q = \mu_Q z^Q$. Thus, we can chose $\mu_Q > 0$ such that f_r has $|\mathcal{P}(K)| - 1 = m - 1$ distinct eigenvalues. □

If $-\infty < r < 0$, the function f we constructed in Theorem 5.2.3 is infinitely differentiable on $\text{int}(K)$. One could ask whether there exists a continuous homogeneous order-preserving map on K as in Theorem 5.2.3 which is C^1 on $\text{int}(K)$ and for which $Df(x)$ extends continuously to 0. For $\dim K > 1$ the answer is trivially no: If L denotes the limit of $Df(x)$ as $x \to 0$, L would necessarily have $m - 1$ distinct eigenvalues.

The following result shows that the cone spectrum need not be finite if the cone is not polyhedral. To prove it we need Straszewicz's theorem [210] (see also [186, p. 167]), which says that the exposed points of a closed convex set S in a finite-dimensional vector space V are dense in the extreme points of S.

Recall that a point x in a closed convex set $C \subseteq V$ is said to be *exposed* if there exists a hyperplane $H \subseteq V$ with $\dim(H) = \dim(V) - 1$ such that $H \cap C = \{x\}$.

Theorem 5.2.4 *If $K \subseteq V$ is a solid closed cone, then there exists a continuous order-preserving homogeneous map $f: K \to K$ with infinitely many distinct eigenvalues if and only if K is nonpolyhedral.*

Proof By Theorem 5.2.3 it suffices to construct for each solid closed nonpolyhedral cone $K \subseteq V$ a continuous order-preserving homogeneous map $f: K \to K$ with infinitely many distinct eigenvalues.

Let $u \in \text{int}(K)$ and note that if K is nonpolyhedral, then K^* is also nonpolyhedral. Define $\Sigma^* = \{\varphi \in K^*: \varphi(u) = 1\}$, which is a compact convex set in V^* by Lemma 1.2.4. As K^* is nonpolyhedral, Σ^* is not a polyhedron, and hence it has infinitely many extreme points. By Straszewicz's theorem [210], we know that the exposed points of Σ^* are dense in the extreme points of Σ^*. Therefore we can find a sequence $(\varphi_k)_k$ in Σ^* of distinct exposed points such that $\varphi_k \to \psi \in \Sigma^*$ as $k \to \infty$ and $\varphi_k \neq \psi$ for all $k \geq 1$.

As φ_k is an exposed point of Σ^*, $F = \{\lambda\varphi_k : \lambda \geq 0\}$ is an exposed face of K^*. So, as $K^{**} = K$, there exists $x^k \in K$ with $\|x^k\| = 1$ such that $\varphi_k(x^k) = 0$ and $\varphi(x^k) > 0$ for all $\varphi \in \Sigma^* \setminus \{\varphi_k\}$. In particular,

$$\psi(x^k) > 0 \quad \text{and} \quad \varphi_m(x^k) > 0 \quad \text{for all } m \neq k. \tag{5.3}$$

Now define $f: K \to K$ as follows:

$$f(x) = \sum_{k=1}^{\infty} \left(\frac{1}{2}\right)^k \lambda_k \inf_{m \neq k} \varphi_m(x) x^k \qquad \text{for } x \in K,$$

where $(\lambda_k)_k$ is a bounded sequence of strictly positive reals. Note that for each $k \neq q$ we have that

$$\lambda_k \inf_{m \neq k} \varphi_m(x^q) = 0,$$

as $\varphi_q(x^q) = 0$. On the other hand, if $k = q$, then

$$\lambda_q \inf_{m \neq q} \varphi_m(x^q) > 0,$$

by (5.3) and the fact that $\varphi_k \to \psi$ as $k \to \infty$, $\varphi_k \neq \varphi_q$ for all $k \neq q$, and $\varphi_q \neq \psi$.

Thus for each $q \geq 1$ we have that

$$f(x^q) = \left(\frac{1}{2}\right)^q \lambda_q \inf_{m \neq q} \varphi_m(x^q) x^q,$$

which shows that x^q is an eigenvector of f. We can select $0 < \lambda_q \leq 1$ in such a way as to ensure that f has infinitely many distinct eigenvalues. □

The cone spectrum of the map f constructed in the proof of Theorem 5.2.4 is a countably infinite set. The following example shows that the cone spectrum of a continuous order-preserving homogeneous map may contain a continuum, even if the cone is finite-dimensional.

Example 5.2.5 The example we give is defined on the cone of positive-semidefinite symmetric 2×2 matrices, $\Pi_2(\mathbb{R})$, but similar examples exist on $\Pi_n(\mathbb{R})$ for $n > 2$. Recall from Section 2.4 that $\Pi_2(\mathbb{R})$ is a self-dual cone with respect to the inner-product $\langle R, S\rangle = \mathrm{tr}(RS)$. Let

$$A = \begin{pmatrix} 1 & 0 \\ 0 & 0 \end{pmatrix} \in \partial \Pi_2(\mathbb{R})$$

and define $f\colon \Pi_2(\mathbb{R}) \to \Pi_2(\mathbb{R})$ by

$$f(X) = \Big(\mathrm{tr}(XA)X\Big)^{1/2} \quad \text{for } X \in \Pi_2(\mathbb{R}).$$

Clearly if $X \leq Y$, then $\mathrm{tr}(XA) \leq \mathrm{tr}(YA)$, as $A \in \Pi_2(\mathbb{R})$. It follows that

$$\mathrm{tr}(XA)X \leq \mathrm{tr}(YA)X \leq \mathrm{tr}(YA)Y,$$

which implies that $f(X) \leq f(Y)$, since the map $R \mapsto R^{1/2}$ is order-preserving on $\Pi_2(\mathbb{R})$ (see Section 1.4). Note also that f is homogeneous.

For $0 \leq \alpha \leq 1$ let

$$X_\alpha = \begin{pmatrix} 1-\alpha & 0 \\ 0 & \alpha \end{pmatrix} \in \Pi_2(\mathbb{R}).$$

So, $\mathrm{tr}(X_\alpha A) = 1 - \alpha$ and

$$f(X_\alpha) = \begin{pmatrix} (1-\alpha)^2 & 0 \\ 0 & (1-\alpha)\alpha \end{pmatrix}^{1/2} = \begin{pmatrix} (1-\alpha) & 0 \\ 0 & \sqrt{(1-\alpha)\alpha} \end{pmatrix}.$$

We see that if $f(X_\alpha) = \lambda X_\alpha$, then $\lambda = 1$ and $\sqrt{(1-\alpha)\alpha} = \alpha$, which is equivalent to $\alpha = 0$ or $\alpha = 1/2$. Thus, X_0 and $X_{1/2}$ are eigenvectors of f with eigenvalue 1. As $X_{1/2}$ lies in the interior of the cone, it follows from Lemma 5.2.1 that the maximum eigenvalue of f is equal to 1.

Now consider $Z \in \partial\Pi_2(\mathbb{R})$ with $\mathrm{tr}(Z) = 1$. Note that Z can be written as

$$Z = \begin{pmatrix} \cos\vartheta & -\sin\vartheta \\ \sin\vartheta & \cos\vartheta \end{pmatrix} X_0 \begin{pmatrix} \cos\vartheta & \sin\vartheta \\ -\sin\vartheta & \cos\vartheta \end{pmatrix}$$
$$= \begin{pmatrix} \cos^2\vartheta & \cos\vartheta\sin\vartheta \\ \cos\vartheta\sin\vartheta & \sin^2\vartheta \end{pmatrix}$$

for some $0 \leq \vartheta \leq 2\pi$. Thus,

$$f(Z) = \big(\mathrm{tr}(ZA)Z\big)^{1/2} = |\cos\vartheta|Z^{1/2} = |\cos\vartheta|Z,$$

which shows that $Z \in \partial\Pi_2(\mathbb{R})$ is an eigenvector with eigenvalue $|\cos\vartheta|$, and hence the cone spectrum of f is equal to $[0, 1]$.

With regard to Example 5.2.5 we mention the following open problem.

Problem 5.2.6 *Which finite-dimensional closed cones K admit a continuous order-preserving homogeneous map $f\colon K \to K$ with a continuum in $\sigma_K(f)$?*

5.3 The cone spectral radius

The main part of the Perron–Frobenius theorem concerns the spectral radius and its corresponding eigenvector. In this section we discuss generalizations of the spectral radius to nonlinear cone maps. Initially we shall only assume that $f\colon K \to K$ is a continuous homogeneous map on a solid closed cone K in a finite-dimensional vector space V. Two natural definitions for the cone spectral radius, denoted $r_K(f)$, are considered, and it will be shown that they coincide. Unlike the linear case, it may happen that there exists no $x \in K \setminus \{0\}$ such that

$$f(x) = r_K(f)x. \tag{5.4}$$

In the next section, however, it will be shown that if $f\colon K \to K$ is also assumed to be order-preserving, then $r_K(f)$ is an eigenvalue of f. It must be remarked that the issue does not arise if $f\colon K \to K$ is linear, as f is necessarily order-preserving in that case.

Let $K \subseteq V$ be a closed clone and $f\colon K \to K$ be a continuous homogeneous map. As f is continuous at 0 and homogeneous, we can define

$$\|f\|_K = \sup\{\|f(x)\|\colon x \in K \text{ and } \|x\| \leq 1\}, \tag{5.5}$$

where $\|\cdot\|$ is a fixed norm on V. Since f is homogeneous, $\|f(x)\| \leq \|f\|_K\|x\|$ for all $x \in K$. This implies that

$$\|f^{k+m}(x)\| \leq \|f^k\|_K\|f^m\|_K\|x\|$$

for all $k, m \geq 1$ and $x \in K$, so that

$$\|f^{k+m}\|_K \leq \|f^k\|_K \|f^m\|_K \qquad \text{for all } k, m \geq 1.$$

By Fekete's sub-additive lemma [64], we find that

$$\lim_{m\to\infty} \|f^m\|_K^{1/m} = \inf_{m\geq 1} \|f^m\|_K^{1/m}. \tag{5.6}$$

For linear maps f, Bonsall [32] called $\lim_{m\to\infty} \|f^m\|_K^{1/m}$ the cone spectral radius of f. For a continuous homogeneous map $f: K \to K$ we therefore define the *Bonsall cone spectral radius* of f to be

$$\hat{r}_K(f) = \lim_{m\to\infty} \|f^m\|_K^{1/m} = \inf_{m\geq 1} \|f^m\|_K^{1/m}. \tag{5.7}$$

Note that $\hat{r}_K(f)$ is independent of norm, as all norms on a finite-dimensional vector space are equivalent.

Another natural definition of the cone spectral radius of $f: K \to K$ is the following. For each $x \in K$ with $x \neq 0$ let

$$\mu_K(x, f) = \limsup_{m\to\infty} \|f^m(x)\|^{1/m}, \tag{5.8}$$

and define the *cone spectral radius* of f by

$$r_K(f) = \sup\{\mu_K(x, f): x \in K \text{ and } x \neq 0\}. \tag{5.9}$$

Note that $\|f^m(x)\|^{1/m} \leq \|f^m\|_K^{1/m} \|x\|^{1/m}$ for all $x \in K$ with $x \neq 0$ and $m \geq 1$. As $\lim_{m\to\infty} \|f^m\|_K^{1/m} = \hat{r}_K(f)$, we see that $\mu_K(x, f) \leq \hat{r}_K(f)$ for all $x \in K$ with $x \neq 0$. Thus,

$$r_K(f) \leq \hat{r}_K(f). \tag{5.10}$$

By making essential use of the assumption that V is finite-dimensional, it will be shown that $r_K(f) = \hat{r}_K(f)$. In fact, the equality does not necessarily hold in infinite-dimensional Banach spaces: see [138]. The following result is a special case of [138, theorem 2.3].

Theorem 5.3.1 *If $f: K \to K$ is a continuous homogeneous map on a closed cone $K \subseteq V$, then $r_K(f) = \hat{r}_K(f)$.*

Proof By inequality (5.10) it suffices to prove that $\hat{r}_K(f) \leq r_K(f)$. Suppose that $r_K(f) < \hat{r}_K(f)$. For $\lambda \geq 0$ fixed, define $f_\lambda(x) = f(\lambda x) = \lambda f(x)$. Clearly $r_K(f_\lambda) = \lambda r_K(f)$ and $\hat{r}_K(f_\lambda) = \lambda \hat{r}_K(f)$. Thus, replacing f by f_λ, we may assume that $r_K(f) < 1$ and $\hat{r}_K(f) > 1$.

For $\alpha > 0$ let $B_\alpha = \{x \in K: \|x\| \leq \alpha\}$. As $r_K(f) < 1$, there exists for each $x \in K$ an integer $m_x \geq 1$ such that $f^{m_x}(x) \in B_{1/3}$. By continuity, there exists a relatively open neighborhood V_x of x in K such that $\|f^{m_x}(y)\| < 1/2$ for all $y \in V_x$.

The collection $\{V_x : x \in B_1\}$ is an open covering of the compact set B_1, and hence there exist finitely many $x_1, \ldots, x_p \in B_1$ such that $B_1 \subseteq \bigcup_{j=1}^{p} V_{x_j}$. For notational convenience write $V_j = B_1 \cap V_{x_j}$ and $m_j = m_{x_j}$. So, $B_1 = \cup_{j=1}^{p} V_j$ and $\|f^{m_j}(y)\| < 1/2$ for all $y \in V_j$.

Let $M = \max_j m_j$ and $W = \cup_{j=0}^{M} f^j(B_1)$. We claim that $f(W) \subseteq W$. If $w \in W$, then $w = f^j(x)$ for some $x \in B_1$ and $0 \leq j \leq M$. Clearly if $j < M$, then $f(w) = f^{j+1}(x) \in W$. On the other hand, if $w = f^M(x)$, then we can find $1 \leq k \leq p$ such that $x \in V_k$. Note that $f^{m_k}(x) \in B_{1/2} \subseteq B_1$ and $1 \leq m_k \leq M$. Putting $\xi = f^{m_k}(x) \in B_1$ and using the fact that $B_1 \subseteq W$, we find that $f(w) = f^{M-m_k+1}(\xi) \in W$.

Since W is compact, there exists a constant $R > 0$ such that $\|w\| < R$ for all $w \in W$. Thus, for each $x \in B_1$, we have that $\|f^m(x)\| \leq R$ for all $m \geq 1$. This implies that $\hat{r}_K(f) \leq 1$, which is a contradiction. □

The cone spectral radius $r_K(f)$, and hence also $\hat{r}_K(f)$, has the following basic properties.

Proposition 5.3.2 *If $K \subseteq V$ is a closed cone and $g\colon K \to K$ is a continuous homogeneous map, then the following hold:*

(i) For each $k \geq 1$ we have that $r_K(g^k) = r_K(g)^k$.
(ii) If $g^k(x) = \lambda^k x$ for some $x \in K \setminus \{0\}$ and $k \geq 1$, then $\lambda \leq r_K(g)$.

Moreover, if $K_1 \subseteq V_1$ and $K_2 \subseteq V_2$ are closed cones, and $f_1\colon K_1 \to K_2$ and $f_2\colon K_2 \to K_1$ are continuous homogeneous maps, then

$$r_{K_1}(f_2 \circ f_1) = r_{K_2}(f_1 \circ f_2).$$

Proof By Theorem 5.3.1 it suffices to prove the statements for $\hat{r}_K(g)$. Remark that

$$\hat{r}_K(g^k) = \lim_{m\to\infty} \|g^{km}\|_K^{1/m} = \lim_{m\to\infty} \left(\|g^{km}\|_K^{1/(mk)}\right)^k = \hat{r}_K(g)^k.$$

To prove the second assertion let $x \in K \setminus \{0\}$ and $g^k(x) = \lambda^k x$. Clearly $g^{km}(x) = \lambda^{km} x$ for all $m \geq 1$, so that

$$\lambda = \lim_{m\to\infty} \|g^{km}(x)\|^{1/(mk)} \leq \limsup_{j\to\infty} \|g^j(x)\|^{1/j} \leq r_K(g).$$

To prove the last assertion let $\|\cdot\|$ be a norm on V_1 and $\|\cdot\|'$ be a norm on V_2. Define $\|f_1\| = \sup\{\|f_1(x)\|' \colon x \in K_1 \text{ and } \|x\| \leq 1\}$. Likewise let $\|f_2\| = \sup\{\|f_2(y)\| \colon y \in K_2 \text{ and } \|y\|' \leq 1\}$. For every integer $m \geq 1$,

$$\|(f_2 \circ f_1)^{m+1}(x)\| = \|f_2((f_1 \circ f_2)^m(f_1(x)))\| \leq \|f_2\| \|(f_1 \circ f_2)^m\|_{K_2} \|f_1\| \|x\|,$$

so that

$$\|(f_2 \circ f_1)^{m+1}\|_{K_1} \leq \|f_2\| \|(f_1 \circ f_2)^m\|_{K_2} \|f_1\|.$$

If $\|f_1\| = 0$, or if $\|f_2\| = 0$, then

$$\hat{r}_{K_1}(f_2 \circ f_1) = 0 \leq \hat{r}_{K_2}(f_1 \circ f_2).$$

On the other hand, if both are non-zero, then

$$\begin{aligned}\hat{r}_{K_1}(f_2 \circ f_1) &= \lim_{m \to \infty} \|(f_2 \circ f_1)^{m+1}\|_{K_1}^{1/(m+1)} \\ &\leq \lim_{m \to \infty} \|f_2\|^{1/(m+1)} \left(\|(f_1 \circ f_2)^m\|_{K_2}^{1/m}\right)^{m/(m+1)} \|f_1\|^{1/(m+1)} \\ &= \hat{r}_{K_2}(f_1 \circ f_2).\end{aligned}$$

From Theorem 5.3.1, we conclude that $r_{K_1}(f_2 \circ f_1) \leq r_{K_2}(f_1 \circ f_2)$. Interchanging the roles of K_1 and K_2 shows that $r_{K_1}(f_2 \circ f_1) = r_{K_2}(f_1 \circ f_2)$. □

It is natural to ask if the cone spectral radius of a continuous homogeneous map $f: K \to K$ is an eigenvalue. Define

$$\tilde{r}_K(f) = \sup\{\lambda \geq 0 \colon g(x) = \lambda x \text{ for some } x \in K \setminus \{0\}\}.$$

One may suspect that $\tilde{r}_K(f) = r_K(f)$, particularly as $\tilde{r}_K(f) \leq r_K(f)$ by Proposition 5.3.2(ii). The following example, however, shows that the inequality can be strict, even in finite dimensions.

Example 5.3.3 Consider the Lorentz cone $\Lambda_3 = \{(x_1, x_2, x_3) \in \mathbb{R}^3 \colon x_1^2 - x_2^2 - x_3^2 \geq 0 \text{ and } x_1 \geq 0\}$. Let $U_\vartheta \colon \mathbb{R}^2 \to \mathbb{R}^2$ denote the rotation through ϑ about the origin, and assume that ϑ/π is irrational. Define $f_\vartheta \colon \Lambda_3 \to \Lambda_3$ by

$$f_\vartheta(x) = (\frac{1}{2}x_1 + \sqrt{x_2^2 + x_3^2}, U_\vartheta(x_2, x_3)) \quad \text{for } x \in \Lambda_3.$$

It is easy to verify that $f_\vartheta(\Lambda_3) \subseteq \Lambda_3$, and that f_ϑ is a continuous homogeneous map. Furthermore, for each $k \geq 1$ we have that

$$f_\vartheta^k(x) = ((1/2)^k x_1 + \sqrt{x_2^2 + x_3^2} \sum_{j=0}^{k-1} (1/2)^j, U_\vartheta^k(x_2, x_3)), \tag{5.11}$$

since $\|(x_2, x_3)\|_2 = \|U_\vartheta(x_2, x_3)\|_2$. As ϑ/π is irrational, we know that $U_\vartheta(x_2, x_3) = \lambda(x_2, x_3)$ for some $\lambda \geq 0$ if and only if $(x_2, x_3) = (0, 0)$. Thus, $x \in \Lambda_3$ with $x \neq 0$ satisfies $f_\vartheta(x) = \lambda x$ for some $\lambda \geq 0$ if and only if $x_1 > 0$, $x_2 = x_3 = 0$, and $\lambda = 1/2$. This shows that $\tilde{r}_{\Lambda_3}(f_\vartheta) = 1/2$. By taking $x = (x_1, x_2, x_3)$ with $x_1 \geq 0$ and $x_1^2 = 1/2 = x_2^2 + x_3^2$ and using (5.11), we see that $\|f_\vartheta^k\|_2 \geq 1$. So, $r_{\Lambda_3}(f_\vartheta) \geq 1$, which shows that $\tilde{r}_{\Lambda_3}(f_\vartheta) < r_{\Lambda_3}(f_\vartheta)$.

To ensure that the cone spectral radius is an eigenvalue, extra assumptions are required. In the next section it will be shown that it is sufficient to assume, in addition, that f is order-preserving. In fact, the following slightly more general condition suffices. There exists a closed cone $C \subseteq V$ with $K \subseteq C$ such that $f\colon K \to K$ is a continuous homogeneous map that is order-preserving with respect to C. The assumption that f is order-preserving with respect to C instead of K may seem artificial; indeed, in many natural examples f is order-preserving with respect to K. It may, however, happen that it is easier to prove that f preserves $\leq_C$ even if f is K-order-preserving. To illustrate this, consider the following example.

Example 5.3.4 For $1 \leq i, j \leq n$ let $\alpha_{ij} \geq 0$ be such that $\sum_{j=1}^n \alpha_{ij} > 0$ for each i. Define $f\colon \text{int}(\mathbb{R}^n_+) \to \mathbb{R}^n_+$ by

$$f_i(x) = \Big(\sum_{j=1}^n \alpha_{ij} x_j^{-1}\Big)^{-1}$$

for all $1 \leq i \leq n$ and $x \in \text{int}(\mathbb{R}^n_+)$. Obviously f is continuous, homogeneous, and order-preserving with respect to $\mathbb{R}^n_+$. By Theorem 5.1.5 f has a continuous homogeneous order-preserving extension to $\mathbb{R}^n_+$.

Let $K = \{x \in \mathbb{R}^n : 0 \leq x_1 \leq x_2 \leq \ldots \leq x_n\}$. We claim that $f(K) \subseteq K$ if and only if

$$\sum_{j=1}^k \alpha_{ij} \geq \sum_{j=1}^k \alpha_{(i+1)j} \quad \text{for } 1 \leq i < n \text{ and } 1 \leq k \leq n. \tag{5.12}$$

To show this we note that, by continuity of f, $f(K)$ is contained in K if and only if $f(x) \in K$ for all $x \in K$ with $0 < x_1 \leq x_2 \leq \ldots \leq x_n$. For such $x \in K$, $f(x) \in K$ is equivalent to

$$\sum_{j=1}^n \alpha_{ij} x_j^{-1} \geq \sum_{i=1}^n \alpha_{(i+1)j} x_j^{-1} \quad \text{for } 1 \leq i < n. \tag{5.13}$$

Write $y_j = x_j^{-1}$, and note that $y_1 \geq y_2 \geq \ldots \geq y_n > 0$ and $y_j \to 0$ as $x_j \to \infty$. Thus, (5.13) is equivalent to

$$\sum_{j=1}^n \alpha_{ij} y_j \geq \sum_{j=1}^n \alpha_{(i+1)j} y_j \tag{5.14}$$

for all $1 \leq i \leq n$ and $y_1 \geq y_2 \geq \ldots \geq y_n \geq 0$.

For i fixed let $\beta_j = \alpha_{ij} - \alpha_{(i+1)j}$ and define $S_k = \sum_{j=1}^{k} \beta_j$ and $S_0 = 0$. Summation by parts shows that

$$\sum_{j=1}^{n} \beta_j y_j = \sum_{k=1}^{n-1} S_k(y_k - y_{k+1}) + S_n y_n. \tag{5.15}$$

As $y_1 \geq y_2 \geq \ldots \geq y_n > 0$ and, by (5.12), $S_k \geq 0$ for all $1 \leq k \leq n$, we see that (5.14) is satisfied.

To show the equivalence we fix $1 \leq k \leq n$ and consider the vector y given by $y_j = 1$ for $1 \leq j \leq k$ and $y_j = 0$ for $j > k$. In that case (5.14) gives $\sum_{j=1}^{k} \alpha_{ij} \geq \sum_{j=1}^{k} \alpha_{(i+1)j}$. This shows that $f(K) \subseteq K$ if and only if (5.12) holds.

Even though $f(K)$ is contained in K it may happen that f is not order-preserving with respect to K. For instance one can use Theorem 1.3.1 to show that if

$$\alpha_{nn}(\alpha_{(n-1)n})^2 - \alpha_{(n-1)n}(\alpha_{n1})^2 < 0$$

then f is not order-preserving with respect to K.

We conclude this section with several useful inequalities for the cone spectral radius.

Proposition 5.3.5 *Let $K \subseteq C$ be closed cones in V, and $f\colon K \to K$ and $g\colon K \to K$ be continuous homogeneous maps such that*

$$f(x) \leq_C g(x) \quad \textit{for all } x \in K. \tag{5.16}$$

If either f or g is order-preserving with respect to C, then $r_K(f) \leq r_K(g)$.

Proof We first show by induction that $f^k(x) \leq_C g^k(x)$ for all $x \in K$. The case $k = 1$ is our assumption. Suppose that $f^m(x) \leq_C g^m(x)$ for all $x \in K$ and $m \leq k$. If f is order-preserving with respect to C, then it follows from (5.16) that

$$f^{k+1}(x) = f(f^k(x)) \leq_C f(g^k(x)) \leq_C g(g^k(x)) = g^{k+1}(x).$$

On the other hand, if g is order-preserving with respect to C, then

$$f^{k+1}(x) = f(f^k(x)) \leq_C g(f^k(x)) \leq_C g(g^k(x)) = g^{k+1}(x).$$

As C is a closed cone in a finite-dimensional vector space, C is normal by Lemma 1.2.5. Thus, there exists $M > 0$ such that $\|u\| \leq M\|v\|$ for all $u, v \in C$ with $u \leq_C v$. it follows that

$$\|f^k(x)\| \leq M\|g^k(x)\| \quad \text{for all } x \in K \text{ and } k \geq 1.$$

So, $\limsup_{k\to\infty} \|f^k(x)\|^{1/k} \leq \limsup_{k\to\infty} \|g^k(x)\|^{1/k}$ for all $x \in K$, and hence $r_K(f) \leq r_K(g)$. □

Proposition 5.3.6 *Let $K \subseteq C$ be closed cones in V and $f: K \to K$ be a continuous homogeneous map. If f is order-preserving with respect to C and there exist $u \in K \setminus \{0\}$ and $\rho > 0$ such that*

$$\rho u \leq_C f(u),$$

then $r_K(f) \geq \rho$. Moreover, if K has a non-empty interior, then for each $x \in$ $\mathrm{int}(K)$ we have that

$$r_K(f) = \lim_{k\to\infty} \|f^k(x)\|^{1/k}.$$

Proof If $\rho u \leq_C f(u)$, then $\rho^k u \leq_C f^k(u)$ for all $k \geq 1$. As C is normal, we deduce that $\rho^k\|u\| \leq M\|f^k(u)\|$ for some constant $M > 0$. This implies that

$$\rho \leq \limsup_{k\to\infty} \|f^k(u)\|^{1/k} \leq r_K(f).$$

If $x \in \mathrm{int}(K)$, then there exists a constant $M' > 0$ such that $y \leq_C M'x$ for all $y \in K$ with $\|y\| \leq 1$. Again using normality of C, we find that $\|f^k(y)\| \leq M'M\|f^k(x)\|$ for all $y \in K$ with $\|y\| \leq 1$ and $k \geq 1$. This implies that

$$\|f^k\|_K \leq M'M\|f^k(x)\| \leq M'M\|x\|\|f^k\|_K.$$

As $\lim_{k\to\infty} \|f^k\|_K^{1/k} = r_K(f)$ by Theorem 5.3.1, we conclude that

$$\lim_{k\to\infty} \|f^k(x)\|^{1/k} = r_K(f).$$

□

5.4 Eigenvectors corresponding to the cone spectral radius

As in classical Perron–Frobenius theory the cone spectral radius of a continuous homogeneous order-preserving map is an eigenvalue with eigenvector in the cone. The proof relies on the Brouwer fixed-point theorem. The idea that fixed-point theory can be a powerful tool to show existence of eigenvectors is an old one, and can be traced back to the works of Alexandroff and Hopf [8, pp. 480–1] and Kreĭn and Rutman [117].

Theorem 5.4.1 *Let $K \subseteq C$ be closed cones in V, where K has non-empty interior, and let $f: \mathrm{int}(K) \to K$ be a continuous homogeneous map which is order-preserving with respect to C. Suppose that $\vartheta \in \mathrm{int}(C^*)$ and $u \in \mathrm{int}(K)$. For $\varepsilon > 0$ define $f_\varepsilon: \mathrm{int}(K) \to \mathrm{int}(K)$ by*

$$f_\varepsilon(x) = f(x) + \varepsilon\vartheta(x)u \quad \textit{for } x \in \mathrm{int}(K). \tag{5.17}$$

The following assertions hold:

(i) The map f_ε is continuous, homogeneous, and order-preserving with respect to C.
(ii) There exists $x_\varepsilon \in \mathrm{int}(K)$ with $\vartheta(x_\varepsilon) = 1$ such that $f_\varepsilon(x_\varepsilon) = r_K(f_\varepsilon)x_\varepsilon$.
(iii) If $0 < \eta < \varepsilon$, then $r_K(f_\eta) < r_K(f_\varepsilon)$ and hence $\lim_{\varepsilon\to 0} r_K(f_\varepsilon) = r$ exists.
(iv) There exists a convergent sequence $(x_{\varepsilon_k})_k$ in $\{x_\varepsilon : 0 < \varepsilon \leq 1\}$ with $x_{\varepsilon_k} \to x$ and $\varepsilon_k \to 0$ as $k \to \infty$, such that $f(x_{\varepsilon_k}) \to y$ as $k \to \infty$ and $y = rx$.

Proof Let $\Sigma = \{x \in K : \vartheta(x) = 1\}$. So, Σ is a convex compact set by Lemma 1.2.4. Clearly f_ε is a continuous homogeneous map which preserves $\leq_C$, as $\vartheta \in C^*$. Moreover, $f_\varepsilon(K) \subseteq \mathrm{int}(K)$, since $u \in \mathrm{int}(K)$.

As $u \in \mathrm{int}(K) \subseteq \mathrm{int}(C)$ and $\Sigma \subseteq K$ is compact, there exists $M_1 > 0$ such that

$$x \leq_C M_1 u \quad \text{for all } x \in \Sigma. \tag{5.18}$$

This implies that $f_\varepsilon(x) = f(x) + \varepsilon u \leq_C M_1 f(u) + \varepsilon u$ for all $x \in \Sigma \cap \mathrm{int}(K)$. As $u \in \mathrm{int}(C)$, there also exists $M_2 > 0$ such that

$$M_1 f(u) \leq_C M_2 u. \tag{5.19}$$

Thus,

$$f_\varepsilon(x) \leq_C (M_2 + \varepsilon)u \quad \text{for all } x \in \Sigma \cap \mathrm{int}(K). \tag{5.20}$$

It follows that

$$\varepsilon\vartheta(u) \leq \vartheta(f_\varepsilon(x)) \leq (M_2 + \varepsilon)\vartheta(u) \quad \text{for all } x \in \Sigma \cap \mathrm{int}(K). \tag{5.21}$$

Now define $g_\varepsilon \colon \Sigma \cap \mathrm{int}(K) \to \Sigma \cap \mathrm{int}(K)$ by

$$g_\varepsilon(x) = \frac{f_\varepsilon(x)}{\vartheta(f_\varepsilon(x))} \quad \text{for all } x \in \Sigma \cap \mathrm{int}(K).$$

By (5.21) we know that

$$\frac{\varepsilon u}{(M_2 + \varepsilon)\vartheta(u)} \leq_K \frac{f_\varepsilon(x)}{(M_2 + \varepsilon)\vartheta(u)} \leq_K g_\varepsilon(x).$$

As C is a closed cone in a finite-dimensional vector space V, it is normal and hence there exists a constant $M_3 > 0$ such that $\|f(x)\| \leq M_3\|u\|$ for all $x \in \Sigma \cap \mathrm{int}(K)$ by (5.18) and (5.19). Since $f(x) \in K$ for all $x \in \Sigma \cap \mathrm{int}(K)$

and $u \in \text{int}(K)$, there exists a constant $M > 0$ such that $f(x) \leq_K Mu$ for all $x \in \Sigma \cap \text{int}(K)$. By using (5.21) we see that

$$g_\varepsilon(x) \leq_K \frac{f(x) + \varepsilon u}{\varepsilon\vartheta(u)} \leq_K \frac{(M+\varepsilon)u}{\varepsilon\vartheta(u)} \quad \text{for all } x \in \Sigma \cap \text{int}(K).$$

Thus, there exist constants $0 < A \leq B$ such that

$$g_\varepsilon(\Sigma \cap \text{int}(K)) \subseteq \{x \in \Sigma \cap \text{int}(K) \colon Au \leq_K x \leq_K Bu\}. \tag{5.22}$$

Write $\Omega_\varepsilon = \{x \in \Sigma \cap \text{int}(K) \colon d_H(x,u) \leq \log(B/A)\}$, where the Hilbert metric is determined by K. So, Ω_ε is a compact convex subset of $\Sigma \cap \text{int}(K)$ by Lemma 2.6.1. The inclusion (5.22) ensures that $g_\varepsilon(\Omega_\varepsilon) \subseteq \Omega_\varepsilon$. It follows from the Brouwer fixed-point theorem that $g_\varepsilon(x_\varepsilon) = x_\varepsilon$ for some $x_\varepsilon \in \Omega_\varepsilon \subseteq \Sigma \cap \text{int}(K)$. Write $\lambda_\varepsilon = \vartheta(f_\varepsilon(x_\varepsilon))$ and note that $f_\varepsilon(x_\varepsilon) = \lambda_\varepsilon x_\varepsilon$ and $\vartheta(x_\varepsilon) = 1$. Proposition 5.3.6 implies that $r_K(f_\varepsilon) = \lim_{k\to\infty} \|f_\varepsilon^k(x_\varepsilon)\|^{1/k} = \lambda_\varepsilon$. This completes the proof of part (ii).

Now suppose that $0 < \eta < \varepsilon$. As $f_\eta(x_\eta) = r_K(f_\eta)x_\eta$, we have that $f_\varepsilon(x_\eta) = r_K(f_\eta)x_\eta + (\varepsilon - \eta)u$. As $u \in \text{int}(C)$, there exists $\delta > 0$ such that $\delta x_\eta \leq_C (\varepsilon - \eta)u$, so that

$$(r_K(f_\eta) + \delta)x_\eta \leq_C f_\varepsilon(x_\eta). \tag{5.23}$$

Recall that $x_\varepsilon \in \text{int}(C)$, and hence there exists $\alpha > 0$ such that $\alpha x_\eta \leq_C x_\varepsilon$. Combining this inequality with (5.23) and iterating f_ε gives

$$\alpha(r_K(f_\eta) + \delta)^k x_\eta \leq_C \alpha f_\varepsilon^k(x_\eta) \leq_C f_\varepsilon^k(x_\varepsilon) = r_K(f_\varepsilon)^k x_\varepsilon.$$

This implies that

$$\left(\frac{r_K(f_\varepsilon)}{r_K(f_\eta) + \delta}\right)^k x_\varepsilon - \alpha x_\eta \in C \quad \text{for all } k \geq 1. \tag{5.24}$$

If $r_K(f_\varepsilon) < r_K(f_\eta) + \delta$, we can let $k \to \infty$ in (5.24) and conclude that $-\alpha x_\eta \in C$, which is impossible. Thus, $r_K(f_\varepsilon) \geq r_K(f_\eta) + \delta$, which shows that $r_K(f_\eta) < r_K(f_\varepsilon)$ for $0 < \eta < \varepsilon$ and hence $\lim_{\varepsilon\to 0} r_K(f_\varepsilon) = r$ exists.

As $\{x_\varepsilon \colon 0 < \varepsilon \leq 1\}$ and $\{f(x_\varepsilon) \colon 0 < \varepsilon \leq 1\}$ are bounded subsets of the finite-dimensional closed cone K, there exists a sequence $(\varepsilon_k)_k$ with $\varepsilon_k \to 0$ as $k \to \infty$ such that $(x_{\varepsilon_k})_k$ converges to some $x \in \Sigma$ and $(f(x_{\varepsilon_k}))_k$ converges to some $y \in K$. As $f_{\varepsilon_k}(x_{\varepsilon_k}) = r_K(f_{\varepsilon_k})x_{\varepsilon_k} \to rx$ as $k \to \infty$, we find that $y = rx$. □

If the map $f \colon \text{int}(K) \to K$ in Theorem 5.4.1 has a continuous homogeneous C-order-preserving extension $F \colon K \to K$, then F has an eigenvector in K with eigenvalue r. In that case we will see that $r = r_K(F)$. The example discussed in Section 5.1 shows that in general f may fail to have such a continuous extension. However, in case K is polyhedral essentially the same arguments as in Section 5.1 can be used to show that f always admits a continuous homogeneous extension $F: K \to K$ which is order-preserving with respect to C. We refer the reader to [41, theorem 7.2] for details.

Corollary 5.4.2 *Let $K \subseteq C$ be closed cones in V with $\mathrm{int}(K)$ non-empty. If $f: K \to K$ is a continuous homogeneous map which is order-preserving with respect to C, then there exists $x \in K \setminus \{0\}$ with $f(x) = r_K(f)x$.*

Proof We use the notation from Theorem 5.4.1. For each $x \in K$, $f(x) \leq_C f_\varepsilon(x)$, so Proposition 5.3.5 implies that $r_K(f) \leq r_K(f_\varepsilon)$. This shows that $r \geq r_K(f)$, where $r = \lim_{\varepsilon \to 0} r_K(f_\varepsilon)$. By part (iv) of Theorem 5.4.1 there exists a sequence $(x_{\varepsilon_k})_k$ in $\Sigma \cap \mathrm{int}(K)$ with $x_{\varepsilon_k} \to x$ and $\varepsilon_k \to 0$ as $k \to \infty$ such that $f(x_{\varepsilon_k}) \to y$ as $k \to \infty$ and $y = rx$. Since f is continuous on K, we get that $rx = y = f(x)$, so that $r \leq r_K(f)$ by definition of $r_K(f)$. Thus, $r = r_K(f)$ and $f(x) = r_K(f)x$. □

Corollary 5.4.2 is closely related to a classical result by Kreĭn and Rutman [117, theorem 9.1], which deals with compact, nonlinear cone maps in a Banach space. Direct generalizations to noncompact cone maps have been given by Nussbaum and Mallet-Paret [138, 139], where more sophisticated ideas from fixed-point theory, such as the fixed-point index, are used.

5.5 Continuity of the cone spectral radius

Given a continuous homogeneous order-preserving map $f: K \to K$ on a solid closed cone $K \subseteq V$, and a sequence $(f_k)_k$ of continuous homogeneous order-preserving maps on K such that

$$\lim_{k \to \infty} \sup_{x \in \Sigma} \|f(x) - f_k(x)\| = 0, \tag{5.25}$$

where $\Sigma = \{x \in K : \varphi(x) = 1\}$ and $\varphi \in \mathrm{int}(K^*)$, it is natural to ask if

$$\lim_{k \to \infty} r_K(f_k) = r_K(f). \tag{5.26}$$

We say that $f: K \to K$ has a *continuous cone spectral radius* if (5.26) holds for every sequence $(f_k)_k$ of continuous homogeneous order-preserving maps on K satisfying (5.25). There exist examples [120] in infinite-dimensional Banach spaces of linear maps that are compact on a closed cone, but not compact as a linear map on the whole space, which do not have a continuous cone spectral radius. In finite dimensions, however, no such examples are known. In fact, the following problem is open.

Problem 5.5.1 *Is the cone spectral radius continuous for every continuous homogeneous order-preserving map $f: K \to K$ on a solid closed cone K in a finite-dimensional vector space V?*

The cone spectral radius is always upper-semicontinuous, as the following lemma shows.

Lemma 5.5.2 *If $f: K \to K$ is a continuous homogeneous order-preserving map on a solid closed cone $K \subseteq V$ and $(f_k)_k$ is a sequence of continuous homogeneous order-preserving maps on K satisfying (5.25), then*

$$\limsup_{k\to\infty} r_K(f_k) \leq r_K(f).$$

Proof For $k \geq 1$ write $\lambda_k = r_K(f_k)$ and let $u_k \in K$ be such that $f_k(u_k) = \lambda_k u_k$. There exists a subsequence $(\lambda_{k_i})_i$ such that

$$\lim_{i\to\infty} \lambda_{k_i} = \limsup_{k\to\infty} \lambda_k = \lambda'.$$

As $\Sigma_\varphi = \{x \in K : \varphi(x) = 1\}$, with $\varphi \in \text{int}(K^*)$, is compact by Lemma 1.2.4, we may after taking a further subsequence assume that $(u_{k_i})_i$ converges in Σ_φ. Denote its limit by v. As f is continuous, $f(v) = \lambda' v$ by (5.25). Now the definition of $r_K(f)$ implies that $\lambda' \leq r_K(f)$, and we are done. □

If the cone is polyhedral, the cone spectral radius is always continuous. This is a consequence of the following result.

Theorem 5.5.3 *If $f: K \to K$ is a continuous homogeneous order-preserving map on a solid closed cone $K \subseteq V$, and there exists $u \in K \setminus \{0\}$ at which condition* G *is satisfied with $f(u) = r_K(f)u$, then the cone spectral radius is continuous.*

Proof For each $k \geq 1$ write $u_k = f_k(u)$. By (5.25), $\lim_{k\to\infty} u_k = r_K(f)u$. As condition G is satisfied at u, there exists $0 < \mu_k < 1$ with $\mu_k \to 1$ such that

$$\mu_k r_K(f)u \leq_K u_k \quad \text{for all } k \geq 1.$$

This implies that $\mu_k r_K(f)u \leq_K f_k(u)$ for all $k \geq 1$. It follows from Proposition 5.3.6 that $\mu_k r_K(f) \leq r_K(f_k)$ for all $k \geq 1$. Letting $k \to \infty$ we deduce that

$$r_K(f) \leq \liminf_{k\to\infty} r_K(f_k).$$

Combining this inequality with Lemma 5.5.2 shows that $\lim_{k\to\infty} r_K(f_k) = r_K(f)$. □

Recall from Lemma 5.1.4 that if K is a polyhedral cone, then condition G is satisfied at every point $x \in K$. Thus, Theorem 5.5.3 has the following direct consequence.

Corollary 5.5.4 *The cone spectral radius is continuous for every continuous homogeneous order-preserving map on a polyhedral cone.*

Of course, to apply Theorem 5.5.3 we only need to know that condition G is satisfied at an eigenvector $u \in K$ with $f(u) = r_K(f)u$. An interesting situation occurs when $u \in \text{int}(K)$. In that case condition G is satisfied at u regardless of the geometry of the cone K.

Corollary 5.5.5 *If $f: K \to K$ is a continuous homogeneous order-preserving map and there exists $u \in \text{int}(K)$ with $f(u) = r_K(f)u$, then f has a continuous cone spectral radius.*

Further results concerning Problem 5.5.1 were obtained in [120] using the fixed-point index. It was shown there that if $f: K \to K$ is a continuous homogeneous order-preserving map and there exists a sequence $(\mu_k)_k$ in $(0, r_K(f)]$ such that $\lim_{k\to\infty} \mu_k = r_K(f)$ and each μ_k is not an eigenvalue of f, then $r_K(f)$ is continuous. We know from Example 5.2.5 that such a sequence $(\mu_k)_k$ does not always exist. In that specific example, however, the homogeneous order-preserving map has an eigenvector in the interior of the cone, and hence it has a continuous spectral radius by Corollary 5.5.5.

We end this section with an elementary result, which we will need later.

Lemma 5.5.6 *If $f: K \to K$ is a continuous homogeneous order-preserving map on a solid closed cone K and $(f_k)_k$ is a sequence of continuous homogeneous order-preserving maps on K satisfying (5.25) and*

$$f(x) \le_K f_k(x) \quad \text{for all } x \in K \text{ and } k \ge 1,$$

then $\lim_{k\to\infty} r_K(f_k) = r_K(f)$.

Proof Let $u \in K$ be such that $f(u) = r_K(f)u$ and note that $r_K(f)u = f(u) \le_K f_k(u)$. So, by Proposition 5.3.6 we know that $r_K(f) \le r_K(f_k)$ for all $k \ge 1$, so that $r_K(f) \le \liminf_{k\to\infty} r_K(f_k)$. Using Lemma 5.5.2 we obtain the equality. □

5.6 A Collatz–Wielandt formula

A classical result going back to Collatz [52] and Wielandt [225] asserts that, for every nonnegative matrix A, the spectral radius $r(A)$ satisfies

$$r(A) = \inf_{x \in \mathrm{int}(\mathbb{R}^n_+)} \max_{1 \le i \le n} \frac{(Ax)_i}{x_i}. \tag{5.27}$$

Such a "minimax" variational formula for the cone spectral radius exists for general continuous order-preserving homogeneous maps.

Let $K \subseteq V$ be a solid closed cone and $u \in \mathrm{int}(K)$. Following the notation from Section 2.1 let $\Sigma_u^* = \{\varphi \in K^* : \varphi(u) = 1\}$ and denote the set of extreme points of Σ_u^* by $\mathcal{E}_u^*$. According to (2.6),

$$M(x/y) = \max_{\varphi \in \mathrm{cl}(\mathcal{E}_u^*)} \frac{\varphi(x)}{\varphi(y)}$$

for $x, y \in K$ with $y \in \mathrm{int}(K)$. Given a continuous order-preserving homogeneous map $f : K \to K$, define the *Collatz–Wielandt number* by

$$\mathrm{cw}(f) = \inf_{x \in \mathrm{int}(K)} M(f(x)/x) = \inf_{x \in \mathrm{int}(K)} \max_{\varphi \in \mathrm{cl}(\mathcal{E}_u^*)} \frac{\varphi(f(x))}{\varphi(x)}. \tag{5.28}$$

Note that if $K = \mathbb{R}^n_+$, then

$$\mathrm{cw}(f) = \inf_{x \in \mathrm{int}(\mathbb{R}^n_+)} \max_{1 \le i \le n} \frac{f_i(x)}{x_i}.$$

Theorem 5.6.1 *If $f : K \to K$ is a continuous order-preserving homogeneous map on a solid closed cone $K \subseteq V$, then*

$$r_K(f) = \mathrm{cw}(f).$$

Proof If $x \in \mathrm{int}(K)$, then $f(x) \le_K M(f(x)/x)x$, as K is closed. It follows from Proposition 5.3.6 that

$$r_K(f) = \lim_{k \to \infty} \|f^k(x)\|^{1/k} \le M(f(x)/x),$$

and hence $r_K(f) \le \inf_{x \in \mathrm{int}(K)} M(f(x)/x) = \mathrm{cw}(f)$.

To prove equality we first consider the case where f has an eigenvector $x_0 \in \mathrm{int}(K)$, so $f(x_0) = r_K(f)x_0$. Clearly, in that case, $M(f(x_0)/x_0) = r_K(f)$ and hence $\mathrm{cw}(f) \le r_K(f)$, which gives the equality.

If f does not have an eigenvector in $\mathrm{int}(K)$, we consider for each integer $k \ge 1$ the continuous order-preserving homogeneous map $f_k : K \to \mathrm{int}(K)$ given by

$$f_k(x) = f(x) + \frac{\varphi(x)}{k} u,$$

where $\varphi \in \operatorname{int}(K^*)$ is fixed. Note that

$$\lim_{k\to\infty} \sup_{x\in\Sigma} \|f(x) - f_k(x)\| = 0,$$

and $f(x) \leq_K f_k(x)$ for all $x \in K$. From Lemma 5.5.6 we know that $r_K(f) = \lim_{k\to\infty} r_K(f_k)$. As $f_k(K) \subseteq \operatorname{int}(K)$, each f_k has an eigenvector in $\operatorname{int}(K)$. Thus, by the previous case, $\operatorname{cw}(f_k) = r_K(f_k)$. It is clear that $M(f(x)/x) \leq M(f_k(x)/x)$ for all $k \geq 1$ and $x \in \operatorname{int}(K)$. Thus, $\operatorname{cw}(f) \leq \operatorname{cw}(f_k)$ for all $k \geq 1$. Now taking limits as $k \to \infty$ gives

$$r_K(f) = \lim_{k\to\infty} r_K(f_k) = \lim_{k\to\infty} \operatorname{cw}(f_k) \geq \operatorname{cw}(f).$$

This shows that $r_K(f) = \operatorname{cw}(f)$. □

6

Eigenvectors in the interior of the cone

It is frequently important to know whether a homogeneous order-preserving map $f\colon \operatorname{int}(K) \to \operatorname{int}(K)$ on a solid closed cone $K \subseteq V$ has an eigenvector $u \in \operatorname{int}(K)$. For example, to prove that a topical map $g\colon \mathbb{R}^n \to \mathbb{R}^n$ has an *additive eigenvector* $v \in \mathbb{R}^n$, i.e.,

$$g(v) = \lambda \mathbb{1} + v \quad \text{for some } \lambda \in \mathbb{R}, \tag{6.1}$$

we can consider the log-exp transform $f\colon \operatorname{int}(\mathbb{R}^n_+) \to \operatorname{int}(\mathbb{R}^n_+)$ of g. It is easy to verify that g has an additive eigenvector $v \in \mathbb{R}^n$ satisfying $g(v) = \lambda \mathbb{1} + v$ if and only if f has an eigenvector $u \in \operatorname{int}(\mathbb{R}^n_+)$ with $f(u) = e^{\lambda} u$.

The problem of the existence of an eigenvector in the interior of the cone appears to be subtle and difficult, even for simple examples on $\mathbb{R}^n_+$, and there seems to be no clear-cut solution. In this chapter several general principles are given, which may be useful in proving existence of eigenvectors in the interior. These principles, and their limitations, are illustrated in Section 6.6 by analyzing homogeneous order-preserving maps on $\mathbb{R}^n_+$ involving means. In Chapter 7 further applications to the solution of matrix scaling problems are discussed.

6.1 First principles

To appreciate the subtlety of the problem of finding an eigenvector in the interior of the cone we consider the following basic example.

Example 6.1.1 Let $\alpha = (a, b, c) \in \operatorname{int}(\mathbb{R}^3_+)$ and define the order-preserving homogeneous map $f_\alpha\colon \mathbb{R}^3_+ \to \mathbb{R}^3_+$ by

$$f_\alpha(x) = \begin{pmatrix} ax_1 + a\min\{x_2, x_3\} \\ bx_2 + b\min\{x_1, x_3\} \\ cx_3 + c\min\{x_1, x_2\} \end{pmatrix} \quad \text{for } x \in \mathbb{R}^3_+.$$

Given a coordinate permutation $\sigma\colon (x_1, x_2, x_3) \mapsto (x_{\sigma(1)}, x_{\sigma(2)}, x_{\sigma(3)})$, it is easy to verify that

$$f_{\sigma(\alpha)}(\sigma(x)) = \sigma(f_\alpha(x)) \quad \text{for all } x \in \mathbb{R}^3_+.$$

So, f_α has an eigenvector in $\mathrm{int}(\mathbb{R}^3_+)$ if and only if $f_{\sigma(\alpha)}$ has one, and $f^k_\alpha(x)/\|f^k_\alpha(x)\|_\infty \to v$ is equivalent to $f^k_{\sigma(\alpha)}(\sigma(x))/\|f^k_{\sigma(\alpha)}(\sigma(x))\|_\infty \to \sigma(v)$. Thus, in the analysis of the iterates of f_α and the eigenvectors of f_α in the interior of $\mathbb{R}^3_+$, we can always assume that $a \leq b \leq c$ by replacing α by $\sigma(\alpha)$ for a suitable permutation σ, if necessary.

Lemma 6.1.2 *If $\alpha = (a, b, c) \in \mathrm{int}(\mathbb{R}^3_+)$ and $c \geq a + b$, then*

$$\lim_{k\to\infty} \frac{f^k_\alpha(x)}{\|f^k_\alpha(x)\|_\infty} = \begin{pmatrix} 0 \\ 0 \\ 1 \end{pmatrix} \tag{6.2}$$

for all $x \in \mathrm{int}(\mathbb{R}^3_+)$.

Proof First note that if $t > 0$ and $g(x) = tf_\alpha(x)$, then (6.2) holds if and only if

$$\lim_{k\to\infty} \frac{g^k(x)}{\|g^k(x)\|_\infty} = \begin{pmatrix} 0 \\ 0 \\ 1 \end{pmatrix} \quad \text{for all } x \in \mathrm{int}(\mathbb{R}^3_+).$$

Taking $t = 1/c$, we may as well assume that $a, b > 0$, $c = 1$, and $a + b \leq 1$. For $(x_1, x_2, x_3) \in \mathrm{int}(\mathbb{R}^3_+)$ and integers $k \geq 1$, define $(x^k_1, x^k_2, x^k_3) = f^k((x_1, x_2, x_3))$ and $m_k = \min\{x^k_1, x^k_2\}$. Because $x^{k+1}_1 = ax^k_1 + a\min\{x^k_2, x^k_3\} \leq a(x^k_1 + x^k_2)$ and $x^{k+1}_2 = bx^k_2 + b\min\{x^k_1, x^k_3\} \leq b(x^k_1 + x^k_2)$, we know that $x^{k+1}_1 + x^{k+1}_2 \leq (a+b)(x^k_1 + x^k_2)$, so that $x^k_1 + x^k_2 \leq (a+b)^k(x_1 + x_2)$ for all $k \geq 1$.

If $a + b < 1$, it follows that $x^k_1 + x^k_2 \to 0$ as $k \to \infty$. If $a + b = 1$, then $x^{k+1}_1 + x^{k+1}_2 \leq x^k_1 + x^k_2$ for all $k \geq 1$, and we find that $x^k_1 + x^k_2 \to r$ for some $r \geq 0$ as $k \to \infty$. On the other hand, it is easy to see that $x^k_3 = x_3 + \sum_{j=1}^k m_j$, so that $x^{k+1}_3 \geq x^k_3$ and $x^k_3 \geq x_3 > 0$ for all $k \geq 1$. Thus, if $x^k_1 + x^k_2 \to 0$, then (6.2) holds.

To complete the proof it remains to consider the case where $a + b = 1$ and $x^k_1 + x^k_2 \to r > 0$ as $k \to \infty$. Let $\delta = \limsup_{k\to\infty} m_k$. If $\delta > 0$, then $x^k_3 = x_3 + \sum_{j=1}^k m_j \to \infty$ as $k \to \infty$. In that case (6.2) holds, because $x^k_1 + x^k_2 \to r < \infty$. Finally assume that $\delta = 0$; so, $\lim_{k\to\infty} m_k = 0$. In that case we will derive a contradiction from the assumptions that $x^k_1 + x^k_2 \to r > 0$ and $\min\{x^k_1, x^k_2\} \to 0$. Indeed, select $\varepsilon > 0$ and $m \geq 1$ such that

$$\max\{x_1^j, x_2^j\} \geq r - \varepsilon \quad \text{and} \quad \min\{x_1^j, x_2^j\} \leq \varepsilon \quad \text{for all } j \geq m.$$

Such an integer m exists, since $\max\{x_1^j, x_2^j\} + \min\{x_1^j, x_2^j\} = x_1^j + x_2^j$. If $x_1^j = \max\{x_1^j, x_2^j\}$ and $j \geq m$, then $x_1^{j+1} \geq ax_1^j \geq a(r - \varepsilon)$ and $x_2^{j+1} \geq b\min\{x_3, r - \varepsilon\}$. Similarly, if $x_2^j = \max\{x_1^j, x_2^j\}$ and $j \geq m$, then $x_2^{j+1} \geq bx_2^j \geq b(r - \varepsilon)$ and $x_1^{j+1} \geq a\min\{x_3, r - \varepsilon\}$. It follows that if $j \geq m$, then $x_1^{j+1} \geq a\min\{x_3, r - \varepsilon\}$ and $x_2^{j+1} \geq b\min\{x_3, r - \varepsilon\}$. Thus, for $\varepsilon > 0$ sufficiently small, $x_1^{j+1} > \varepsilon$ and $x_2^{j+1} > \varepsilon$ for all $j \geq m$, which contradicts $\min\{x_1^k, x_2^k\} \to 0$ as $k \to \infty$. □

We can use Lemma 6.1.2 to prove the following result for f_α.

Proposition 6.1.3 *The map f_α has an eigenvector in* $\text{int}(\mathbb{R}^3_+)$ *if and only if $a < b+c$ and $b < a+c$ and $c < a+b$. Furthermore, if f_α has an eigenvector in* $\text{int}(\mathbb{R}^3_+)$, *it is unique up to positive scalar multiples.*

Proof By virtue of Lemma 6.1.2 and the remarks preceding it, we see that f_α has no eigenvector in the interior of $\mathbb{R}^3_+$ if $a \geq b+c$ or $b \geq a+c$ or $c \geq a+b$, which shows that the conditions are necessary.

To prove that they are sufficient, we may assume without loss of generality that $0 < a \leq b \leq c$. As $c < a + b$ and $a \leq b \leq c$, we get that

$$1 \leq \frac{b}{a} \leq \frac{c}{a+b-c}.$$

A simple calculation shows that $u = (1, b/a, c/(a + b - c))$ is an eigenvector of f_α in $\text{int}(\mathbb{R}^3_+)$ with eigenvalue $a + b$. Now suppose that $v = (v_1, v_2, v_3)$ is also an eigenvector of f_α in the interior of $\mathbb{R}^3_+$. It follows from Corollary 5.2.2 that v has eigenvalue $a + b$, and hence

$$\begin{aligned} av_1 + a\min\{v_2, v_3\} &= (a+b)v_1, \\ bv_2 + b\min\{v_1, v_3\} &= (a+b)v_2, \\ cv_3 + c\min\{v_1, v_2\} &= (a+b)v_3. \end{aligned}$$

The first equation implies that $v_1 = (a/b)\min\{v_2, v_3\}$. As $a \leq b$, we deduce that $v_1 \leq v_2$ and $v_1 \leq v_3$. It now follows from the second equation that $v_2 = (b/a)v_1$, and the third equation gives $v_3 = (c/(a + b - c))v_1$. Thus, $v = v_1 u$, which shows that u is the unique, up to positive scalar multiples, eigenvector of f_α in the interior of $\mathbb{R}^3_+$ if $a \leq b \leq c$ and $c < a + b$. □

The following basic principle to check for eigenvectors in the interior of the cone is a direct generalization of the irreducibility condition for linear maps.

Proposition 6.1.4 *If $f: K \to K$ is a continuous homogeneous order-preserving map on a solid closed cone $K \subseteq V$ and for each $x \in K \setminus \{0\}$ there exists an integer $k_x \geq 1$ such that*

$$\sum_{k=0}^{k_x} f^k(x) \in \text{int}(K),$$

then each eigenvector of f lies in $\text{int}(K)$.

Proof If $u \in K \setminus \{0\}$ with $f(u) = \lambda u$, then

$$\sum_{k=0}^{k_x} f^k(u) = \Big(\sum_{k=0}^{k_x} \lambda^k\Big) u \in \text{int}(K),$$

This implies that $u \in \text{int}(K)$. □

The hypotheses in Proposition 6.1.4 are often too restrictive, as homogeneous order-preserving maps frequently have eigenvectors in the boundary of the cone as well as in the interior. Given a continuous homogeneous order-preserving map $f: K \to K$ on a solid closed cone K, define

$$r_{\partial K}(f) = \sup\{\lambda \geq 0 \colon f(x) = \lambda x \text{ for some } x \in \partial K \setminus \{0\}\}.$$

If the set is empty, we put $r_{\partial K}(f) = -\infty$.

Proposition 6.1.5 *If $f: K \to K$ is a continuous homogeneous order-preserving map on a solid closed cone K, and there exist $u \in K \setminus \{0\}$ and $\rho > 0$ such that $\rho u \leq_K f(u)$ and $r_{\partial K}(f) < \rho$, then f has an eigenvector in* $\text{int}(K)$.

Proof By Proposition 5.3.6 it follows that $\rho \leq r_K(f)$, as $\rho u \leq_K f(u)$, and hence $r_{\partial K}(f) < r_K(f)$. By Corollary 5.4.2 there exists $x \in K$ with $x \neq 0$ such that $f(x) = r_K(f)x$. By the definition of $r_{\partial K}(f)$, $x \notin \partial K$, and hence $x \in \text{int}(K)$. □

Of course, Proposition 6.1.5 is only useful if it can be shown that $r_{\partial K}(f) < r_K(f)$, which is often hard for specific examples. In theory, however, one frequently has that $r_{\partial K}(f) < r_K(f)$ if f has an eigenvector in $\text{int}(K)$. To prove this we need to recall the idea of semi-differentiability ([7] which extends the notion of semi-differentiability as discussed in [187]).

Let $\xi \in V$, where V is a finite-dimensional normed vector space. For $r > 0$ write $B_r(\xi) = \{v \in V \colon \|v - \xi\| < r\}$. If $D \subseteq V$ and $\xi \in D$, we say that D is *locally convex at* ξ if there exists $r > 0$ such that $B_r(\xi) \cap D$ is convex. Given $D \subseteq V$ locally convex at $\xi \in D$, define

$$S_\xi = \{v \in V : \text{there exists } t_v > 0 \text{ such that } \xi + tv \in D \text{ for all } 0 \leq t \leq t_v\}. \tag{6.3}$$

Using local convexity at ξ, it is straightforward to check that S_ξ is convex and $\lambda S_\xi = S_\xi$ for all $\lambda > 0$. In general, however, S_ξ is not closed. For example, if D is the Lorentz cone Λ_3 and $\xi \in \partial\Lambda_3$ with $\xi \neq 0$, then S_ξ is an open half-space. However, it is easy to show that if $D \subseteq V$ is a polyhedral cone, S_ξ is closed for all $\xi \in D$.

If $D \subseteq V$ is locally convex at $\xi \in D$ and $f : D \to V$, it may happen that the one-sided Gateaux derivative

$$\lim_{t \to 0^+} \frac{f(\xi + tv) - f(\xi)}{t} \tag{6.4}$$

exists for all $v \in S_\xi$. Note that the map $f'_\xi : S_\xi \to V$, where $f'_\xi(v)$ is the limit in (6.4), is positively homogeneous. In that case we can write, for all $v \in S_\xi$ with $\xi + v \in D$,

$$f(\xi + v) - f(\xi) - f'_\xi(v) = R(v). \tag{6.5}$$

If $f'_\xi : S_\xi \to V$ is continuous and

$$\lim_{\|v\| \to 0} \frac{\|R(v)\|}{\|v\|} = 0, \tag{6.6}$$

we say that $f : D \to V$ is *semi-differentiable* at $\xi \in D$. Here Equation (6.6) must be interpreted as follows. For each $\varepsilon > 0$ there exists $\delta > 0$ such that for $v \in S_\xi$ with $\xi + v \in D$ and $\|v\| \leq \delta$,

$$\|R(v)\| \leq \varepsilon \|v\|.$$

The map $f'_\xi : S_\xi \to V$ is called the *semi-derivative of* f *at* ξ with respect to D. Although the map f'_ξ need not be linear, the chain rule does hold for semi-differentiable maps.

Lemma 6.1.6 *Let V_1, V_2, and V_3 be finite-dimensional normed vector spaces, and let $D_1 \subseteq V_1$ and $D_2 \subseteq V_2$ be open sets. If $f : D_1 \to V_2$ is semi-differentiable at $x \in D_1$ and $g : D_2 \to V_3$ is semi-differentiable at $y = f(x) \in D_2$, then $g \circ f$ is semi-differentiable at x and*

$$(g \circ f)'_x = g'_y \circ f'_x.$$

Furthermore, if there exists an open neighborhood U of x such that $f_{|U}$ is Lipschitz with Lipschitz constant $C > 0$, then f'_x is Lipschitz on V_1 with Lipschitz constant C.

Proof Since there will be no confusion, we simply write $\|\cdot\|$ to denote the norm on V_i for $i = 1, 2, 3$. As f is semi-differentiable at ξ,

$$f(x+v) = f(x) + f_x'(v) + \|v\|\varepsilon_1(v)$$

for all $v \in V_1$ with $x + v \in D_1$. Here $f_x'\colon V_1 \to V_2$ is a continuous positively homogeneous map and $\lim_{\|v\|\to 0} \varepsilon_1(v) = 0$. There exists $\delta > 0$ so that $f(x + v) \in D_2$ for all $\|v\| < \delta$. As g is semi-differentiable at $y = f(x) \in D_2$, we have for $\|v\| < \delta$,

$$\begin{aligned} g(f(x+v)) = g(f(x)) &+ g_y'(f_x'(v) + \|v\|\varepsilon_1(v)) \\ &+ \big\| f_x'(v) + \|v\|\varepsilon_1(v) \big\| \varepsilon_2(f_x'(v) + \|v\|\varepsilon_1(v)), \end{aligned}$$

where $\lim_{\|w\|\to 0} \varepsilon_2(w) = 0$.

Since f_x' is continuous and positively homogeneous, there exists $M > 0$ with $\|f_x'(v)\| \le M\|v\|$ for all $v \in V_1$. Using this estimate we see that

$$\big\| f_x'(v) + \|v\|\varepsilon_1(v) \big\| \varepsilon_2(f_x'(v) + \|v\|\varepsilon_1(v)) = \|v\|\varepsilon_3(v),$$

where $\lim_{\|v\|\to 0} \varepsilon_3(v) = 0$.

As both g_y' and f_x' are positively homogeneous, we can write

$$\begin{aligned} g_y'(f_x'(v) + \|v\|\varepsilon_1(v)) - g_y'(f_x'(v)) =& \|v\| \big(g_y'(f_x'(v/\|v\|) + \varepsilon_1(v)) \\ &- g_y'(f_x'(v/\|v\|)) \big). \end{aligned}$$

Using uniform continuity of g_y' on pre-compact subsets of V_2, we see that

$$\|v\| \big(g_y'(f_x'(v/\|v\|) + \varepsilon_1(v)) - g_y'(f_x'(v/\|v\|)) \big) = \|v\|\varepsilon_4(v),$$

where $\lim_{\|v\|\to 0} \varepsilon_4(v) = 0$. For $\|v\| < \delta$ it follows that

$$g(f(x+v)) = g(f(x)) + g_y'(f_x'(v)) + \|v\|(\varepsilon_3(v) + \varepsilon_4(v)),$$

which shows that $g \circ f$ is semi-differentiable at x with semi-derivative $g_y' \circ f_x'$.

If $U \subseteq D_1$ is an open neighborhood of x and $f_{|U}$ is Lipschitz with Lipschitz constant $C > 0$, it remains to show that f_x' is Lipschitz with Lipschitz constant C. For $u, v \in V_1$ there exists $\varepsilon > 0$ so that $x + tu \in U$ and $x + tv \in U$ for all $0 \le t \le \varepsilon$. Remark that, for $0 \le t \le \varepsilon$,

$$\begin{aligned} \left\| \frac{f(x+tu) - f(x)}{t} - \frac{f(x+tv) - f(x)}{t} \right\| &= \left\| \frac{f(x+tu) - f(x+tv)}{t} \right\| \\ &\le C\|u - v\|. \end{aligned}$$

Letting $t \to 0$ in the above inequality shows that

$$\|f_x'(u) - f_x'(v)\| \le C\|u - v\| \quad \text{for all } u, v \in V_1,$$

which completes the proof. $\square$

The next theorem shows that often $r_{\partial K}(f) < r_K(f)$, whenever f has an eigenvector in $\text{int}(K)$.

Theorem 6.1.7 *Let $f: K \to K$ be a continuous homogeneous order-preserving map on a solid closed cone $K \subseteq V$ with eigenvector $u \in \text{int}(K)$. If f is semi-differentiable at u and the semi-derivative $f'_u: V \to V$ satisfies the condition that for each $w \in K$ with $w < u$ there exists an integer $N \geq 1$ such that*

$$-\sum_{k=1}^{N} (f'_u)^k (w-u) \in \text{int}(K), \tag{6.7}$$

then $r_{\partial k}(f) < r_K(f)$.

Proof If $f(u) = \lambda u$ we must have that $\lambda > 0$, as otherwise $f(x) = 0$ for all $x \in K$ and f'_u cannot satisfy (6.7). Note that if we replace f by $\lambda^{-1} f$, then all the hypotheses remain valid. So we may as well assume from the start that $\lambda = 1$ and $f(u) = u$. By the chain rule for semi-differentiable maps, Lemma 6.1.6, we know that for each $k \geq 1$, f^k is semi-differentiable at u and

$$(f^k)'_u = (f'_u)^k.$$

For the sake of contradiction suppose that there exists $v \in \partial K \setminus \{0\}$ with $f(v) = v$. Write $a = m(u/v)$, so $av < u$ and $u - av \notin \text{int}(K)$. By (6.7) there exists an integer $N \geq 1$ such that

$$-\sum_{k=1}^{n} (f'_u)^k (av-u) \in \text{int}(K).$$

For $0 < t \leq 1$ define $u_t = (1-t)u + tav < u$ and note that

$$f^k(u_t) = f^k(u) + t(f'_u)^k(av-u) + t\|av-u\|\varepsilon_k(t(av-u)),$$

where $\lim_{\|w\| \to 0} \varepsilon_k(w) = 0$. It follows that

$$\sum_{k=1}^{N} (f^k(u) - f^k(u_t)) = t\Big(-\sum_{k=1}^{N} (f'_u)^k (av-u) - \|av-u\| \sum_{k=1}^{N} \varepsilon_k(t(av-u))\Big).$$

Since $-\sum_{k=1}^{n} (f'_u)^k (av-u) \in \text{int}(K)$ and $\lim_{\|w\| \to 0} \sum_{k=1}^{N} \varepsilon_k(w) = 0$, there exists $s > 0$ such that for $0 < t \leq s$

$$\sum_{k=1}^{N} (f^k(u) - f^k(u_t)) \in \text{int}(K).$$

Note that $av < u_t$ for all $0 \leq t < 1$, so that $\sum_{k=1}^{N}(f^k(u_t) - f^k(av)) \in K$. Thus, we find that

$$Nu = \sum_{k=1}^{N} f^k(u) \gg \sum_{k=1}^{N} f^k(av) = Nav$$

and hence there exists $a' > a$ such that $a'v \leq u$, which contradicts the fact that $a = m(u/v)$. □

Of course if in Theorem 6.1.7 f is Fréchet differentiable at u, then the condition (6.7) is equivalent to assuming that the derivative $f'(u)$ is irreducible. A careful inspection of the proof of Theorem 6.1.7 also shows that instead of assuming (6.7) it suffices to demand that, for each $w \in \partial K \setminus \{0\}$ with $w < u$, there exist an integer $N \geq 1$ and a constant $s > 0$ such that

$$\Big(\sum_{k=1}^{N} f^k(u) - f^k((1-t)u + tw)\Big) \in \operatorname{int}(K) \quad \text{for all } 0 < t \leq s.$$

6.2 Perturbation method

A natural approach to finding eigenvectors of a homogeneous order-preserving map $f\colon \operatorname{int}(K) \to \operatorname{int}(K)$ in the interior of K is to consider perturbations $f_\varepsilon\colon \operatorname{int}(K) \to \operatorname{int}(K)$ for $\varepsilon > 0$ such that each f_ε has an eigenvector $x^\varepsilon \in \operatorname{int}(K)$. If it can be arranged that the set $\{x^\varepsilon : \varepsilon > 0\}$ has a limit point $x^* \in \operatorname{int}(K)$, then under suitable conditions on the perturbations f_ε we may expect x^* to be an eigenvector of f in $\operatorname{int}(K)$. As we shall see in this section this idea works in a variety of circumstances.

There exists the following interesting choice for the perturbation f_ε of a homogeneous order-preserving map $f\colon \operatorname{int}(K) \to \operatorname{int}(K)$. A map $f_\varepsilon\colon \operatorname{int}(K) \to \operatorname{int}(K)$ is said to be a *contractive perturbation* of f if it is of the form

$$f_\varepsilon(x) = f(x) + \varepsilon g(x) \quad \text{for } x \in \operatorname{int}(K), \tag{6.8}$$

where $\varepsilon > 0$ and $g\colon K \to K$ is a continuous homogeneous strongly order-preserving map such that $g(K \setminus \{0\}) \subseteq \operatorname{int}(K)$.

Note that as g is strongly order-preserving, f_ε is also strongly order-preserving. It thus follows from Lemma 2.1.6 that f_ε is contractive under Hilbert's metric. Specific examples of maps $g\colon K \to K$ are provided by

$$g(x) = \vartheta(x)u \quad \text{for } x \in K, \tag{6.9}$$

where $\vartheta \in \operatorname{int}(K^*)$ and $u \in \operatorname{int}(K)$.

Lemma 6.2.1 *Let $K \subseteq V$ be a solid closed cone, $\varphi \in \text{int}(K^*)$, and $\Sigma^\circ = \{x \in \text{int}(K) : \varphi(x) = 1\}$. If $f : \text{int}(K) \to \text{int}(K)$ is a homogeneous order-preserving map and f_ε is a contractive perturbation of f, then, for each $\varepsilon > 0$, f_ε has a unique eigenvector $x^\varepsilon \in \Sigma^\circ$ and*

$$\lim_{k\to\infty} \frac{f_\varepsilon^k(x)}{\varphi(f_\varepsilon^k(x))} = x^\varepsilon \quad \textit{for all } x \in \Sigma^\circ. \tag{6.10}$$

Proof Let $f_\varepsilon : \text{int}(K) \to \text{int}(K)$ be given by

$$f_\varepsilon(x) = f(x) + \varepsilon g(x) \quad \text{for } x \in \text{int}(K),$$

where $\varepsilon > 0$ and $g : K \to K$ is a continuous homogeneous strongly order-preserving map such that $g(K \setminus \{0\}) \subseteq \text{int}(K)$. Recall that $\Sigma = \{x \in K : \varphi(x) = 1\}$ is compact, so that $g(\Sigma)$ is a compact subset of $\text{int}(K)$, as g is continuous and $g(K \setminus \{0\}) \subseteq \text{int}(K)$. Select $w \in \Sigma^\circ$ and note that as Σ is compact, there exists a constant $\beta > 0$ such that $y \leq \beta w$ for all $y \in \Sigma$. It follows that $f(y) \leq \beta f(w)$ for all $y \in \Sigma^\circ$, and hence $\text{cl}(\{f(y) : y \in \Sigma^\circ\})$ is a compact subset of K.

A simple compactness argument now shows that there exists a constant $\mu > 0$ such that $u \leq \mu v$ for all $u \in \text{cl}(\{f(y) : y \in \Sigma^\circ\})$ and all $v \in g(\Sigma)$. In particular, we have that

$$f(y) \leq \mu g(x) \quad \text{for all } x \in \Sigma \text{ and } y \in \Sigma^\circ. \tag{6.11}$$

Define $h_\varepsilon : \Sigma^\circ \to \Sigma^\circ$ by

$$h_\varepsilon(x) = \frac{f_\varepsilon(x)}{\varphi(f_\varepsilon(x))} \quad \text{for all } x \in \Sigma^\circ.$$

Select constants $0 < c_1 \leq c_2$ such that $c_1 \leq \varphi(g(x)) \leq c_2$ for all $x \in \Sigma^\circ$. From (6.11) it follows that $f(x) + \varepsilon g(x) \leq (\mu + \varepsilon) g(x)$, so that

$$\varphi(f(x) + \varepsilon g(x)) \leq (\mu + \varepsilon)\varphi(g(x)) \leq (\mu + \varepsilon)c_2.$$

This implies that

$$h_\varepsilon(x) \geq \frac{\varepsilon}{(\mu + \varepsilon)c_2} g(x) \quad \text{for all } x \in \Sigma^\circ. \tag{6.12}$$

Similarly, we see that

$$h_\varepsilon(x) \leq \frac{(\mu + \varepsilon)}{\varepsilon c_1} g(x) \quad \text{for all } x \in \Sigma^\circ, \tag{6.13}$$

because $\varphi(f(x) + \varepsilon g(x)) \geq \varepsilon c_1$ and $f(x) + \varepsilon g(x) \leq (\mu + \varepsilon) g(x)$.

Again, as $g(\Sigma)$ is compact, there exist constants $0 < \alpha_1 \leq \alpha_2$ such that $\alpha_1 w \leq g(x) \leq \alpha_2 w$ for all $x \in \Sigma^\circ$. Using (6.12) and (6.13) we deduce that

$$\frac{\varepsilon\alpha_1}{(\mu+\varepsilon)c_2} w \leq h_\varepsilon(x) \leq \frac{(\mu+\varepsilon)\alpha_2}{\varepsilon c_1} w \quad \text{for all } x \in \Sigma^\circ. \tag{6.14}$$

Note that the set

$$\begin{aligned} T_\varepsilon &= \{z \in \Sigma^\circ \colon \frac{\varepsilon\alpha_1}{(\mu+\varepsilon)c_2} w \leq z \leq \frac{(\mu+\varepsilon)\alpha_2}{\varepsilon c_1} w\} \\ &= \{z \in \Sigma^\circ \colon d_H(w,z) \leq \log\left(\frac{(\mu+\varepsilon)^2 c_2\alpha_2}{\varepsilon^2 c_1\alpha_1}\right)\} \end{aligned}$$

is compact. By (6.14) we see that $h_\varepsilon(\Sigma^\circ) \subseteq T_\varepsilon$ and also that $h_\varepsilon(T_\varepsilon) \subseteq T_\varepsilon$.

As f_ε is homogeneous and strongly order-preserving, h_ε is contractive under Hilbert's metric by Lemma 2.1.6. It follows from Theorem 3.2.2 that h_ε has a unique fixed point $x^\varepsilon \in T_\varepsilon$ and $\lim_{k\to\infty} h_\varepsilon^k(x) = x^\varepsilon$ for all $x \in T_\varepsilon$. Thus, x^ε is a locally attracting fixed point of h_ε in Σ°. It now follows from Proposition 3.2.3 that

$$\lim_{k\to\infty} h_\varepsilon^k(x) = x^\varepsilon \quad \text{for all } x \in \Sigma^\circ,$$

which completes the proof. □

The eigenvector $x^\varepsilon \in \Sigma$ of a contractive perturbation f_ε can be useful in detecting eigenvectors of f in the interior of the cone.

Theorem 6.2.2 *Let $K \subseteq V$ be a solid closed cone, $\varphi \in \text{int}(K^*)$, and $\Sigma^\circ = \{x \in \text{int}(K) \colon \varphi(x) = 1\}$. Suppose that $f \colon \text{int}(K) \to \text{int}(K)$ is a homogeneous order-preserving map and $f_\varepsilon \colon \text{int}(K) \to \text{int}(K)$ is a contractive perturbation of f with eigenvector $x^\varepsilon \in \Sigma^\circ$ for $\varepsilon > 0$. Then f has an eigenvector $x^* \in \text{int}(K)$ if and only if $\text{cl}(\{x^\varepsilon \in \Sigma^\circ \colon \varepsilon > 0\})$ is a compact subset of Σ°.*

Proof Suppose that $v \in \text{int}(K)$ is an eigenvector of f. Note that as $g(K \setminus \{0\}) \subseteq \text{int}(K)$ and g is continuous, $g(\Sigma)$ is a compact subset of $\text{int}(K)$, and hence there exists $0 \leq R < \infty$ such that

$$\sup\{d_H(g(x), v) \colon x \in \text{int}(K)\} \leq R.$$

Write $B_R(v) = \{z \in \text{int}(K) \colon d_H(z, v) \leq R\}$. So, for each $\varepsilon > 0$ we have that $\varepsilon g(x) \in B_R(v)$. Furthermore, for $x \in \text{int}(K)$ with $d_H(x, v) \leq R$,

$$d_H(f(x), v) = d_H(f(x), f(v)) \leq d_H(x, v) \leq R.$$

We know that $B_R(v)$ is convex and $\lambda z \in B_R(v)$ for all $\lambda > 0$ and $z \in B_R(v)$. Since $f(x) \in B_R(v)$ for all $x \in B_R(v)$ and $\varepsilon g(x) \in B_R(v)$ for all $x \in \text{int}(K)$, it follows that

$$f_\varepsilon(x) = f(x) + \varepsilon g(x) \in B_R(v) \quad \text{for all } x \in B_R(v) \text{ and } \varepsilon > 0.$$

Let $h_\varepsilon : \Sigma^\circ \to \Sigma^\circ$ be given by

$$h_\varepsilon(x) = \frac{f_\varepsilon(x)}{\varphi(f_\varepsilon(x))} \quad \text{for all } x \in \Sigma^\circ.$$

So $h_\varepsilon(B_R(v) \cap \Sigma^\circ) \subseteq B_R(v) \cap \Sigma^\circ$. As $B_R(v)$ is compact and h_ε is contractive under Hilbert's metric by Lemma 2.1.6, we know from Theorem 3.2.2 that h_ε has a unique fixed point in $B_R(v) \cap \Sigma^\circ$. From Lemma 6.2.1 we conclude that

$$\mathrm{cl}(\{x^\varepsilon \in \Sigma^\circ : \varepsilon > 0\}) \subseteq B_R(v) \cap \Sigma^\circ,$$

and hence $\mathrm{cl}(\{x^\varepsilon \in \Sigma^\circ : \varepsilon > 0\})$ is a compact subset of Σ°.

Now suppose that x^* is a limit point of $\{x^\varepsilon \in \Sigma^\circ : \varepsilon > 0\}$ in Σ°. Let $\lambda_\varepsilon \geq 0$ be such that $f_\varepsilon(x^\varepsilon) = \lambda_\varepsilon x^\varepsilon$. There exists a sequence $(\varepsilon_k)_k$ with $\lim_{k\to\infty} \varepsilon_k = 0$ such that $\lim_{k\to\infty} x^{\varepsilon_k} = x^*$ and $\lim_{k\to\infty} \lambda_{\varepsilon_k} = \lambda^*$. Note that

$$\lim_{k\to\infty} \sup_{x\in\Sigma^\circ} \|f(x) - f_\varepsilon(x)\| = \lim_{k\to\infty} \sup_{x\in\Sigma^\circ} \varepsilon_k \|g(x)\| = 0.$$

As f is continuous on $\mathrm{int}(K)$ we find that $\lim_{k\to\infty} f_{\varepsilon_k}(x^{\varepsilon_k}) = f(x^*)$ and $\lim_{k\to\infty} f_{\varepsilon_k}(x^{\varepsilon_k}) = \lim_{k\to\infty} \lambda_{\varepsilon_k} x^{\varepsilon_k} = \lambda^* x^*$, so that $f(x^*) = \lambda^* x^*$. As $x^* \in \mathrm{int}(K)$ and $f(x^*) \in \mathrm{int}(K)$, we also see that $\lambda^* > 0$. □

If in (6.8) we take $g(x) = \vartheta(x)u$, where $\vartheta \in \mathrm{int}(K^*)$ and $u \in \mathrm{int}(K)$, then the arguments in the proof of Theorem 6.2.2 provide information about the location of possible eigenvectors of f in the interior of K. More precisely, if for some $\varepsilon > 0$ we have found an eigenvector $x^\varepsilon \in \Sigma^\circ$ of f_ε and $d_H(x^\varepsilon, u) = 2R$, then each eigenvector $v \in \Sigma^\circ$ of f satisfies $d_H(u, v) \geq R$. Indeed, if $d_H(x^\varepsilon, u) = 2R$ and $v \in \Sigma^\circ$ is an eigenvector of f with $d_H(u, v) < R$, then

$$R < d_H(x^\varepsilon, u) - d_H(u, v) < d_H(x^\varepsilon, v),$$

which is impossible, as it follows from the proof of Theorem 6.2.2 that $d_H(x^\varepsilon, v) \leq R$.

We also wish to point out that there are many examples of homogeneous order-preserving maps $f : \mathrm{int}(K) \to \mathrm{int}(K)$ for which the set of normalized eigenvectors, $\{x \in \mathrm{int}(K) : f(x) = r_f(K)x \text{ and } \|x\| = 1\}$, is a compact subset of $\mathrm{int}(K)$, but not a single point. A simple example is provided by the following map. Let $0 < a, b < 1$ and $C = \{x \in \mathbb{R}^2_+ : ax_1 \leq x_2 \text{ and } bx_2 \leq x_1\}$. Define a retraction g of $\mathbb{R}^2_+$ onto the closed cone C inside $\mathbb{R}^2_+$ by

$$g(x) = \begin{cases} \left(\frac{x_1+x_2}{1+a}, \frac{a(x_1+x_2)}{1+a}\right) & \text{for } x_2 \leq ax_1, \\ (x_1, x_2) & \text{for } ax_1 \leq x_2 \text{ and } bx_2 \leq x_1, \\ \left(\frac{b(x_1+x_2)}{1+b}, \frac{x_1+x_2}{1+b}\right) & \text{for } x_1 \leq bx_2. \end{cases}$$

Obviously g is continuous, homogeneous, and a retraction onto C. A simple calculation shows that g satisfies

$$g(x) = \begin{pmatrix} \max\left(\min\{x_1, \frac{x_1+x_2}{1+a}\}, \frac{b(x_1+x_2)}{1+b}\right) \\ \max\left(\min\{x_2, \frac{x_1+x_2}{1+b}\}, \frac{a(x_1+x_2)}{1+a}\right) \end{pmatrix}$$

for each $x \in \mathbb{R}^2_+$. From this identity we see that g preserves the ordering induced by $\mathbb{R}^2_+$.

In the remainder of the section we will see how Theorem 6.2.2 can be used to prove a variety of sufficient conditions for the existence of eigenvectors of homogeneous order-preserving maps in the interior of $\mathbb{R}^n_+$. These conditions were obtained by Gaubert and Gunawardena [74] and involve a directed graph, which is a direct generalization of the usual directed graph associated with a nonnegative matrix.

For $J \subseteq \{1, \ldots, n\}$ and a positive real $u > 0$, define $u^J \in \mathbb{R}^n_+$ by $u^J_j = u$ if $j \in J$, and $u^J_j = 1$ otherwise. We associate a digraph $G_f = (V, A)$ with a homogeneous order-preserving map $f\colon \operatorname{int}(\mathbb{R}^n_+) \to \operatorname{int}(\mathbb{R}^n_+)$ as follows. The vertex set $V = \{1, \ldots, n\}$ and there exists an arc $(i, j) \in A$ if

$$\lim_{u\to\infty} f_i(u^{\{j\}}) = \infty. \tag{6.15}$$

Recall that a digraph $G = (V, A)$ is *strongly connected* if for each $i, j \in V$ there exists a directed path from i to j in G. For $r > 0$ define

$$\Psi(r) = \sup\{u \geq 0\colon \inf_{(i,j)\in A} f_i(u^{\{j\}}) \leq r\}.$$

So, if $(i, j) \in A$ and $f_i(u^{\{j\}}) \leq r$, then $u \leq \Psi(r)$. We also remark that Ψ is an increasing function of r, and $\Psi(r) < \infty$ for all $r > 0$ by (6.15).

Theorem 6.2.3 *Let* $\Delta_n^\circ = \{x \in \operatorname{int}(\mathbb{R}^n_+)\colon \sum_i x_i = 1\}$ *and suppose that* $f\colon \operatorname{int}(\mathbb{R}^n_+) \to \operatorname{int}(\mathbb{R}^n_+)$ *is a homogeneous order-preserving map. If* G_f *is strongly connected, then the following assertions hold:*

(i) If $f_\varepsilon\colon \operatorname{int}(\mathbb{R}^n_+) \to \operatorname{int}(\mathbb{R}^n_+)$ *is a contractive perturbation of* f *with eigenvector* $x^\varepsilon \in \Delta_n^\circ$ *for* $\varepsilon > 0$*, then* $\operatorname{cl}(\{x^\varepsilon \in \Delta_n^\circ\colon \varepsilon > 0\})$ *is a compact subset of* Δ_n°.

(ii) There exists an eigenvector of f *in* $\operatorname{int}(\mathbb{R}^n_+)$.

(iii) The set $\{v \in \Delta_n^\circ\colon f(v) = r_{\mathbb{R}^n_+}(f)v\}$ *is a compact subset of* Δ_n°.

Proof As we are dealing with $\mathbb{R}^n_+$ both f and f_ε can be extended continuously to $\partial\mathbb{R}^n_+$ by Theorem 5.1.5. So we may as well assume from the beginning that f and f_ε are continuous homogeneous order-preserving maps on $\mathbb{R}^n_+$.

To prove the first assertion we argue by contradiction. Suppose that $\mathrm{cl}(\{x^\varepsilon \in \Delta_n^\circ : \varepsilon > 0\})$ is not a compact subset of Δ_n°. This implies that there exists a sequence $(x^{\varepsilon_k})_k$ with $\lim_{k\to\infty} x^{\varepsilon_k} = x^* \in \partial\Delta_n$, where $\Delta_n = \{x \in \mathbb{R}_+^n : \sum_i x_i = 1\}$ and $\lim_{k\to\infty} \varepsilon_k = 0$. Denote $\lambda_k = r_{\mathbb{R}_+^n}(f_{\varepsilon_k})$ for all $k \geq 1$, so $f_{\varepsilon_k}(x^{\varepsilon_k}) = \lambda_k x^{\varepsilon_k}$. As $f(x) \leq f_{\varepsilon_k}(x)$ for all $x \in \mathbb{R}_+^n$ and $k \geq 1$, and

$$\lim_{k\to\infty} \sup_{x\in\Delta_n} \|f(x) - f_{\varepsilon_k}(x)\| = \lim_{k\to\infty} \sup_{x\in\Delta_n} \varepsilon_k \|g(x)\| = 0,$$

it follows from Lemma 5.5.6 that

$$\lim_{k\to\infty} \lambda_k = r_{\mathbb{R}_+^n}(f). \tag{6.16}$$

By taking a further subsequence we may assume for some $1 \leq i \leq n$ that

$$x_i^{\varepsilon_k} = \mathsf{b}(x^{\varepsilon_k}) \quad \text{for all } k \geq 1, \tag{6.17}$$

and $\lim_{k\to\infty} x_i^{\varepsilon_k} = 0$.

Now the idea is to use the fact that G_f is strongly connected to show that $x_j^* = 0$ for all j, which is impossible, as $x^* \in \Delta_n$. Take $j \in \{1, \ldots, n\}$ and let $i = i_1 \to i_2 \to i_3 \to \cdots \to i_m = j$ be a directed path from i to j in G_f where $m \leq n$. (Such a path always exists as G_f is strongly connected.) It follows that

$$\lim_{u\to\infty} f_{i_1}(u^{\{i_2\}}) = \infty.$$

From (6.16) there exists a constant $M_0 > 0$ such that $\lambda_k \leq M_0$ for all $k \geq 1$. Using (6.17) we find for $u = x_{i_2}^{\varepsilon_k}/x_{i_1}^{\varepsilon_k}$ and $K \geq 1$ that

$$x_{i_1}^{\varepsilon_k} f_{i_1}(u^{\{i_2\}}) \leq x_{i_1}^{\varepsilon_k} f_{i_1}(x^{\varepsilon_k}/x_{i_1}^{\varepsilon_k}) \leq x_{i_1}^{\varepsilon_k} (f_{\varepsilon_k})_{i_1}(x^{\varepsilon_k}/x_{i_1}^{\varepsilon_k}) = \lambda_k x_{i_1}^{\varepsilon_k} \leq M_0 x_{i_1}^{\varepsilon_k},$$

as $f(x) \leq f_{\varepsilon_k}(x)$ for all $x \in \mathrm{int}(\mathbb{R}_+^n)$.

Put $M_1 = \Psi(M_0)$ and note that for $u = x_{i_2}^{\varepsilon_k}/x_{i_1}^{\varepsilon_k}$ we have that $u \leq M_1$; so,

$$x_{i_2}^{\varepsilon_k} \leq M_1 x_{i_1}^{\varepsilon_k} \text{ for all } k \geq 1.$$

We can repeat this argument. Indeed, for $u = x_{i_3}^{\varepsilon_k}/x_{i_1}^{\varepsilon_k}$ we have that

$$x_{i_1}^{\varepsilon_k} f_{i_2}(u^{\{i_3\}}) \leq x_{i_1}^{\varepsilon_k} f_{i_2}(x^{\varepsilon_k}/x_{i_1}^{\varepsilon_k}) \leq x_{i_1}^{\varepsilon_k} (f_{\varepsilon_k})_{i_2}(x^{\varepsilon_k}/x_{i_1}^{\varepsilon_k}) = \lambda_k x_{i_2}^{\varepsilon_k} \leq M_0 M_1 x_{i_1}^{\varepsilon_k},$$

and $u \leq \Psi(M_0 M_1) = M_2$. Thus,

$$x_{i_3}^{\varepsilon_k} \leq M_2 x_{i_1}^{\varepsilon_k} \quad \text{for all } k \geq 1.$$

Iterating this argument yields, after finitely many steps, a constant $M > 0$ such that

$$x_j^{\varepsilon_k} \leq M x_{i_1}^{\varepsilon_k} \quad \text{for all } k \geq 1.$$

Now letting $k \to \infty$ we deduce that $x_j^* = \lim_{k\to\infty} x_j^{\varepsilon_k} = 0$, which shows that $x^* = 0$. Thus, $\mathrm{cl}(\{x^\varepsilon \in \Delta_n^\circ : \varepsilon > 0\})$ is a compact subset of Δ_n°.

The second assertion is a direct consequence of part (i) and Theorem 6.2.2. To show that $\{v \in \Delta_n^\circ : f(v) = r_{\mathbb{R}_+^n}(f)v\}$ is a compact subset of Δ_n° we observe that the same argument as in the proof of part (i) can be applied to f instead of f_{ε_k}. □

A simple example to which Theorem 6.2.3 applies is the map $f : \mathbb{R}_+^3 \to \mathbb{R}_+^3$ given by

$$f(x) = \begin{pmatrix} (x_1 \vee x_2) \wedge 2\sqrt{x_2x_3} \\ \sqrt{x_2^2 + x_3^2} \wedge 3\sqrt{x_3x_1} \\ x_1 \vee (x_2 \wedge x_3) \end{pmatrix}$$

for $x \in \mathbb{R}_+^3$. It is easy to verify that G_f is the digraph given in Figure 6.1. On the other hand, the reader can check that Theorem 6.2.3 does not apply to Example 6.1.1.

There exists a dual version of Theorem 6.2.3. Let $\tau : \mathrm{int}(\mathbb{R}_+^n) \to \mathrm{int}(\mathbb{R}_+^n)$ be given by

$$\tau(x) = (x_1^{-1}, \ldots, x_n^{-1}) \quad \text{for } x \in \mathrm{int}(\mathbb{R}_+^n).$$

Given a homogeneous order-preserving map $f : \mathrm{int}(\mathbb{R}_+^n) \to \mathrm{int}(\mathbb{R}_+^n)$, define $f^- : \mathrm{int}(\mathbb{R}_+^n) \to \mathrm{int}(\mathbb{R}_+^n)$ by $f^-(x) = (\tau \circ f \circ \tau)(x)$ for all $x \in \mathrm{int}(\mathbb{R}_+^n)$. Clearly f^- is a homogeneous order-preserving map, and f has an eigenvector in $\mathrm{int}(\mathbb{R}_+^n)$ if and only if f^- does.

Corollary 6.2.4 *If $f : \mathrm{int}(\mathbb{R}_+^n) \to \mathrm{int}(\mathbb{R}_+^n)$ is a homogeneous order-preserving map and G_{f^-} is strongly connected, then f has an eigenvector in $\mathrm{int}(\mathbb{R}_+^n)$.*

The digraph G_{f^-} can be easily constructed from f by noting that there exists an arrow from i to j in G_{f^-} if and only if $\lim_{u\to 0^+} f_i(u^{\{j\}}) = 0$. It turns out that if G_{f^-} is strongly connected, then $r_{\partial\mathbb{R}_+^n}(f) < r_{\mathbb{R}_+^n}(f)$.

Corollary 6.2.5 *If $f : \mathbb{R}_+^n \to \mathbb{R}_+^n$ is a continuous homogeneous order-preserving map such that $f(\mathrm{int}(\mathbb{R}_+^n)) \subseteq \mathrm{int}(\mathbb{R}_+^n)$ and G_{f^-} is strongly connected, then $r_{\partial\mathbb{R}_+^n}(f) \leq 0 < r_{\mathbb{R}_+^n}(f)$.*

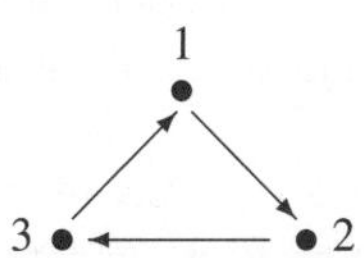

Figure 6.1 The digraph G_f.

Proof For the sake of contradiction suppose that there exists $w \in \partial\mathbb{R}^n_+ \setminus \{0\}$ such that $f(w) = \lambda w$ and $\lambda > 0$. By scaling w we may assume that $w \leq \mathbb{1}$. Let $J = \{j : w_j = 0\}$ and $I = \{1, \dots, n\} \setminus J$. Note that $I \neq \emptyset$, as $w \neq 0$. Recall that G_{f^-} is strongly connected, and hence there exists an arc from some $i \in I$ to some $j \in J$ in G_{f^-}. As $w \leq u^{\{j\}}$ for all $u > 0$, we find that

$$0 \leq \lambda w_i = f_i(w) \leq \lim_{u \to 0^+} f_i(u^{\{j\}}) = 0.$$

But $w_i > 0$ and $\lambda > 0$, which is impossible. Thus, either f has no eigenvector in $\partial\mathbb{R}^n_+$, in which case $r_{\partial\mathbb{R}^n_+}(f) = -\infty$, or $r_{\partial\mathbb{R}^n_+}(f) = 0$.

As $f(\mathrm{int}(\mathbb{R}^n_+)) \subseteq \mathrm{int}(\mathbb{R}^n_+)$, there exists $\delta > 0$ such that $\delta\mathbb{1} \leq f(\mathbb{1})$. So, by Proposition 5.3.6, we also know that $r_{\mathbb{R}^n_+}(f) \geq \delta > 0$. □

6.3 Bounded invariant sets

Another useful strategy for proving the existence of eigenvectors in the interior of the cone is given by the following result.

Proposition 6.3.1 *If* $f : \mathrm{int}(K) \to \mathrm{int}(K)$ *is a homogeneous order-preserving map on the interior of a solid closed cone* $K \subseteq V$, *then there exists* $W \subseteq \mathrm{int}(K)$ *with* $\mathrm{diam}(W) = \sup\{d_H(x, y) : x, y \in W\} < \infty$ *and* $f(W) \subseteq W$ *if and only if* f *has an eigenvector in* $\mathrm{int}(K)$.

Proof Clearly f has an eigenvector $u \in \mathrm{int}(K)$; then for each $\varepsilon > 0$ the closed ball $B_\varepsilon(u) = \{x \in \mathrm{int}(K) : d_H(x, u) \leq \varepsilon\}$ satisfies the conditions of the theorem. On the other hand, if there exists $W \subseteq \mathrm{int}(K)$ with $\mathrm{diam}(W) < \infty$ and $f(W) \subseteq W$, we can assume that $tW = W$ for all $t > 0$ by replacing W by $W' = \{tx : t > 0 \text{ and } x \in W\}$. Now we can consider the scaled map $g : \Sigma^\circ \to \Sigma^\circ$ given by

$$g(x) = \frac{f(x)}{\varphi(f(x))}$$

for $x \in \Sigma^\circ$, where $\Sigma^\circ = \{x \in \mathrm{int}(K) : \varphi(x) = 1\}$ and $\varphi \in \mathrm{int}(K^*)$. Clearly $g(W \cap \Sigma^\circ) \subseteq W \cap \Sigma^\circ$. Moreover, as $\mathrm{diam}(W) < \infty$, the orbit of each $w \in W \cap \Sigma^\circ$ under g has a limit point in Σ°. It therefore follows from Corollary 3.2.5 that g has a fixed point in Σ°, and hence f has an eigenvector in $\mathrm{int}(K)$. □

There are various natural invariant sets of homogeneous order-preserving maps $f : \mathrm{int}(K) \to \mathrm{int}(K)$. For $\beta > 0$ define the *super-eigenspace of* f *corresponding to* β by

$$S^\beta(f) = \{x \in \mathrm{int}(K)\colon f(x) \leq \beta x\}.$$

Similarly, for $\alpha > 0$ define the *sub-eigenspace of f corresponding to α* by

$$S_\alpha(f) = \{x \in \mathrm{int}(K)\colon \alpha x \leq f(x)\}.$$

Note that $S^\beta(f)$ and $S_\alpha(f)$ are non-empty for $\beta > 0$ large and $\alpha > 0$ small. The intersection of a sub-eigenspace and a super-eigenspace is called a *slice space*, and is denoted by

$$S_\alpha^\beta(f) = S_\alpha(f) \cap S^\beta(f).$$

Clearly, sub-eigenspaces, super-eigenspaces, and slice spaces are invariant under f.

In case $K = \mathbb{R}^n_+$ it was shown by Gaubert and Gunawardena [74] that each super-eigenspace of f is bounded in Hilbert's metric if G_f is strongly connected, and that each sub-eigenspace of f is bounded in Hilbert's metric if G_{f^-} is strongly connected. Sometimes it can be shown that all slice spaces of a homogeneous order-preserving map $f\colon \mathrm{int}(\mathbb{R}^n_+) \to \mathrm{int}(\mathbb{R}^n_+)$ are bounded in Hilbert's metric by using the so-called *recession map* $\hat{f}\colon \mathrm{int}(\mathbb{R}^n_+) \to \mathrm{int}(\mathbb{R}^n_+)$, which is defined by

$$\hat{f}(x) = \lim_{r\to\infty} f(x_1^r, x_2^r, \ldots, x_n^r)^{1/r} \quad \text{for } x \in \mathrm{int}(\mathbb{R}^n_+).$$

Note that $\hat{f}(\mathbb{1}) = \mathbb{1}$, so $\mathbb{1}$ is always an eigenvector of $\hat{f}$ in $\mathrm{int}(\mathbb{R}^n_+)$. The recession map $\hat{f}$ need not always exist, but if it does it can be a useful tool for proving the existence of eigenvectors in the interior of $\mathbb{R}^n_+$, as the following result by Gaubert and Gunawardena [74] shows.

Theorem 6.3.2 *If $f\colon \mathrm{int}(\mathbb{R}^n_+) \to \mathrm{int}(\mathbb{R}^n_+)$ is a homogeneous order-preserving map for which $\hat{f}$ exists, and $\mathbb{1}$ is the only eigenvector of $\hat{f}$ in $\mathrm{int}(\mathbb{R}^n_+)$, up to a positive scalar multiple, then all slice spaces of f are bounded under Hilbert's metric and f has an eigenvector in $\mathrm{int}(\mathbb{R}^n_+)$.*

Proof Suppose, by way of contradiction, that $S_\alpha^\beta(f)$ is unbounded under Hilbert's metric. Then there exists a sequence $(x^k)_k$ in $S_\alpha^\beta(f)$, with $\mathsf{b}(x^k) = 1$ for all $k \geq 1$, and $d_H(x^k, \mathbb{1}) \to \infty$ as $k \to \infty$. As

$$d_H(x^k, \mathbb{1}) = \log \frac{\mathsf{t}(x^k)}{\mathsf{b}(x^k)} = \log \mathsf{t}(x^k),$$

we see that $\lim_{k\to\infty} \mathsf{t}(x^k) = \infty$. So, we may assume that $t_k = \log \mathsf{t}(x^k) > 0$ for all $k \geq 1$.

Let L be the coordinatewise log function, and E be the coordinatewise exponential function. For each $k \geq 1$ write $y^k = L(x^k)$, and note that, as $x^k \in S_\alpha^\beta(f)$,

$$\alpha E(y^k) \leq f(E(y^k)) \leq \beta E(y^k). \tag{6.18}$$

Remark that $\mathsf{b}(E(y^k/t_k)) = e^0 = 1$ and $\mathsf{t}(E(y^k/t_k)) = e$, so that

$$\mathbb{1} \leq E(y^k/t_k) \leq e\mathbb{1} \quad \text{for all } k \geq 1.$$

Thus, after taking a subsequence we may assume that $\lim_{k\to\infty} E(y^k/t_k) = E(y)$ for some $y \in \mathbb{R}^n$. Also note that $\mathsf{b}(E(y)) = 1$ and $\mathsf{t}(E(y)) = e$; so, $E(y) \neq \mathbb{1}$.

To obtain a contradiction we show that $\hat{f}(E(y)) = E(y)$. Remark that

$$\begin{aligned}\log M(f(E(y^k))^{1/t_k}/f(E(yt_k))^{1/t_k}) &= \frac{1}{t_k}\log M(f(E(y^k))/f(E(yt_k)))\\ &\leq \frac{1}{t_k}\log M(E(y^k)/E(yt_k))\\ &= \log M(E(y^k/t_k)/E(y)),\end{aligned}$$

as f is homogeneous and order-preserving. Likewise,

$$\log M(f(E(yt_k))^{1/t_k}/f(E(y^k)^{1/t_k}) \leq \log M(E(y)/E(y^k/t_k))$$

for all $k \geq 1$. Thus,

$$\lim_{k\to\infty} d_H(f(E(y^k))^{1/t_k}, f(E(yt_k))^{1/t_k}) = 0. \tag{6.19}$$

From (6.18) we deduce that

$$\left(\alpha E(y^k)\right)^{1/t_k} \leq f(E(y^k))^{1/t_k} \leq (\beta E(y^k))^{1/t_k},$$

where the left-hand side and the right-hand side converge to $E(y)$ as $k \to \infty$. It follows from (6.19) that $\lim_{k\to\infty} f(E(yt_k))^{1/t_k} = E(y)$, which shows that $\hat{f}(E(y)) = E(y)$. This is a contradiction, as $E(y) \neq \mathbb{1}$. Thus, all slice spaces of f are bounded, and f has an eigenvector in $\mathrm{int}(\mathbb{R}^n_+)$ by Proposition 6.3.1. □

Note that if $x = (x_1, \ldots, x_n)$ is an eigenvector of $\hat{f}$ in $\mathrm{int}(\mathbb{R}^n_+)$, then, for each $s > 0$,

$$\lim_{r\to\infty} f(x_1^{sr}, \ldots, x_n^{sr})^{1/r} = \hat{f}(x)^s = (x_1^s, \ldots, x_n^s).$$

So, for each $s > 0$, $x^s = (x_1^s, \ldots, x_n^s)$ is an eigenvector of $\hat{f}$ in $\mathrm{int}(\mathbb{R}^n_+)$. Note that if $x \neq \mathbb{1}$, then $d_H(x^s, \mathbb{1}) \to \infty$ as $s \to \infty$. We conclude that either each eigenvector of $\hat{f}$ in $\mathrm{int}(\mathbb{R}^n_+)$ is a multiple of $\mathbb{1}$, or the set of eigenvectors of $\hat{f}$ in $\mathrm{int}(\mathbb{R}^n_+)$ is unbounded under d_H.

To illustrate the use of the recession map consider the following example:

$$g(x) = \begin{pmatrix} x_1 + 2x_2 + 3x_3 \\ 2(x_1^{-1} + x_2^{-1})^{-1} + 3x_3 \\ (2x_2^{-1} + 3x_3^{-1})^{-1} \end{pmatrix}$$

for $x \in \operatorname{int}(\mathbb{R}^3_+)$. It is easy to verify that

$$\hat{g}(x) = \begin{pmatrix} x_1 \vee x_2 \vee x_3 \\ (x_1 \wedge x_2) \vee x_3 \\ x_2 \wedge x_3 \end{pmatrix},$$

and $\hat{g}(x) = x$ for $x \in \operatorname{int}(\mathbb{R}^3_+)$ if and only if x is a positive multiple of $\mathbb{1}$. Thus, g has an eigenvector in the interior of $\mathbb{R}^3_+$ by Theorem 6.3.2.

On the other hand, for the map f in Example 6.1.1 the recession map is given by

$$\hat{f}(x) = \begin{pmatrix} x_1 \vee (x_2 \wedge x_3) \\ x_2 \vee (x_1 \wedge x_3) \\ x_3 \vee (x_1 \wedge x_2) \end{pmatrix}.$$

Notice that $\hat{f}$ is independent of the parameters a, b, and c. However, if $c \geq a + b$, then f has no eigenvector in the interior of $\mathbb{R}^3_+$. So, we can deduce from Theorem 6.3.2 that $\mathbb{1}$ is not the only eigenvector $\hat{f}$ in the interior of $\mathbb{R}^3_+$. Indeed, it is easy to see that $\hat{f}(2, 1, 1) = (2, 1, 1)$.

We shall see in Section 6.6 that it is often hard to decide if a continuous homogeneous order-preserving map $f \colon \mathbb{R}^n_+ \to \mathbb{R}^n_+$ has an eigenvector in the interior of $\mathbb{R}^n_+$ if the map is constructed from means $M_{r,\sigma}(\cdot)$ with $-\infty \leq r < 0$.

6.4 Uniqueness of the eigenvector

If a homogeneous order-preserving map $f \colon \operatorname{int}(K) \to \operatorname{int}(K)$ has an eigenvector $u \in \operatorname{int}(K)$, it is interesting to understand when u is the unique, up to positive scalar multiples, eigenvector of f in the interior of K. In this section several conditions for uniqueness are discussed which involve the semi-derivative of f at the eigenvector $u \in \operatorname{int}(K)$. We start with the following result.

Theorem 6.4.1 *Let $K \subseteq V$ be a solid closed cone, $\varphi \in \operatorname{int}(K^*)$, and $\Sigma^\circ = \{x \in \operatorname{int}(K) \colon \varphi(x) = 1\}$. Suppose that $g \colon \Sigma^\circ \to \Sigma^\circ$ is non-expansive under Hilbert's metric and $g(u) = u$ for some $u \in \Sigma^\circ$. If g is semi-differentiable on a neighborhood $U \subseteq \Sigma^\circ$ of u and*

$$x - g'_u(x) \neq 0 \tag{6.20}$$

for all $x \in V_\varphi = \{x \in V \colon \varphi(x) = 0\}$ with $x \neq 0$, then u is the only fixed point of g in Σ°.

Proof It is easy to verify that $S_u = \{x \in V : \varphi(x) = 0\}$ in (6.3). For the sake of contradiction suppose that $w \in \Sigma^\circ$ with $w \neq u$ and $g(w) = w$. It follows from Theorem 2.6.3 that $\gamma(t) = (1-t)u + tw$, where $0 \leq t \leq 1$, is, after re-parametrization, a geodesic path connecting u and w in (Σ°, d_H). Let $\delta = d_H(u, w)$; so,

$$d_H(u, \gamma(t)) + d_H(\gamma(t), w) = \delta$$

for all $0 \leq t \leq 1$. Moreover, for each $0 < \lambda < \delta$ there exists $0 < t < 1$ with

$$d_H(u, \gamma(t)) = \lambda \quad \text{and} \quad d_H(\gamma(t), w) = \delta - \lambda. \tag{6.21}$$

For $0 < \lambda < \delta$ define $\Gamma_\lambda \subseteq \Sigma^\circ$ by

$$\Gamma_\lambda = \{x \in \Sigma^\circ : d_H(u, x) \leq \lambda \text{ and } d_H(x, w) \leq \delta - \lambda\}.$$

It follows from (6.21) that Γ_λ is non-empty. By Corollary 2.5.6 and Lemma 2.6.1 we know that closed Hilbert metric balls are closed, bounded, convex sets in the norm topology on Σ°. As g is non-expansive under d_H, and $g(u) = u$ and $g(w) = w$, we find that $g(\Gamma_\lambda) \subseteq \Gamma_\lambda$. Thus, by the Brouwer fixed-point theorem, g has a fixed point $x_\lambda \in \Gamma_\lambda$ for $0 < \lambda < \delta$.

By using the assumption that g is semi-differentiable at u, we see that

$$x_\lambda = g(x_\lambda) = g(u) + g'_u(x_\lambda - u) + R(x_\lambda - u) = u + g'_u(x_\lambda - u) + R(x_\lambda - u), \tag{6.22}$$

where

$$\lim_{\lambda \to 0^+} \frac{R(x_\lambda - u)}{\|x_\lambda - u\|} = 0.$$

Now select a sequence $\lambda_k \to 0^+$ such that

$$\lim_{k \to \infty} \frac{x_{\lambda_k} - u}{\|x_{\lambda_k} - u\|} = \xi \in V_\varphi$$

exists. Recall that g'_u is positively homogeneous and note that $\xi \neq 0$, as $\|\xi\| = 1$. Now take $\lambda = \lambda_k$ in (6.22) and bring u to the other side. Subsequently divide both sides by $\|x_\lambda - u\|$ and let $k \to \infty$ to deduce that $\xi = g'_u(\xi)$, which contradicts (6.20). □

To fully exploit Theorem 6.4.1 the following lemma will be useful.

Lemma 6.4.2 *Let $K \subseteq V$ be a solid closed cone. Suppose that $f : \operatorname{int}(K) \to \operatorname{int}(K)$ is a map with $f(u) = ru$ for some $u \in \operatorname{int}(K)$ and f is semi-differentiable at u. If $\varphi \in \operatorname{int}(K^*)$ with $\varphi(u) = 1$, and $h : \operatorname{int}(K) \to \operatorname{int}(K)$ is given by*

$$h(x) = \frac{f(x)}{\varphi(f(x))} \quad \text{for } x \in \operatorname{int}(K),$$

then h is semi-differentiable at u and

$$h'_u(v) = \frac{1}{r}(f'_u(v) - \varphi(f'_u(v))u) \quad \textit{for } v \in V. \tag{6.23}$$

Proof Let $P\colon \mathrm{int}(K) \to \mathrm{int}(K)$ be given by $P(y) = y/\varphi(y)$ for $y \in \mathrm{int}(K)$. The reader can verify that P is continuously Fréchet differentiable on $\mathrm{int}(K)$ with derivative

$$DP(y)w = \frac{\varphi(y)w - \varphi(w)y}{\varphi(y)^2} \quad \text{for } w \in V.$$

It follows from the chain rule for semi-derivatives (see Lemma 6.1.6) that h is semi-differentiable at u with semi-derivative

$$h'_u(v) = DP(f(u))(f'_u(v)) = \frac{1}{r}(f'_u(v) - \varphi(f'_u(v))u) \quad \text{for } v \in V.$$

□

By combining Theorem 6.4.1 and Lemma 6.4.2 we now derive a sufficient condition for uniqueness of the eigenvector in the interior of the cone. The result will be formulated, and proved, in the more general context of subhomogeneous maps, as it requires little extra effort and provides interesting additional information.

Theorem 6.4.3 *Let $K \subseteq V$ be a solid closed cone, $\varphi \in \mathrm{int}(K^*)$, and $\Sigma^\circ = \{x \in \mathrm{int}(K)\colon \varphi(x) = 1\}$. Suppose that $f\colon \mathrm{int}(K) \to \mathrm{int}(K)$ is a subhomogeneous order-preserving map, with $f(u) = ru$ for some $u \in \Sigma^\circ$. If f is semi-differentiable at u and*

$$rv - f'_u(v) + \varphi(f'_u(v))u \neq 0 \quad \textit{for all } v \in \{v \in V \setminus \{0\}\colon \varphi(v) = 0\}, \tag{6.24}$$

then u is the only eigenvector of f in Σ°.

Proof Let $V_\varphi = \{v \in V : \varphi(v) = 0\}$ and define $h\colon \mathrm{int}(K) \to \mathrm{int}(K)$ by

$$h(x) = \frac{f(x)}{\varphi(f(x))} \quad \text{for } x \in \mathrm{int}(K).$$

By Lemma 6.4.2 we know that h is semi-differentiable at $u \in \mathrm{int}(K)$ and $h'_u(v) = \frac{1}{r}(f'_u(v) - \varphi(f'_u(v)u)$ for $v \in V$. Restricting h to Σ° gives a map $g\colon \Sigma^\circ \to \Sigma^\circ$ with semi-derivative $g'_u\colon V_\varphi \to V_\varphi$ at u, which is the restriction of h'_u to V_φ.

A similar argument as in Lemma 2.1.6 shows that g is non-expansive under Hilbert's metric. Indeed, if $\alpha = m(x/y)$ and $\beta = M(x/y)$ for $x, y \in \Sigma^\circ$, then $\alpha y \le x$ and $x \le \beta y$, as K is closed. Applying φ to the inequalities shows that

$\alpha \leq 1$ and $1 \leq \beta$. As f is subhomogeneous and order-preserving we deduce that $\alpha f(y) \leq f(\alpha y) \leq f(x)$ and $f(x) \leq f(\beta y) \leq \beta f(y)$. Thus,

$$\alpha \frac{\varphi(f(y))}{\varphi(f(x))} g(y) \leq g(x) \quad \text{and} \quad g(x) \leq \beta \frac{\varphi(f(y))}{\varphi(f(x))} g(y),$$

which shows that $d_H(g(x), g(y)) \leq \log \beta/\alpha = d_H(x, y)$.

Thus, we can apply Theorem 6.4.1 to deduce from (6.24) that u is the only fixed point of g in Σ°. This implies that u is the only eigenvector of f in Σ°. □

In many interesting cases the map f is semi-differentiable at each point in the interior of the cone. In this regard we mention the following results concerning super-additive maps. A map $f\colon D \to K$, with $D \subseteq K$, is *super-additive* with respect to K if

$$f(x) + f(y) \leq_K f(x+y) \quad \text{for all } x, y \in D \text{ with } x + y \in D.$$

Lemma 6.4.4 *Let $K \subseteq V$ be a solid closed cone and $f\colon K \to K$ be a continuous homogeneous map. If f is super-additive and for $\xi \in K$ there exists $\delta > 0$ such that f restricted to $B_\delta(\xi) = \{x \in K : \|x - \xi\| < \delta\}$ is Lipschitz, then the following assertions hold:*

(i) $f'_\xi(v)$ exists for all $v \in S_\xi$.
(ii) $f'_\xi\colon S_\xi \to V$ is Lipschitz and extends as a Lipschitz map to $\mathrm{cl}(S_\xi)$.
(iii) If $\Gamma \subseteq S_\xi$ is compact, then

$$\lim_{t \to 0^+} \frac{f(\xi + tv) - f(\xi)}{t} = f'_\xi(v)$$

uniformly for $v \in \Gamma$.
(iv) If S_ξ is closed, f is semi-differentiable at ξ.

Proof To show that

$$f'_\xi(v) = \lim_{t \to 0^+} \frac{f(\xi + tv) - f(\xi)}{t}$$

exists for all $v \in S_\xi$, let $\tau > 0$ be such that $\xi + tv \in B_\delta(\xi) \cap K$ for $0 \leq t \leq \tau$. Now let $\varphi \in K^* \setminus \{0\}$. As f is homogeneous and super-additive, $(f(x) + f(y))/2 \leq f((x+y)/2)$ for all $x, y \in K$.

Let $\psi\colon [0, \tau] \to \mathbb{R}$ be given by

$$\psi(t) = \varphi(f(\xi + tv)) \quad \text{for } 0 \leq t \leq \tau.$$

Note that $\psi((s+t)/2) \geq (\psi(s)+\psi(t))/2$, as $\varphi \in K^*$, so ψ is concave. It follows that the quotient $(\psi(t)-\psi(0))/t$ is increasing for $t \to 0^+$. Moreover, the quotient is bounded, since ψ is Lipschitz. Thus,

$$\lim_{t\to 0^+} \frac{\varphi(f(\xi+tv))-\varphi(f(\xi))}{t}$$

exists. Since K^* has non-empty interior, each $\vartheta \in V^*$ can be written as $\vartheta = \varphi_1 - \varphi_2$ with $\varphi_1, \varphi_2 \in K^*$. We conclude that

$$\lim_{t\to 0^+} \frac{\vartheta(f(\xi+tv))-\vartheta(f(\xi))}{t} \tag{6.25}$$

exists for all $\vartheta \in V^*$.

Since V is finite-dimensional, there exists a basis $e_1, e_2, \ldots, e_n$ for V and a dual basis $\vartheta_1, \vartheta_2, \ldots, \vartheta_n$ for V^* with $\vartheta_i(e_j) = \delta_{ij}$, the Kronecker delta. It follows that $f(\xi+tv) = \sum_i \vartheta_i(f(\xi+tv))e_i$, so that, by (6.25),

$$f'_\xi(v) = \lim_{t\to 0^+} \frac{f(\xi+tv)-f(\xi)}{t}$$

exists for all $v \in S_\xi$.

Exactly the same argument as in the proof of Lemma 6.1.6 shows that f'_ξ is Lipschitz with constant C if f is Lipschitz in $B_\delta(\xi)\cap K$ with Lipschitz constant C. If $v \in \mathrm{cl}(S_\xi)$ and $(v_k)_k$ is a sequence in S_ξ converging to v, then $(f'_\xi(v_k))_k$ is Cauchy, as f'_ξ is Lipschitz on S_ξ. Thus, $f'_\xi(v_k) \to w$ as $k \to \infty$, and the limit is independent of the choice of $(v_k)_k$. So, we can define $f'_\xi(v) = w$. If $u, v \in \mathrm{cl}(S_\xi)$, and $(u_k)_k$ and $(v_k)_k$ in S_ξ are such that $u_k \to u$ and $v_k \to v$ as $k \to \infty$, then

$$\|f'_\xi(u) - f'_\xi(w)\| = \lim_{k\to\infty} \|f'_\xi(u_k) - f'_\xi(v_k)\| \leq C \lim_{k\to\infty} \|u_k - v_k\| = C\|u-v\|,$$

which shows that f'_ξ extends as a Lipschitz map to $\mathrm{cl}(S_\xi)$.

If $\Gamma \subseteq S_\xi$ is compact and $\varepsilon > 0$, we can find points $v_1, \ldots, v_m \in \Gamma$ such that $\cup_i B_{\varepsilon/3C}(v_i)$ is an open cover of Γ. Here C is the Lipschitz constant of f on $B_\delta(\xi)$. It follows from the first assertion that there exists $0 < \eta < \delta$ such that, for all $1 \leq i \leq m$ and $0 < t < \eta$,

$$\left\| \frac{f(\xi+tv_i)-f(\xi)}{t} - f'_\xi(v_i) \right\| < \varepsilon/3.$$

If $v \in \Gamma$, select j with $v \in B_{\varepsilon/3C}(v_j)$. If $0 < t < \eta$ and $\xi + tv \in K$, it follows that $\xi + tv \in B_\delta(\xi) \cap K$. Using the fact that f is Lipschitz on $B_\delta(\xi) \cap K$ with Lipschitz constant C, and f'_ξ is Lipschitz on S_ξ with the same constant as f, we see that

$$\left\|\frac{f(\xi+tv_i)-f(\xi)}{t}-\frac{f(\xi+tv)-f(\xi)}{t}\right\|<C\|v_i-v\|<\varepsilon/3$$

and $\|f'_\xi(v_i)-f'_\xi(v)\|\leq C\|v_i-v\|<\varepsilon/3$

Combining these inequalities shows that if $0<t<\eta$, $v\in\Gamma$, and $\xi+tv\in K$, then

$$\left\|\frac{f(\xi+tv)-f(\xi)}{t}-f'_\xi(v)\right\|<\varepsilon.$$

Finally, assume that S_ξ is closed. To prove that f is semi-differentiable at ξ we must show that for each $\varepsilon>0$ there exists $\mu>0$ such that if $\|v\|<\mu$ and $\xi+v\in K$, then $\|f(\xi+v)-f(\xi)-f'_\xi(v)\|\leq\varepsilon\|v\|$. Arguing by contradiction, we suppose that there exists $\varepsilon>0$ and a sequence $(u_k)_k$ in S_ξ such that $\|u_k\|\to 0$ as $k\to\infty$, $\xi+u_k\in K$, and $\|f(\xi+u_k)-f(\xi)-f'_\xi(u_k)\|>\varepsilon\|u_k\|$ for all k. Define $w_k=u_k/\|u_k\|$ and $t_k=\|u_k\|>0$, so

$$\left\|\frac{f(\xi+t_kw_k)-f(\xi)}{t_k}-f'_\xi(w_k)\right\|>\varepsilon \tag{6.26}$$

for all k. As S_ξ is closed, we may assume, after taking a subsequence, that $w_k\to w\in S_\xi$ as $k\to\infty$. Note that $\Gamma=\{w_k\colon k\geq 1\}\cup\{w\}$ is a compact subset of S_ξ. It follows from the third assertion that there exists $\eta>0$ such that if $0<t<\eta$, $z\in\Gamma$, and $\xi+tz\in K$, then

$$\left\|\frac{f(\xi+t_kz)-f(\xi)}{t_k}-f'_\xi(z)\right\|<\varepsilon. \tag{6.27}$$

Since $t_k\to 0$ as $k\to\infty$, equations (6.26) and (6.27) are contradictory, and we are done. □

We note that if $K=\mathbb{R}^n_+$, then S_ξ is closed for all $\xi\in K$. Therefore if $f\colon\mathbb{R}^n_+\to\mathbb{R}^n_+$ is a continuous homogeneous order-preserving map which is super-additive and locally Lipschitz, then f is semi-differentiable at every $\xi\in\mathbb{R}^n_+$. Nussbaum [159, section 3] analyzed a class $\mathcal{M}_-$ of maps on $\mathbb{R}^n_+$ satisfying these conditions. In fact, each map in $\mathcal{M}_-$ is C^∞ on the interior of $\mathbb{R}^n_+$, but in general only semi-differentiable on $\partial\mathbb{R}^n_+$. The class $\mathcal{M}_-$ concerns maps that are constructed from means $M_{r\sigma}(x)=(\sum_i\sigma_ix_i^r)^{1/r}$, where $r<0$. It turns out that for maps in the class $\mathcal{M}_-$ it is particularly difficult to decide if there exists an eigenvector in the interior of $\mathbb{R}^n_+$. We shall briefly discuss this issue in Section 6.6, but for a more detailed account the reader is referred to [159].

If in Theorem 6.4.3 the map f is Fréchet differentiable at $u\in\operatorname{int}(K)$, weaker conditions on the Fréchet derivative $Df(u)$ exist, which ensure uniqueness of the eigenvector in the interior. In particular, it suffices to assume that the linear map $Df(u)\colon V\to V$ is irreducible. In fact, the following slightly weaker assumption works.

Definition 6.4.5 Let $L \colon V \to V$ be a linear map leaving a closed solid cone $K \subseteq V$ invariant, and ρ be the spectral radius of L. We say that L satisfies the *Kreĭn–Rutman condition* if either $\rho = 0$, or $\rho > 0$ and (1) $\dim(\ker(\rho I - L)) = 1$ and (2) there exist $w \in K \setminus \{0\}$ with $Lw = \rho w$ and $w^* \in K^* \setminus \{0\}$ with $L^* w^* = \rho w^*$ such that $w^*(w) > 0$.

Obviously, every irreducible linear map $L \colon V \to V$ satisfies the Kreĭn–Rutman condition by the Perron–Frobenius theorem. However, it may easily happen that L satisfies the Kreĭn–Rutman condition even though L is not irreducible. For example, consider $\mathbb{R}^4_+$ and the linear map L defined by the matrix

$$M = \begin{bmatrix} 0 & 0 & 0 & \alpha \\ \beta & 0 & 0 & 0 \\ 0 & \gamma & 0 & 0 \\ \delta & 0 & 0 & 0 \end{bmatrix},$$

where $\alpha, \beta, \gamma, \delta > 0$.

Theorem 6.4.6 *Suppose that $f \colon \mathrm{int}(K) \to \mathrm{int}(K)$ is a subhomogeneous order-preserving map and $u \in \mathrm{int}(K)$ with $f(u) = ru$. If f is Fréchet differentiable at u, then the following holds:*

(i) The spectral radius ρ of $Df(u)$ satisfies $\rho \leq r$.
(ii) If $\rho < r$, then for each $\vartheta \in K^*$ with $\vartheta(u) = 1$, u is the only eigenvector of f in $\Sigma_\vartheta = \{x \in \mathrm{int}(K) \colon \vartheta(x) = 1\}$.
(iii) If $\rho = r$ and $Df(u)$ satisfies the Kreĭn–Rutman condition, where $w \in K \setminus \{0\}$ and $w^* \in K^* \setminus \{0\}$ are such that $Df(u)w = \rho w$ and $Df(u)^* w^* = \rho w^*$, then u is the only eigenvector of f in Σ_ϑ for all $\vartheta \in K^*$ with $\vartheta(u) = 1$ and $\vartheta(w) > 0$.

Proof For simplicity write $L = Df(u)$. To prove the first assertion note that as f is subhomogeneous

$$Lu = \lim_{t \to 1^+} \frac{f(tu) - f(u)}{t - 1} \leq \lim_{t \to 1^+} \frac{tf(u) - f(u)}{t - 1} = f(u) = ru. \qquad (6.28)$$

Since f is order-preserving, $L \colon V \to V$ leaves K invariant. As $u \in \mathrm{int}(K)$, (6.28) and Proposition 5.3.6 imply that $\rho \leq r$.

Before continuing the proof we remark that L has an eigenvector $w \in K \setminus \{0\}$ with $Lw = \rho w$, and likewise L^* has an eigenvector $w^* \in K^* \setminus \{0\}$ with $L^* w^* = \rho w^*$. If $\rho = r$, the Kreĭn–Rutman condition ensures that w and w^* are unique up to positive scalar multiples, and $w^*(w) > 0$. As $u \in \mathrm{int}(K)$, $w^*(u) > 0$. Thus, if we let $\vartheta = tw^*$ we can choose $t > 0$ such that $\vartheta(u) = 1$ and $\vartheta(w) > 0$. It follows that if $\rho = r$, there exists $\vartheta \in K^* \setminus \{0\}$ as in the third assertion. Furthermore, by replacing f by $r^{-1} f$, we may assume that $r = 1$ and $f(u) = u$.

To prove the last two assertions it suffices by Theorem 6.4.3 to show that if $\rho < 1$ or $\rho = 1$, then

$$v - Lv + \vartheta(Lv)u \neq 0$$

for all $v \in V_\vartheta \setminus \{0\}$, where $V_\vartheta = \{v \in V : \vartheta(v) = 1\}$.

For the sake of contradiction suppose that there exists $v \in V_\vartheta \setminus \{0\}$ such that

$$v - Lv = -\vartheta(Lv)u. \tag{6.29}$$

If $\rho < 1$, we find that

$$v = -(I - L)^{-1}(\vartheta(Lv)u)) = -\vartheta(Lv)\sum_{k=0}^{\infty} L^k u. \tag{6.30}$$

Since $u \in \text{int}(K)$ and $v \neq 0$, it follows that $\vartheta(Lv) \neq 0$, and $v \in \text{int}(K)$ or $-v \in \text{int}(K)$. If $v \in \text{int}(K)$, we must have that $\vartheta(Lv) \geq 0$, as $Lv \in K$, so that $\vartheta(Lv) > 0$. But then (6.30) implies $-v \in \text{int}(K)$, which is impossible, as $v \neq 0$. A similar argument gives a contradiction if $-v \in \text{int}(K)$. Thus, if $\rho < 1$, then Equation (6.29) cannot hold for any non-zero $v \in V_\vartheta$. Notice that in this case it is only required that $\vartheta \in K^*$ satisfy $\vartheta(u) = 1$, but not that $\vartheta(w) > 0$.

To treat the case $\rho = 1$ we again argue by contradiction. So, suppose there exists a non-zero $v \in V_\vartheta$ satisfying (6.29). It follows that for each integer $N \geq 0$

$$\sum_{k=0}^{N} L^k(v - Lv) = v - L^{N+1}v = -\vartheta(Lv)\sum_{k=0}^{N} L^k u. \tag{6.31}$$

Note that if $\vartheta(Lv) = 0$, then (6.29) implies that $v - Lv = 0$. As L satisfies the Kreĭn–Rutman condition, we find that $v = tw$ for some $t \neq 0$. Thus, $\vartheta(Lv) = t\vartheta(w) = 0$, which contradicts $\vartheta(w) > 0$. So, $\vartheta(Lv) \neq 0$.

To derive a contradiction we show that $(\|v - L^{N+1}v\|)_N$ remains bounded, while $\lim_{N\to\infty} \|\sum_{k=0}^{N} L^k u\| = \infty$, which violates (6.31). As K is a finite-dimensional closed cone, K is normal by Lemma 1.2.5. Let $\kappa \geq 1$ be the normality constant. Note that if $x \in K$ and $y \in V$ are such that $-bx \leq y \leq bx$, then $0 \leq y + bx \leq 2bx$, and hence

$$\|y\| \leq (2\kappa + 1)b\|x\|. \tag{6.32}$$

As $u \in \text{int}(K)$ there exists $a > 0$ such that $-au \leq v \leq au$. Recall that $r = 1$, so (6.28) implies $-au \leq -aL^{N+1}u \leq L^{N+1}v \leq aL^{N+1}u \leq au$. Thus, for each $N \geq 0$,

$$-2au \leq v - L^{N+1}v \leq 2au. \tag{6.33}$$

We conclude from (6.32) and (6.33) that

$$\|v - L^{N+1}v\| \leq 2a(2\kappa + 1)\|u\|. \tag{6.34}$$

As $u \in \text{int}(K)$, there exists $\alpha > 0$ with $\alpha w \leq u$. Recall that $Lw = w$. Thus,

$$\sum_{k=0}^{N} L^k(u) \geq \alpha \sum_{k=0}^{N} L^k w = \alpha(N+1)w.$$

Using the normality of K, we find that

$$\left\| \vartheta(Lv) \sum_{k=0}^{N} L^k u \right\| \geq \frac{\alpha}{\kappa}(N+1)|\vartheta(Lv)| \left\| w \right\|. \tag{6.35}$$

As $\vartheta(Lv) \neq 0$, the inequalities (6.34) and (6.35) contradict (6.31) for N large. □

If in Theorem 6.4.6 it is assumed that f is homogeneous, the assumptions simplify considerably and the following sharper results hold.

Corollary 6.4.7 *If $f\colon \text{int}(K) \to \text{int}(K)$ is a homogeneous order-preserving map on a solid closed cone $K \subseteq V$, and f is Fréchet differentiable at $u \in \text{int}(K)$ with $f(u) = ru$, then the following hold:*

(i) The spectral radius of $Df(u)$ is equal to r.
(ii) If, in addition, $\dim(\ker(rI - Df(u))) = 1$, then u is the unique, up to positive scalar multiples, eigenvector of f in $\text{int}(K)$.

Proof Write $L = Df(u)$; so,

$$Lu = \lim_{t \to 1^+} \frac{f(tu) - f(u)}{t - 1} = f(u) = ru.$$

By the Perron–Frobenius theorem there exists $u^* \in K^* \setminus \{0\}$ with $L^*u^* = ru^*$. As $u \in \text{int}(K)$, we know that $u^*(u) > 0$. Under the assumption that $\dim(\ker(rI - L)) = 1$, we conclude that L satisfies the Kreĭn–Rutman condition. Also note that $\vartheta(u) > 0$ for all non-zero $\vartheta \in K^*$. Thus, if $\vartheta \in K^* \setminus \{0\}$ and $\vartheta(u) = 1$, then u is the unique eigenvector of f in $\Sigma_\vartheta = \{x \in \text{int}(K)\colon \vartheta(x) = 1\}$ by Theorem 6.4.6. But if y is an eigenvector of f in the interior of K, then $ty \in \Sigma_\vartheta$ for some $t > 0$, and hence $ty = u$, which completes the proof. □

Another interesting refinement of Theorem 6.4.6 is obtained by assuming that the Fréchet derivative $Df(u)$ is irreducible.

Corollary 6.4.8 *Suppose $f\colon \mathrm{int}(K) \to \mathrm{int}(K)$ is a subhomogeneous order-preserving map on a solid closed cone $K \subseteq V$, and f is Fréchet differentiable at $u \in \mathrm{int}(K)$ with $f(u) = ru$. If the Fréchet derivative $Df(u)$ is irreducible, then for each $\vartheta \in K^* \setminus \{0\}$ with $\vartheta(u) = 1$, u is the only eigenvector of f in $\Sigma_\vartheta = \{x \in \mathrm{int}(K)\colon \vartheta(x) = 1\}$. Moreover, if $y \in \mathrm{int}(K)$ is an eigenvector of f, then $y \le u$ or $u \le y$.*

Proof As before, denote $L = Df(u)$. To show that u is the only eigenvector of f in Σ_ϑ for all $\vartheta \in K^* \setminus \{0\}$ with $\vartheta(u) = 1$, it suffices, by Theorem 6.4.6, to prove that if the spectral radius ρ of L is equal to r, then $Lu = ru$. From the inequality (6.28) we know that $Lu \le ru$. Now if $Lu \ne ru$, then $z = u - r^{-1}Lu \in K \setminus \{0\}$, and irreducibility of L implies that there exists an integer $N \ge 0$ such that

$$\sum_{k=0}^{N} (r^{-1}L)^k z = u - (r^{-1}L)^{N+1}u \in \mathrm{int}(K).$$

This implies that there exists $s > 0$ with $(1 - s)u \ge (r^{-1}L)^{N+1}u$, so that $\rho < r$, which is a contradiction.

To prove the second assertion we argue by contradiction. Assume that $y \in \mathrm{int}(K)$ is an eigenvector of f and $z = y - u \notin K$ and $-z = u - y \notin K$. As K is closed, there exists $\delta > 0$ such that $B_\delta(z) \cap K$ and $B_\delta(-z) \cap K$ are both empty. Here $B_\delta(z) = \{v \in V\colon \|z - v\| < \delta\}$. The Hahn–Banach separation theorem [186, theorem 11.4] implies that there exist $\varphi_+ \in K^* \setminus \{0\}$ with $\varphi_+(v) < 0$ for all $v \in B_\delta(z)$, and $\varphi_- \in K^* \setminus \{0\}$ with $\varphi_-(v) < 0$ for all $v \in B_\delta(-z)$. In particular, we have that $\varphi_+(z) < 0$ and $\varphi_-(-z) < 0$. Rescaling φ_+ by a positive scalar we may assume that $\varphi_+(z) = \varphi_-(-z)$. Putting $\psi = \varphi_+ + \varphi_- \in K^* \setminus \{0\}$ we deduce that

$$\psi(u) = \psi_+(u) + \psi_-(u) = -\varphi_+(z) + \varphi_-(-z) + \varphi_+(y) + \varphi_-(y) = \psi(y).$$

As $\psi(u) > 0$, we can rescale ψ so that $\psi(u) = \psi(y) = 1$. It follows that u and y are distinct eigenvectors of f in Σ_ψ, which is impossible. □

6.5 Convergence to a unique eigenvector

If $f\colon \mathrm{int}(K) \to \mathrm{int}(K)$ is a homogeneous order-preserving map on a solid closed cone $K \subseteq V$ and f has a unique eigenvector $u \in \mathrm{int}(K)$ with $f(u) = ru$, it is interesting to understand when the ray through u is globally attracting. In particular, it is natural to ask whether

$$\lim_{k\to\infty} \frac{f^k(x)}{\varphi(f^k(x))} = u \quad \text{for all } x \in \text{int}(K),$$

where $\varphi \in K^*$ with $\varphi(u) = 1$, or whether for each $x \in \text{int}(K)$ there exists $\lambda_x > 0$ such that

$$\lim_{k\to\infty} \frac{f^k(x)}{r^k} = \lambda_x u.$$

Surprisingly, a slight strengthening of the conditions that ensure uniqueness of the eigenvector in the interior of the cone give global convergence results of this type. The underlying reason for this phenomenon is Proposition 3.2.3, which says that a locally attracting fixed point of non-expansive maps on a geodesic metric space is globally attracting. A simple, although often too restrictive, condition on the eigenvector $u \in \text{int}(K)$ which ensures that u is a locally attracting fixed point of the normalized map,

$$g(x) = \frac{f(x)}{\varphi(f(x))} \quad \text{for } x \in \Sigma^\circ,$$

where $\Sigma^\circ = \{x \in \text{int}(K) : \varphi(x) = 1\}$ and $\varphi \in K^*$ with $\varphi(u) = 1$, is the following. If $f : \text{int}(K) \to \text{int}(K)$ and $u \in \text{int}(K)$, we say that f is *locally strongly order-preserving at u* if there exists an open neighborhood $U \subseteq \text{int}(K)$ of u such that for $x, y \in U$ with $x < y$ we have that $f(x) \ll f(y)$.

Theorem 6.5.1 *Let $K \subseteq V$ be a solid closed cone and $\varphi \in K^*$. If $f : \text{int}(K) \to \text{int}(K)$ is a homogeneous order-preserving map and there exists $u \in K$, with $\varphi(u) = 1$, such that $f(u) = ru$ and f is strongly order-preserving at u, then*

$$\lim_{k\to\infty} \frac{f^k(x)}{\varphi(f^k(x))} = u \quad \textit{for all } x \in \text{int}(K).$$

Proof Let $U \subseteq \text{int}(K)$ be an open neighborhood of u such that $x, y \in U$ and $x < y$ implies $f(x) \ll f(y)$. As the norm topology on $\text{int}(K)$ coincides with the topology induced by Thompson's metric, there exists $R > 1$ such that the closed Thompson's metric ball, $B_T(u) = \{x \in \text{int}(K) : R^{-1}u \leq x \leq Ru\}$, with radius $\log R$ and center u is contained in U. Let $h : \text{int}(K) \to \text{int}(K)$ be given by

$$h(x) = \frac{f(x)}{r} \quad \text{for } x \in \text{int}(K).$$

Thus, $h(u) = u$. Suppose that $y \in B_T(u)$ and $y \neq \lambda u$ for all $\lambda > 0$. Write $\alpha = m(y/u)$ and $\beta = M(y/u)$. Note that $\alpha u < y < \beta u$, as K is closed, and $R^{-1} \leq \alpha$ and $\beta \leq R$. As f is strongly order-preserving at u,

$$\alpha u = \alpha h(u) \ll h(y) \ll \beta h(u) = \beta u.$$

This implies that there exist $\mu > \alpha$ and $\tau < \beta$ such that $\mu u < h(y) < \tau u$, and hence

$$d_T(h(y), u) \leq \log(\max\{\mu^{-1}, \tau\}) < \log(\max\{\alpha^{-1}, \beta\}) = d_T(y, u). \quad (6.36)$$

for all $y \in B_T(u)$ with $y \neq \lambda u$ for all $\lambda > 0$.

As h is non-expansive under Thompson's metric and $h(u) = u$, we know that $h(B_T(u)) \subseteq B_T(u)$. Now let $w \in B_T(u)$. We can use the compactness of $B_T(u)$ to find an increasing sequence of integers $(k_i)_i$ and $z \in B_T(u)$ such that $\lim_{i\to\infty} h^{k_i}(w) = z$. As

$$\begin{aligned} d_T(z, u) &= \lim_{i\to\infty} d_T(h^{k_i}(w), h^{k_i}(u)) \\ &\leq \lim_{i\to\infty} d_T(h(h^{k_{i-1}}(w)), h(h^{k_{i-1}}(u))) \\ &\leq d_T(h(z), u), \end{aligned}$$

it follows from (6.36) that $z = \lambda_w u$ for some $\lambda_w > 0$. Since f is homogeneous and $\lim_{i\to\infty} h^{k_i}(w) = \lambda_w u$,

$$\lim_{i\to\infty} \frac{f^{k_i}(w)}{\varphi(f^{k_i}(w))} = \lim_{i\to\infty} \frac{h^{k_i}(w)}{\varphi(h^{k_i}(w))} = u. \quad (6.37)$$

Let $\Sigma^\circ = \{x \in \mathrm{int}(K) \colon \varphi(x) = 1\}$ and define $g \colon \Sigma^\circ \to \Sigma^\circ$ by $g(x) = f(x)/\varphi(f(x))$ for $x \in \Sigma^\circ$. Using the non-expansiveness of g under Hilbert's metric we see that, for each $w \in B_T(u) \cap \Sigma^\circ$,

$$\begin{aligned} \lim_{k\to\infty} d_H(g^k(w), u) &= \lim_{k\to\infty} d_H(g^k(w), g^k(u)) \\ &\leq \lim_{i\to\infty} d_H(g^{k_i}(w), g^{k_i}(u)) \\ &= \lim_{i\to\infty} d_H(g^{k_i}(w), u). \end{aligned}$$

The right-hand side is 0 by (6.37), and hence u is a locally attracting fixed point of g in the geodesic metric space (Σ°, d_H). It follows from Proposition 3.2.3 that $\lim_{k\to\infty} g^k(x) = u$ for all $x \in \Sigma^\circ$. □

The assumption that f is strongly order-preserving at u can be considerably weakened. To this end we introduce the following notion.

Definition 6.5.2 Let $f \colon \mathrm{int}(K) \to \mathrm{int}(K)$ where $K \subseteq V$ is a solid closed cone. We say that f satisfies *condition* L *at* $x \in \mathrm{int}(K)$ if there exists an open neighborhood $U \subseteq \mathrm{int}(K)$ of x such that for each $y \in U$ with $y < x$ there exists an integer $k_y \geq 1$ with $f^{k_y}(y) \ll f^{k_y}(x)$. Likewise we say that f satisfies *condition* U *at* $x \in \mathrm{int}(K)$ if there exists an open neighborhood $U \subseteq \mathrm{int}(K)$ of x such that for each $y \in U$ with $x < y$ there exists an integer $k_y \geq 1$ with $f^{k_y}(x) \ll f^{k_y}(y)$.

Similar conditions have been considered by Brualdi, Parter, and Schneider [39] and by Kloeden and Rubinov [103]. The following lemma shows that the neighborhood U in Definition 6.5.2 is irrelevant.

Lemma 6.5.3 *Let $K \subseteq V$ be a solid closed cone. If $f\colon \mathrm{int}(K) \to \mathrm{int}(K)$ is an order-preserving map and f satisfies condition* L *at $x \in \mathrm{int}(K)$, then for each $y \in \mathrm{int}(K)$ with $y < x$ there exists an integer $k_y \geq 1$ such that $f^{k_y}(y) \ll f^{k_y}(x)$. Likewise, if f satisfies condition* U *at $x \in \mathrm{int}(K)$, then for each $y \in \mathrm{int}(K)$ with $x < y$ there exists an integer $k_y \geq 1$ such that $f^{k_y}(x) \ll f^{k_y}(y)$.*

Proof We shall only prove the first assertion and leave the analogous proof of the second assertion to the reader. If $y \in \mathrm{int}(K)$ with $y < x$, define, for $0 < t < 1$, $y_t = (1-t)x + ty$. Note that $x - y_t = t(x-y) \in K \setminus \{0\}$ and $y_t - y = (1-t)(x-y) \in K \setminus \{0\}$ for all $0 < t < 1$. Let U be an open neighborhood of x as in Definition 6.5.2. Let $0 < t < 1$ be so small that $y_t \in U$. Then there exists an integer $k_y \geq 1$ with $f^{k_y}(y_t) \ll f^{k_y}(x)$. As f is order-preserving, $f^{k_y}(y) \leq f^{k_y}(y_t)$, so that $f^{k_y}(y) \leq f^{k_y}(y_t) \ll f^{k_y}(x)$. □

The assumption that f satisfies condition L or condition U at an eigenvector in the interior of the cone turns out to be sufficient to prove global convergence results. We shall discuss this idea in the more general setting of subhomogeneous maps, rather than homogeneous maps, as it requires very little extra effort and yields interesting additional information. We begin by proving two auxiliary lemmas. The first one concerns the ω-limit sets of subhomogeneous order-preserving maps.

Lemma 6.5.4 *If $h\colon \mathrm{int}(K) \to \mathrm{int}(K)$ is a subhomogeneous order-preserving map on the interior of a solid closed cone $K \subseteq V$, and $h(u) = u$ for some $u \in \mathrm{int}(K)$, then for each $x \in \mathrm{int}(K)$ there exists $0 < \alpha \leq 1$ such that $m(y/u) = \alpha$ for all $y \in \omega(x)$, or there exists $\beta \geq 1$ such that $M(y/u) = \beta$ for all $y \in \omega(x)$.*

Proof For $k \geq 1$ define $\alpha_k = m(h^k(x)/h^k(u)) = m(h^k(x)/u)$ and $\beta_k = M(h^k(x)/h^k(u)) = M(h^k(x)/u)$. Clearly $\alpha_k \leq \beta_k$ for all $k \geq 1$. If there exists $m \geq 1$ such that $\beta_m \leq 1$ or, equivalently, $h^m(x) \leq u$, then $h^{k+m}(x) \leq u$ for all $k \geq 0$. So, in that case, $\alpha_k \leq 1$ for all $k \geq m$. A similar argument shows that if $\alpha_m \geq 1$ for some $m \geq 1$, then $1 \leq \alpha_k \leq \beta_k$ for all $k \geq m$.

Thus, there are two cases to consider: (1) $\beta_k > 1$ for all $k \geq 1$, and (2) there exists $m \geq 1$ such that $\alpha_k \leq 1$ for all $k \geq m$. Let us assume that we are in the first case. As f is subhomogeneous and order-preserving, and $h^k(x) \leq \beta_k u$,

$$h^{k+1}(x) \leq h(\beta_k u) \leq \beta_k h(u) = \beta_k u,$$

which shows that $\beta_{k+1} \leq \beta_k$ for all $k \geq 1$. It follows that $\lim_{k\to\infty} \beta_k = \beta \geq 1$ exists. Thus, if $h^{k_i}(x) \to y$ as $i \to \infty$, we find that

$$M(y/u) = \lim_{i\to\infty} M(h^{k_i}(x)/u) = \lim_{i\to\infty} \beta_k = \beta.$$

In the second case, it follows from $\alpha_k u \leq h^k(x)$ that $\alpha_k u = \alpha_k h(u) \leq h(\alpha_k u) \leq h^{k+1}(x)$ for all $m \geq k$, and hence $\alpha_k \leq \alpha_{k+1}$ for all $k \geq m$. So, in this case, $\lim_{k\to\infty} \alpha_k = \alpha \leq 1$ exists. Moreover, if $h^{k_i}(x) \to y$ as $i \to \infty$, we have that

$$m(y/u) = \lim_{i\to\infty} m(h^{k_i}(x)/u) = \lim_{i\to\infty} \alpha_k = \alpha.$$

□

It is useful to look at the proof of Lemma 6.5.4 again when h is homogeneous. In that case α_k is increasing and β_k is decreasing. As $\alpha_k \leq \beta_k$ for all $k \geq 1$, we see that both limits, $\lim_{k\to\infty} \alpha_k = \alpha$ and $\lim_{k\to\infty} \beta_k = \beta$, exist and $0 < \alpha \leq \beta$. Thus, if h in Lemma 6.5.4 is homogeneous, then for each $x \in \text{int}(K)$ there exist $0 < \alpha \leq \beta$ such that

$$m(y/u) = \alpha \quad \text{and} \quad M(y/u) = \beta \tag{6.38}$$

for all $y \in \omega(x)$.

Of course, if h in Lemma 6.5.4 is homogeneous, then $h(tu) = tu$ for all $t > 0$. The next lemma shows that if h is merely subhomogeneous, then $\{t > 0 \colon h(tu) = tu\}$ is still an interval.

Lemma 6.5.5 *Let $h\colon \text{int}(K) \to \text{int}(K)$ be a subhomogeneous order-preserving map on the interior of a solid closed cone with $h(u) = u$ for some $u \in \text{int}(K)$. If $s_0 < t < t_0$, where $s_0 = \inf\{t > 0 \colon h(tu) = tu\}$ and $t_0 = \sup\{t > 0 \colon h(tu) = tu\}$, then $h(tu) = tu$. Moreover, if $s_0 > 0$, then $h(s_0u) = s_0u$. Likewise, $t_0 < \infty$ implies $h(t_0u) = t_0u$. Furthermore, $h(su) > su$ for all $0 < s < s_0$, and $h(tu) < tu$ for all $t > t_0$.*

Proof First remark that as $h(u) = u$, $s_0 \leq 1 \leq t_0$. If $t \geq 1$, $t^{-1}h(tu) \leq h(u)$, so that $h(tu) \leq tu$. Likewise, if $0 < s \leq 1$, then $su = sh(u) \leq h(su)$. We shall now show by contradiction that if $h(\tau u) < \tau u$ for some $\tau \geq 1$, then $h(tu) < tu$ for all $t \geq \tau$. Suppose $h(tu) \geq tu$ for some $t \geq \tau \geq 1$. By the previous observations, $h(tu) = tu$. But this implies that $h(\tau u) \geq (\tau/t)h(tu) = \tau u$, which is impossible. A similar argument shows that if $h(\sigma u) > \sigma u$ for some $0 < \sigma \leq 1$, then $h(su) > su$ for all $0 < s \leq \sigma$.

To complete the proof we remark that the continuity of h implies that $h(s_0u) = s_0u$ if $s_0 > 0$, and $h(t_0u) = t_0u$ if $t_0 < \infty$. □

The next theorem shows that the assumption that f satisfies condition L or condition U at an eigenvector in the interior of the cone ensures global convergence to a unique eigenvector in the interior.

Theorem 6.5.6 *Let $f\colon \mathrm{int}(K) \to \mathrm{int}(K)$ be an order-preserving map on the interior of a solid closed cone $K \subseteq V$ and $f(u) = ru$ for some $u \in \mathrm{int}(K)$. Suppose that $h\colon \mathrm{int}(K) \to \mathrm{int}(K)$ is given by $h(x) = r^{-1}f(x)$ for all $x \in \mathrm{int}(K)$. Then the following assertions hold:*

(i) If f is subhomogeneous and h satisfies conditions L and U at u, then for each $x \in \mathrm{int}(K)$ there exists $\lambda_x > 0$ such that

$$\lim_{k\to\infty} h^k(x) = \lambda_x u.$$

Moreover, $\{\lambda > 0\colon f(\lambda u) = \lambda r u\}$ is an interval.

(ii) If f is homogeneous and f satisfies condition L or U at u, then

$$\lim_{k\to\infty} \frac{f^k(x)}{\varphi(f^k(x))} = u$$

for all $x \in \mathrm{int}(K)$ and $\varphi \in K^*$ with $\varphi(u) = 1$.

Proof To prove (i) we suppose that f is subhomogeneous and consider $h\colon \mathrm{int}(K) \to \mathrm{int}(K)$ to be defined by $h(x) = r^{-1}f(x)$ for $x \in \mathrm{int}(K)$. Clearly $h(u) = u$ and h is subhomogeneous and order-preserving, and hence non-expansive under Thompson's metric. It follows from Lemma 3.1.2 that $h(\omega(x; h)) = \omega(x; h)$. Furthermore we can apply Lemma 6.5.4 to h and conclude that there are two cases: either there exists $0 < \alpha \leq 1$ such that $m(y/u) = \alpha$ for all $y \in \omega(x; h)$, or there exists $\beta \geq 1$ such that $M(y/u) = \beta$ for all $y \in \omega(x; h)$.

We will now show that, in the first case, $\omega(x; h) = \{\alpha u\}$. For the sake of contradiction suppose that there exists $z \in \omega(x; h)$ with $\alpha u \neq z$. Then $\alpha u < z$ and, because h satisfies condition U at u, there exists $m \geq 1$ such that

$$u = h^m(u) \ll h^m(\alpha^{-1}z).$$

As h^m is subhomogeneous, $\alpha h^m(\alpha^{-1}z) \leq h^m(z)$, so that $\alpha u \ll h^m(z)$. It follows that $m(h^m(z)/u) > \alpha$, which is impossible, as $h^m(z) \in \omega(x; h)$.

In a similar way it can be shown that, in the second case, $\omega(x; h) = \{\beta u\}$. Indeed, if there exists $z \in \omega(x; h)$ with $z \neq \beta u$, then $z < \beta u$ and there exists an integer $m \geq 1$ such that

$$h^m(\beta^{-1}z) \ll h^m(u) = u,$$

as h satisfies condition L at u. This implies that $\beta^{-1}h^m(z) \ll u$, since h is subhomogeneous, which contradicts $M(h^m(z)/u) = \beta$.

Thus, we have shown that there exists $\lambda_x > 0$ such that $\omega(x; h) = \{\lambda_x u\}$, and hence

$$\lim_{k\to\infty} h^k(x) = \lambda_x u$$

by Lemma 3.1.3.

If f is homogeneous, then h is homogeneous and hence there exist $0 < \alpha \leq \beta$ such that $m(y/u) = \alpha$ and $M(y/u) = \beta$ for all $y \in \omega(x; h)$ by (6.38). As before we will show that $\omega(x; h) = \{\lambda_x u\}$ for some $\lambda_x > 0$. Suppose that f satisfies condition U at u and there exists $z \in \omega(x; h)$ with $z \neq \alpha u$. Note that as f is homogeneous, h also satisfies condition U at u. Moreover, $\alpha u < z$ and there exists $m \geq 1$ such that $\alpha u = h^m(\alpha u) \ll h^m(z)$. As $h^m(z) \in \omega(x; h)$, this contradicts $m(h^m(z)/u) = \alpha$. In the same way, it can be shown that $\omega(x; h) = \{\beta u\}$ if f satisfies condition L at u.

Thus, $\omega(x; h) = \{\lambda_x u\}$ for some $\lambda_x > 0$, so that $\lim_{k\to\infty} h^k(x) = \lambda_x u$ by Lemma 3.1.3. Now using the homogeneity of f we conclude that

$$\lim_{k\to\infty} \frac{f^k(x)}{\varphi(f^k(x))} = \lim_{k\to\infty} \frac{h^k(x)}{\varphi(h^k(x))} = u$$

for all $x \in \text{int}(K)$ and $\varphi \in K^*$ with $\varphi(u) = 1$. □

Note that if f in Theorem 6.5.6 is homogeneous, then we do not need to know the eigenvalue r explicitly to obtain a convergent scheme. We can also circumvent explicit knowledge of the eigenvector u in the interior of K by assuming that, at each $x \in \text{int}(K)$, f satisfies condition L or U.

Furthermore if f is semi-differentiable at the eigenvector $u \in \text{int}(K)$, then there are simple conditions on the semi-derivative f'_u that ensure that f satisfies condition L or U at u.

Lemma 6.5.7 *Let $K \subseteq V$ be a solid closed cone and $f\colon \text{int}(K) \to \text{int}(K)$ be a subhomogeneous order-preserving map. If $f(u) = ru$ for some $u \in \text{int}(K)$ and f is semi-differentiable at u with semi-derivative $f'_u\colon V \to V$ at u, then the following assertions hold:*

(i) If there exists an integer $p \geq 1$ such that $(f'_u)^p(K \setminus \{0\}) \subseteq \text{int}(K)$, then f satisfies condition U at u.
(ii) If there exists an integer $p \geq 1$ such that $(f'_u)^p(-K \setminus \{0\}) \subseteq -\text{int}(K)$, then f satisfies condition L at u.

Proof We shall only prove the first assertion and leave the second one as an exercise for the reader. Assume there exists an integer $p \geq 1$ such that $(f'_u)^p(K \setminus \{0\}) \subseteq \text{int}(K)$ and define $h\colon \text{int}(K) \to \text{int}(K)$ by $h(x) = r^{-1}f(x)$ for all $x \in \text{int}(K)$. Clearly $h(u) = u$ and h is semi-differentiable at u with semi-derivative $h'_u = r^{-1} f'_u$. So, $(h'_u)^p(K \setminus \{0\}) \subseteq \text{int}(K)$, and f satisfies condition U at u if and only if h does.

By the chain rule for semi-derivatives (see Lemma 6.1.6), we know that

$$h^p(u+v) = h^p(u) + (h'_u)^p(v) + \|v\|\varepsilon_1(v),$$

where $\lim_{\|v\|\to 0} \varepsilon_1(v) = 0$. To show that h satisfies condition U at u, it suffices to show that there exists $\delta > 0$ such that for $v \in K$ with $0 < \|v\| \leq \delta$ we have that $u = h^p(u) \ll h^p(u+v)$. As V is finite-dimensional and K is closed, $S = \{v \in K \colon \|v\| = 1\}$ is compact. This implies that $(h'_u)^p(S)$ is a compact subset of $\text{int}(K)$, since h'_u is continuous. Thus, there exists $\tau > 0$ such that the τ-neighborhood of $(h'_u)^p(S)$, denoted $N_\tau((h'_u)^p(S))$, lies inside $\text{int}(K)$. For $t > 0$ it follows that

$$N_{t\tau}((h'_u)^p(tS)) = tN_\tau((h'_u)^p(S)) \subseteq \text{int}(K).$$

There exists $\delta > 0$ such that $\|\varepsilon_1(v)\| < \tau$ for all $\|v\| \leq \delta$. This implies, for $0 < \|v\| \leq \delta$ with $v \in K$, that $(h'_u)^p(v) + \|v\|\varepsilon_1(v) \in \text{int}(K)$, and hence h satisfies condition U at u. □

A combination of Theorem 6.5.6 and Lemma 6.5.7 immediately gives the following result.

Corollary 6.5.8 *Let $f\colon \text{int}(K) \to \text{int}(K)$ be an order-preserving map on the interior of a solid closed cone $K \subseteq V$, $f(u) = ru$ for some $u \in \text{int}(K)$, and f be semi-differentiable at u with semi-derivative f'_u. Suppose that $h\colon \text{int}(K) \to \text{int}(K)$ is given by $h(x) = r^{-1}f(x)$ for all $x \in \text{int}(K)$. Then the following assertions hold:*

(i) If f is subhomogeneous and there exists an integer $p \geq 1$ such that $(f'_u)^p(K \setminus \{0\}) \subseteq \text{int}(K)$ and $(f'_u)^p(-K \setminus \{0\}) \subseteq -\text{int}(K)$, then for each $x \in \text{int}(K)$ there exists $\lambda_x > 0$ such that

$$\lim_{k\to\infty} h^k(x) = \lambda_x u.$$

(ii) If f is homogeneous and there exists an integer $p \geq 1$ such that $(f'_u)^p(K \setminus \{0\}) \subseteq \text{int}(K)$ or $(f'_u)^p(-K \setminus \{0\}) \subseteq -\text{int}(K)$, then

$$\lim_{k\to\infty} \frac{f^k(x)}{\varphi(f^k(x))} = u$$

for all $x \in \text{int}(K)$ and $\varphi \in K^*$ with $\varphi(u) = 1$.

Note that if in Corollary 6.5.8 f is Fréchet differentiable at u and the linear map $Df(u)$ is *primitive*, that is to say $Df(u)^m(K \setminus \{0\}) \subseteq \text{int}(K)$ for some $m \geq 1$, then the hypotheses are satisfied.

6.6 Means and their eigenvectors

In Section 1.4 we defined the class M of continuous order-preserving homogeneous maps $f\colon \mathbb{R}^n_+ \to \mathbb{R}^n_+$, each of whose coordinate function is of the form

$$f_i(x) = \sum_{(r,\sigma)\in\Gamma_i} c_{ir\sigma} M_{r\sigma}(x) \quad \text{for } x \in \mathbb{R}^n_+, \tag{6.39}$$

where Γ_i is a finite set of pairs (r,σ) with $r \in \mathbb{R}$ and $\sigma \in \mathbb{R}^n_+$ a probability vector, $c_{ir\sigma} > 0$, and $M_{r\sigma}(x)$ is the (r,σ)-mean of x. Let M_+ denote the subclass consisting of those maps $f \in M$ having each coordinate function f_i of the form (6.39) with each $r \geq 0$. Similarly, we say that $f \in M_-$ if each coordinate function f_i is of the form (6.39), but with each $r < 0$.

We denote by $\mathcal{M}$ the smallest class of maps $f\colon \mathbb{R}^n_+ \to \mathbb{R}^n_+$ containing M that is closed under addition and composition. Using, respectively, M_+ and M_-, we similarly define $\mathcal{M}_+$ and $\mathcal{M}_-$ to be the smallest collections of maps $f\colon \mathbb{R}^n_+ \to \mathbb{R}^n_+$ containing, respectively, M_+ and M_- that are closed under addition and composition.

In this section we shall see that the question of whether a map $f \in \mathcal{M}_+$ has an eigenvector in $\text{int}(\mathbb{R}^n_+)$ is quite easy to answer. However, our main objective is to show that the same question for maps in $\mathcal{M}_-$ is often subtle and difficult, and represents a central analytic difficulty in applying the results from this chapter.

Following [156] we say that a nonnegative $n \times n$ matrix $A = (a_{ij})$ is an *incidence matrix* of an order-preserving homogeneous map $f\colon \mathbb{R}^n_+ \to \mathbb{R}^n_+$ if for each pair (i,j) with $a_{ij} > 0$ there exist $c > 0$ and a probability vector $\tau \in \mathbb{R}^n_+$ such that $\tau_j > 0$ and

$$f_i(x) \geq c x^\tau = c\Big(\prod_{k\in\text{supp}(\tau)} x_k^{\tau_k}\Big) \quad \text{for all } x \in \mathbb{R}^n_+.$$

Note that if $A = (a_{ij})$ is an incidence matrix for f, and $B = (b_{ij})$ is a nonnegative matrix with the *same zero pattern*, i.e.,

$$a_{ij} > 0 \quad \text{if and only if} \quad b_{ij} > 0,$$

then B is also an incidence matrix for f. For simplicity we shall write $A \sim B$ if the nonnegative $n \times n$ matrices A and B have the same zero pattern. Also remark that if f has an incidence matrix A and each row of A has a non-zero entry, then $f(\text{int}(\mathbb{R}^n_+)) \subseteq \text{int}(\mathbb{R}^n_+)$.

Clearly if A is an incidence matrix for an order-preserving homogeneous map $f\colon \mathbb{R}^n_+ \to \mathbb{R}^n_+$ and $a_{ij} > 0$, then

$$\lim_{u\to\infty} f_i(u^{\{j\}}) = \infty.$$

Here $u^{\{j\}} \in \mathbb{R}^n_+$ with $u^{\{j\}}_k = 1$ for $k \neq j$ and $u^{\{j\}}_j = u$. Moreover, if A is irreducible, then the directed graph G_f associated with f (see Section 6.2) is strongly connected. This observation yields the following immediate consequence of Theorem 6.2.3.

Proposition 6.6.1 *If $f\colon \mathbb{R}^n_+ \to \mathbb{R}^n_+$ is a continuous order-preserving homogeneous map and f has an irreducible incidence matrix, then f has an eigenvector in* $\mathrm{int}(\mathbb{R}^n_+)$.

The following lemma will be useful in the sequel.

Lemma 6.6.2 *Suppose that $f\colon \mathbb{R}^n_+ \to \mathbb{R}^n_+$ and $g\colon \mathbb{R}^n_+ \to \mathbb{R}^n_+$ are continuous order-preserving homogeneous maps, where f has incidence matrix A and g has incidence matrix B. If each row of A and each row of B has a non-zero entry, then $C = BA$ is an incidence matrix for $g \circ f$ and each row of C contains a non-zero entry. Furthermore, $D = A + B$ is an incidence matrix for $f + g$.*

Proof It is a straightforward exercise to show that $A+B$ is an incidence matrix for $f+g$.

For $1 \leq i \leq n$ select j such that $b_{ij} > 0$. Subsequently select k such that $a_{jk} > 0$. It follows that $c_{ik} \geq b_{ij}a_{jk} > 0$, which shows that each row of C has a non-zero entry. In general, if $c_{ik} > 0$, there exists j such that $b_{ij} > 0$ and $a_{jk} > 0$. Since $b_{ij} > 0$ there exist $d > 0$ and a probability vector $\tau \in \mathbb{R}^n_+$ with $\tau_j > 0$ such that

$$g_i(y) \geq d y^{\tau} \quad \text{for all } y \in \mathbb{R}^n_+.$$

Because each row of A has a positive entry, we know that for each $1 \leq m \leq n$, there exist $c_m > 0$ and a probability vector $\sigma^m \in \mathbb{R}^n_+$ with

$$f_m(x) \geq c_m x^{\sigma^m} \quad \text{for all } x \in \mathbb{R}^n_+.$$

As $a_{jk} > 0$, the k-th coordinate of σ^j can be chosen to be non-zero. It follows that

$$g_i(f(x)) \geq \kappa x^{\gamma} \quad \text{for all } x \in \mathbb{R}^n_+,$$

where $\kappa > 0$ and $\gamma = \sum_{m=1}^n \tau_m \sigma^m$ is a probability vector in $\mathbb{R}^n_+$. Note that the k-th coordinate of γ satisfies

$$\gamma_k = \sum_{m=1}^{n} \tau_m \sigma^m_k \geq \tau_j \sigma^j_k > 0,$$

which shows that $C = BA$ is an incidence matrix for $g \circ f$. □

As usual, if $f\colon \mathbb{R}^n_+ \to \mathbb{R}^n_+$ is C^1 on $\mathrm{int}(\mathbb{R}^n_+)$, we write

$$f'(x) = \left(\frac{\partial f_i}{\partial x_j}(x)\right)$$

to denote the Jacobian matrix of f at x. Since f is order-preserving, it is easy to verify that $\frac{\partial f_i}{\partial x_j}(x) \geq 0$ for all $x \in \mathrm{int}(\mathbb{R}^n_+)$ and $1 \leq i, j \leq n$.

Lemma 6.6.3 *If $f\colon \mathbb{R}^n_+ \to \mathbb{R}^n_+$ is an order-preserving homogeneous map in $\mathcal{M}$, then the following assertions hold:*

(i) f is C^1 on $\mathrm{int}(\mathbb{R}^n_+)$.
(ii) For each $x, y \in \mathrm{int}(\mathbb{R}^n_+)$ we have that $f'(x) \sim f'(y)$.
(iii) For each $x \in \mathrm{int}(\mathbb{R}^n_+)$ each row of $f'(x)$ contains a non-zero entry.
(iv) If $f \in \mathcal{M}_+$ and $x \in \mathrm{int}(\mathbb{R}^n_+)$, then $f'(x)$ is an incidence matrix for f.

Proof Suppose that $f \in M$ and and $f_i\colon \mathbb{R}^n_+ \to [0, \infty)$ is of the form (6.39). Clearly, f is C^1 on $\mathrm{int}(\mathbb{R}^n_+)$ and

$$\frac{\partial f_i}{\partial x_j}(x) = \sum_{(r,\sigma)\in\Gamma_i} c_{ir\sigma} \frac{\partial M_{r\sigma}}{\partial x_j}(x).$$

If $(r, \sigma) \in \Gamma_i$ and $j \notin \mathrm{supp}(\sigma)$, then

$$\frac{\partial M_{r\sigma}}{\partial x_j}(x) = 0.$$

Furthermore, if $\mathrm{supp}(\sigma) = \{j\}$, then

$$\frac{\partial M_{r\sigma}}{\partial x_j}(x) = 1.$$

On the other hand, if $|\mathrm{supp}(\sigma)| > 1$ and $j \in \mathrm{supp}(\sigma)$, then

$$\frac{\partial M_{r\sigma}}{\partial x_j}(x) = \left[\begin{array}{ll} \sigma_j x_j^{r-1}\left(\sum_{k=1}^n \sigma_k x_k^r\right)^{1/r-1} & \text{if } r \neq 0 \\ \sigma_j x_j^{\sigma_j-1}\left(\prod_{k\neq j} x_k^{\sigma_k}\right) & \text{if } r = 0. \end{array}\right.$$

It follows that $\frac{\partial f_i}{\partial x_j}(x) > 0$ if and only if there exists $(r, \sigma) \in \Gamma_i$ such that $j \in \mathrm{supp}(\sigma)$. This implies that $f'(x) \sim f'(y)$ for all $x, y \in \mathrm{int}(\mathbb{R}^n_+)$.

To show the third assertion for $f \in M$, remark that for each i there exists j such that $\frac{\partial f_i}{\partial x_j}(x) > 0$, as Γ_i is non-empty.

Now let $\mathcal{N}$ be the collection of continuous order-preserving homogeneous maps $f\colon \mathbb{R}^n_+ \to \mathbb{R}^n_+$ in $\mathcal{M}$ that satisfy the first three assertions of the lemma. To prove that the first three assertions hold for all $f \in \mathcal{M}$ it suffices to show that $\mathcal{N}$ is closed under addition and composition, as $M \subseteq \mathcal{N}$. So, let $f, g \in \mathcal{N}$ and write $\vartheta = f \circ g$ and $\psi = f + g$. Obviously ϑ and ψ are C^1 on $\mathrm{int}(\mathbb{R}^n_+)$.

Remark that $\psi'(x) = f'(x) + g'(x) \sim f'(y) + g'(y) = \psi'(y)$ for all $x, y \in \mathrm{int}(\mathbb{R}^n_+)$. It is also clear that each row of $\psi'(x)$ contains a non-zero entry.

Let $x, y \in \mathrm{int}(\mathbb{R}^n_+)$ and write $A_1 = f'(g(x))$, $B_1 = g'(x)$, $A_2 = f'(g(y))$, and $B_2 = g'(y)$. As each row of B_1 has a non-zero entry, we know that g maps $\mathrm{int}(\mathbb{R}^n_+)$ into itself, and hence

$$A_1 \sim A_2 \quad \text{and} \quad B_1 \sim B_2.$$

It is a straightforward matrix exercise to show that this implies that $A_1 B_1 \sim A_2 B_2$, and each row of $A_1 B_1$ contains a non-zero entry.

To prove the final assertion we first assume that $f \in M_+$. Suppose that $\frac{\partial f_i}{\partial x_j}(x) > 0$ for $x \in \mathrm{int}(\mathbb{R}^n_+)$. The previous remarks show that there exists $(s, \tau) \in \Gamma_i$ such that $j \in \mathrm{supp}(\tau)$. Since $s \geq 0$, Equation (6.39) gives

$$f_i(x) \geq c_{is\tau} M_{s\tau}(x) \geq c_{is\tau} M_{0\tau}(x) = c_{is\tau} x^\tau,$$

which shows that $f'(x)$ is an incidence matrix for f. Using Lemma 6.6.2 we deduce that $f'(x)$ is an incidence matrix for $f \in \mathcal{M}_+$ and $x \in \mathrm{int}(\mathbb{R}^n_+)$. □

A combination of Proposition 6.6.1 and Lemma 6.6.3 yields the following result, which shows that the question of whether a given map $f \in \mathcal{M}_+$ has an eigenvector is usually quite easy to answer.

Corollary 6.6.4 *If $f : \mathbb{R}^n_+ \to \mathbb{R}^n_+$ is an order-preserving homogeneous map in $\mathcal{M}_+$ and there exists $x \in \mathrm{int}(\mathbb{R}^n_+)$ such that $f'(x)$ is irreducible, then f has a unique normalized eigenvector $u \in \mathrm{int}(\mathbb{R}^n_+)$.*

Proof It follows from Lemma 6.6.3 that $f'(x)$ is an incidence matrix for f. As $f'(x)$ is irreducible we deduce from Proposition 6.6.1 that f has an eigenvector u in $\mathrm{int}(\mathbb{R}^n_+)$. The uniqueness, up to positive scalar multiples, of the eigenvector u follows from Corollary 6.4.8, since $f'(x) \sim f'(u)$ by Lemma 6.6.3. □

Furthermore, remark that if $f \in \mathcal{M}_+$ and there exists $x \in \mathrm{int}(\mathbb{R}^n_+)$ such that $f'(x)$ is primitive, then it follows from Corollary 6.5.8 that not only is the normalized eigenvector $u \in \mathrm{int}(\mathbb{R}^n_+)$ unique, but also

$$\lim_{k\to\infty} \frac{f^k(y)}{\langle \varphi, f^k(y) \rangle} = u \quad \text{for all } y \in \mathrm{int}(\mathbb{R}^n_+)$$

and $\varphi \in \mathbb{R}^n_+$ with $\langle \varphi, u \rangle = 1$.

The following observation can sometimes be used in conjunction with Corollary 6.6.4 to prove that a general continuous order-preserving homogeneous map $g : \mathbb{R}^n_+ \to \mathbb{R}^n_+$ has an eigenvector in $\mathrm{int}(\mathbb{R}^n_+)$.

Lemma 6.6.5 *Suppose that $f: \mathbb{R}^n_+ \to \mathbb{R}^n_+$ is a map in $\mathcal{M}_+$ and $g: \mathbb{R}^n_+ \to \mathbb{R}^n_+$ is a continuous order-preserving homogeneous map such that $g(y) \geq f(y)$ for all $y \in \mathbb{R}^n_+$. If there exists $x \in \mathrm{int}(\mathbb{R}^n_+)$ such that $f'(x)$ is irreducible, then g has an eigenvector in $\mathrm{int}(\mathbb{R}^n_+)$.*

Proof It follows from Lemma 6.6.3 that $f'(x)$ is an incidence matrix for f. As $g(y) \geq f(y)$ for all $y \in \mathbb{R}^n_+$ it follows that $f'(x)$ is also an incidence matrix for g. So, we can apply Proposition 6.6.1 to conclude that g has an eigenvector in $\mathrm{int}(\mathbb{R}^n_+)$. □

For maps $f \in \mathcal{M}_+$ Corollary 6.6.4 provides an effective way to analyze the problem of the existence of an eigenvector in the interior of $\mathbb{R}^n_+$. As mentioned before the same problem is often much harder for maps $f \in \mathcal{M}_-$. This is basically because Theorem 6.2.3 is rarely applicable to maps $f \in \mathcal{M}_-$, as the associated digraph $G_f = (V, A)$ is usually *totally disconnected*, meaning that the set of arcs $A \subseteq \{(i,i): i \in V\}$.

Lemma 6.6.6 *If $f: \mathbb{R}^n_+ \to \mathbb{R}^n_+$ is in M_- and for each $1 \leq i \leq n$ and $(r, \sigma) \in \Gamma_i$ either $\sigma = \{i\}$ or $|\mathrm{supp}(\sigma)| \geq 2$, then G_f is totally disconnected.*

Proof Suppose that $i \neq j$ and for $u > 0$ let $u^{\{j\}} \in \mathbb{R}^n_+$ be given by $u^{\{j\}}_k = 1$ if $k \neq j$, and $u^{\{j\}}_j = u$. If $(r, \sigma) \in \Gamma_i$ and $\sigma = \{i\}$, then clearly $M_{r\sigma}(u^{\{j\}}) = 1$. Also, if $|\mathrm{supp}(\sigma)| \geq 2$ and $j \notin \mathrm{supp}(\sigma)$, then $M_{r\sigma}(u^{\{j\}}) = 1$. Now if $|\mathrm{supp}(\sigma)| \geq 2$ and $j \in \mathrm{supp}(\sigma)$, then

$$\lim_{u \to \infty} M_{r\sigma}(u^{\{j\}}) = \Big(\sum_{k \in \mathrm{supp}(\sigma), k \neq j} \sigma_k \Big)^{1/r} < \infty,$$

as $r < 0$.

It follows that

$$\lim_{u \to \infty} f_i(u^{\{j\}}) < \infty,$$

if $i \neq j$, and hence G_f is totally disconnected. □

The property of having a totally disconnected associated digraph is preserved under addition and composition, as the following proposition shows.

Proposition 6.6.7 *If $f: \mathbb{R}^n_+ \to \mathbb{R}^n_+$ and $g: \mathbb{R}^n_+ \to \mathbb{R}^n_+$ are continuous order-preserving homogeneous maps such that the associated digraphs G_f and G_g are totally disconnected, then G_{f+g} and $G_{g \circ f}$ are also totally disconnected.*

Proof We leave it to the reader as an exercise to show that G_{f+g} is totally disconnected. To prove that $G_{g \circ f}$ is totally disconnected we first remark that, as G_g is totally disconnected and g is order-preserving,

$$\lim_{u\to\infty} g_i(u^{\{j\}}) < \infty \quad \text{for } i \neq j. \tag{6.40}$$

Here g_i is the i-th component of g, and $u_k^{\{j\}} = 1$ if $k \neq j$ and $u_j^{\{j\}} = u$.

Suppose that $(y^m)_m$ is a sequence in $\mathbb{R}^n_+$ with $y^m \leq y^{m+1}$ for all m. Furthermore assume that $\lambda_k = \sup\{y_k^m : m \geq 1\} < \infty$ for all $k \neq j$, and put $\lambda = \max_{k\neq j} \lambda_k + 1 < \infty$. Let u_m be a sequence of positive reals with $u_m \to \infty$ such that

$$u_m > y_j^m/\lambda \quad \text{for all } m \geq 1.$$

As g is order-preserving, it follows that

$$g_i(y^m/\lambda) \leq g_i(u_m^{\{j\}}) \quad \text{for all } m \text{ sufficiently large.}$$

Now using the homogeneity of g, we deduce from (6.40) that

$$\lim_{m\to\infty} g_i(y^m) < \infty \quad \text{for all } i \neq j. \tag{6.41}$$

Now fix i and j with $i \neq j$. Because G_f is totally disconnected, the set of vectors $\{f(u^{\{j\}}) : u > 1\}$ satisfies $f(u_1^{\{j\}}) \leq f(u_2^{\{j\}})$ for $1 < u_1 \leq u_2$ and

$$\sup\{f_k(u^{\{j\}}) : u > 1\} < \infty \quad \text{for all } k \neq j.$$

It follows from (6.41) that

$$\lim_{u\to\infty} g_i(f(u^{\{j\}})) < \infty,$$

which shows that $G_{g\circ f}$ is totally disconnected. □

Lemma 6.6.6 and Proposition 6.6.7 show for $f \in \mathcal{M}_-$ that G_f is usually totally disconnected and rarely of use in proving that f has an eigenvector in $\mathrm{int}(\mathbb{R}^n_+)$.

In the next chapter we shall discuss applications of nonlinear Perron–Frobenius theory to matrix scaling problems. A crucial question in the analysis of matrix scaling problems is to decide whether an order-preserving homogeneous map has an eigenvector in $\mathrm{int}(\mathbb{R}^n_+)$. Among other questions, it will be of interest to determine when, for a given nonnegative $n \times n$ matrix A with no row or column identically zero, the order-preserving homogeneous map $T : \mathrm{int}(\mathbb{R}^n_+) \to \mathrm{int}(\mathbb{R}^n_+)$ given by

$$T(x) = (A^T \circ R \circ A \circ R)(x) \quad \text{for } x \in \mathrm{int}(\mathbb{R}^n_+),$$

where A^T denotes the transpose of A and $R : (x_1, \ldots, x_n) \mapsto (x_1^{-1}, \ldots, x_n^{-1})$, has an eigenvector in $\mathrm{int}(\mathbb{R}^n_+)$. One can verify that the continuous extension of T to $\mathbb{R}^n_+$ belongs to $\mathcal{M}_-$. Moreover, the associated digraph G_T is totally disconnected. Indeed, a simple calculation gives

$$T_i(x) = \sum_{k=1}^{n} a_{ki}\Big(\sum_{j=1}^{n} a_{kj}x_j^{-1}\Big)^{-1} = \sum_{k:\, a_{ki}\neq 0} a_{ki}\Big(\sum_{j=1}^{n} a_{kj}x_j^{-1}\Big)^{-1}.$$

Taking $m \neq i$, we see that

$$T_i(u^{\{m\}}) = \sum_{k:\, a_{ki}\neq 0} a_{ki}\Big(\sum_{j\neq m} a_{kj} + a_{km}u^{-1}\Big)^{-1} \leq \sum_{k:\, a_{ki}\neq 0} a_{ki}(a_{ki})^{-1} \leq n,$$

and hence G_T is totally disconnected.

Even for relatively innocent looking maps $f \in \mathcal{M}_-$ it can be difficult to determine whether f has an eigenvector in the interior of $\mathbb{R}^n_+$. The following simple example has been discussed in detail in [159, chapter 3]. For $s > 0$ and $t > 0$, define

$$\vartheta(s,t) = \frac{1}{2}\Big(\frac{1}{2}s^{-1} + \frac{1}{2}t^{-1}\Big)^{-1}.$$

Let $a_j, b_j, c_j, d_j \geq 0$ for $1 \leq j \leq 4$, and define $f\colon \mathbb{R}^4_+ \to \mathbb{R}^4_+$ by

$$f\begin{pmatrix} x_1 \\ x_2 \\ x_3 \\ x_4 \end{pmatrix} = \begin{pmatrix} a_1x_1 + b_1\vartheta(x_1,x_2) + c_1\vartheta(x_1,x_4) + d_1\vartheta(x_2,x_3) \\ a_2x_2 + b_2\vartheta(x_1,x_2) + c_2\vartheta(x_1,x_4) + d_2\vartheta(x_2,x_3) \\ a_3x_3 + b_3\vartheta(x_3,x_4) + c_3\vartheta(x_1,x_4) + d_3\vartheta(x_2,x_3) \\ a_4x_3 + b_4\vartheta(x_3,x_4) + c_4\vartheta(x_1,x_4) + d_4\vartheta(x_2,x_3) \end{pmatrix}$$

for $x \in \mathbb{R}^4_+$. In [159] complicated conditions on the coefficients in f were given that exactly determine when f has an eigenvector in the interior of $\mathbb{R}^4_+$. The argument, which involves a study of the behavior of f near eigenvectors in $\partial\mathbb{R}^4_+$ and a tedious case-by-case analysis, will be omitted here. The reader should verify that neither the method of recession maps (Section 6.3) nor the consideration of the digraph G_f of f (Section 6.2) provide any information here.

7

Applications to matrix scaling problems

The classic matrix scaling problem as introduced by Sinkhorn [205] asks for which square nonnegative matrices A there exist positive diagonal matrices D and E such that DAE is doubly stochastic. A complete solution to this problem was obtained independently by Sinkhorn and Knopp [206] and by Brualdi, Parter, and Schneider [39]. There exist many natural generalizations of the classic matrix scaling problem, some of which we will discuss in this chapter.

To analyze the matrix scaling problems we will follow a fixed-point approach pioneered by Menon [143]. Given an $m \times n$ nonnegative matrix A and positive vectors $\alpha \in \mathbb{R}^n$ and $\beta \in \mathbb{R}^m$, Menon constructed a nonlinear homogeneous order-preserving map $T = T(A, \alpha, \beta)$ on $\mathbb{R}^n_+$ with the property that T has an eigenvector in $\mathrm{int}(\mathbb{R}^n_+)$ if and only if there exist positive diagonal matrices D and E such that the $m \times n$ matrix DAE has row sums β_i for $1 \leq i \leq m$ and column sums α_j for $1 \leq j \leq n$. The map T allows one to use ideas and methods from nonlinear Perron–Frobenius theory in the analysis of matrix scaling problems. This approach, which was further developed by Menon and Schneider [145] and by Nussbaum [163], will be discussed in detail here.

7.1 Matrix scaling: a fixed-point approach

Recall that a matrix $A = (a_{ij})$ is *positive* if $a_{ij} > 0$ for all i and j, and a vector $x \in \mathbb{R}^n$ is *positive* if $x_i > 0$ for all i. Furthermore we say that $A = (a_{ij})$ has a *positive diagonal* if $a_{ii} > 0$ for all i. Consider the following matrix scaling problem. Given a nonnegative $m \times n$ matrix A and positive vectors $\alpha \in \mathbb{R}^n$ and $\beta \in \mathbb{R}^m$, we ask whether there exist diagonal matrices D and E with positive diagonals such that the matrix $B = DAE$ has row sums β_i for $1 \leq i \leq m$

and column sums α_j for $1 \le j \le n$. By writing $D = \mathrm{diag}\,(d_1, \dots, d_m)$ and $E = \mathrm{diag}\,(e_1, \dots, e_n)$ we see that B has row sums β_i for $1 \le i \le m$ if and only if

$$\sum_{j=1}^{n} d_i a_{ij} e_j = \beta_i \quad \text{for } 1 \le i \le m. \tag{7.1}$$

Similarly, B has column sums α_j for $1 \le j \le n$ if and only if

$$\sum_{i=1}^{m} d_i a_{ij} e_j = \alpha_j \quad \text{for } 1 \le j \le n. \tag{7.2}$$

Clearly, if there exist $d_1, \dots, d_m > 0$ and $e_1, \dots, e_n > 0$ such that (7.1) and (7.2) hold, then no row or column of A can be identically zero. Furthermore, we must have that

$$\sum_{i=1}^{m} \beta_i = \sum_{i=1}^{m} \sum_{j=1}^{n} d_i a_{ij} e_j = \sum_{j=1}^{n} \alpha_j. \tag{7.3}$$

In view of these observations we make the following definition.

Definition 7.1.1 A triple (A, α, β) where A is a nonnegative $m \times n$ matrix with no row or column identically zero, and $\alpha \in \mathrm{int}(\mathbb{R}^n_+)$ and $\beta \in \mathrm{int}(\mathbb{R}^m_+)$ satisfy

$$\sum_{i=1}^{m} \beta_i = \sum_{j=1}^{n} \alpha_j,$$

is called a *DAD problem.*

Of particular interest are DAD problems (A, α, β) where A is square and $\alpha = \beta$, which includes the classic DAD problems where $\alpha_i = \beta_i = 1$ for all $1 \le i \le n$

To set up the fixed-point approach to DAD problems it is convenient to think of $\mathbb{R}^n$ as the space of continuous functions from $\{1, \dots, n\}$ to $\mathbb{R}$. In this way we can view $\mathbb{R}^n$ as a commutative algebra. So, for $x, y \in \mathbb{R}^n$, $z = xy \in \mathbb{R}^n$ is given by $z_i = x_i y_i$ for all i. Clearly $\mathbb{1} \in \mathbb{R}^n$ is the unit in the algebra. For $x \in \mathbb{R}^n$ with $x_i \neq 0$ for all i, we shall write x^{-1} to denote the vector with components x_i^{-1} for all i, so $x^{-1}x = \mathbb{1}$.

By thinking in terms of spaces of continuous functions, we now describe a more general DAD problem. For $1 \le i \le p+1$ and $p \ge 1$ let Q_i be compact Hausdorff spaces with $Q_1 = Q_{p+1}$. Write $X_i = C(Q_i)$, $1 \le i \le p+1$, to denote the space of continuous functions from Q_i to $\mathbb{R}$; so, $X_{p+1} = X_1$.

Furthermore, for $1 \leq i \leq p+1$, let K_i denote the cone of nonnegative functions in X_i, and for $f \in X_i$ let $M_f \colon X_i \to X_i$ denote the *multiplication operator* given by

$$M_f(g)(t) = f(t)g(t) \quad \text{for all } g \in X_i \text{ and } t \in Q_i.$$

If no confusion should arise, we shall simply write fg instead of $M_f(g)$.

A *generalized DAD problem* is a tuple $(L_1, \ldots, L_p, \alpha_1, \ldots, \alpha_p)$, where each $L_i \colon X_i \to X_{i+1}$ is a bounded linear map with $L_i(\text{int}(K_i)) \subseteq \text{int}(K_{i+1})$, and $\alpha_i \in \text{int}(K_i)$. A generalized DAD problem is said to have a *solution* if there exist $\lambda > 0$ and functions $f_i \in \text{int}(K_i)$ for $1 \leq i \leq p$ such that

$$\begin{aligned} f_{i+1}L_i(f_i) &= \alpha_{i+1} \quad \text{for } 1 \leq i < p, \text{ and} \\ f_1 L_p(f_p) &= \lambda \alpha_1. \end{aligned} \tag{7.4}$$

In contrast to our usual convention, we are allowing infinite-dimensional Banach spaces $C(Q_i)$ in this section. For the sequel, however, this generality is not required, and for comfort the reader may wish to assume that each set Q_i is a finite set.

In terms of matrices we see for $p = 2$ that if $L_1 = A$ is an $m \times n$ nonnegative matrix and $L_2 = B$ is an $n \times m$ nonnegative matrix, both with no row or column identically zero, then the generalized DAD problem (A, B, α, β) has a solution if and only if there exist $\lambda > 0$ and diagonal matrices D and E, with positive diagonals, such that DAE has row sums β_j for $1 \leq j \leq m$, and EBD has row sums $\lambda\alpha_i$ for $1 \leq i \leq n$. Moreover, if $B = A^T$ and $\sum_l \beta_l = \sum_j \alpha_j$, then $\lambda = 1$. Thus, the DAD problem (A, α, β) as defined in Definition 7.1.1 corresponds to the case where $p = 2$, $L_1 = A$, $L_2 = A^T$, $\alpha_1 = \alpha$, $\alpha_2 = \beta$, and the parameter λ is equal to 1.

The following theorem shows how solutions of the generalized DAD problem $(L_1, \ldots, L_p, \alpha_1, \ldots, \alpha_p)$ correspond to eigenvectors in the interior of the cone K_1 of a nonlinear map $T \colon \text{int}(K_1) \to \text{int}(K_1)$. For notational convenience define for $1 \leq i \leq p$ the map $R_i \colon \text{int}(K_i) \to \text{int}(K_i)$ by

$$R_i(x)(t) = x^{-1}(t) \quad \text{for all } t \in Q_i,$$

and let $\Lambda_i \colon X_i \to X_{i+1}$ be given by

$$\Lambda_i(g) = L_i(\alpha_i g) \quad \text{for } g \in X_i.$$

Recall that $X_{p+1} = X_1$.

Theorem 7.1.2 *The generalized DAD problem $(L_1, \ldots, L_p, \alpha_1, \ldots, \alpha_p)$ has a solution if and only if the map $T \colon \text{int}(K_1) \to \text{int}(K_1)$ given by*

$$T = \Lambda_p R_p \Lambda_{p-1} R_{p-1} \cdots \Lambda_1 R_1,$$

has an eigenvector $x_1 \in \text{int}(K_1)$ *with eigenvalue* $\lambda > 0$. *Moreover, the eigenvalue* λ *corresponds to the parameter* λ *in (7.4). Furthermore, if p is odd and* $T(x_1) = \lambda x_1$ *with* $\lambda > 0$, *then there exists* $t > 0$ *such that* $T(tx_1) = tx_1$.

Proof First assume that $T(x_1) = \lambda x_1$ for some $x_1 \in \text{int}(K_1)$ and $\lambda > 0$. Define $f_1 = \alpha_1 R_1(x_1)$, $x_{i+1} = \Lambda_i R_i(x_i)$, and $f_{i+1} = \alpha_{i+1} R_{i+1}(x_{i+1})$ for $1 \le i < p$. Note that

$$f_{i+1} L_i(f_i) = \alpha_{i+1} x_{i+1}^{-1} L_i(\alpha_i x_i^{-1}) = \alpha_{i+1} (L_i(\alpha_i x_i^{-1}))^{-1} (L_i(\alpha_i x_i^{-1})) = \alpha_{i+1}$$

for $1 \le i < p$. For $i = p$, we have that

$$\lambda x_1 = (\Lambda_p R_p \Lambda_{p-1} R_{p-1} \cdots \Lambda_1 R_1)(x_1) = L_p(\alpha_p x_p^{-1}),$$

so that $f_1 L_p(f_p) = \alpha_1 x_1^{-1} L_p(\alpha_p x_p^{-1}) = \lambda \alpha_1$.

Conversely, if the generalized DAD problem has a solution, we can define $x_1 = \alpha_1 f_1^{-1}$ and $x_{i+1} = \alpha_{i+1} f_{i+1}^{-1}$ for $1 \le i < p$. Now (7.4) implies that

$$x_{i+1} = \alpha_{i+1} f_{i+1}^{-1} = L_i(f_i) = L_i(\alpha_i x_i^{-1}) = \Lambda_i R_i(x_i)$$

for $1 \le i < p$, and $\lambda x_1 = \lambda \alpha_1 f_1^{-1} = L_p(f_p) = L_p(\alpha_p x_p^{-1}) = \Lambda_p R_p(x_p)$. It follows that $\lambda x_1 = (\Lambda_p R_p \Lambda_{p-1} R_{p-1} \cdots \Lambda_1 R_1)(x_1) = T(x_1)$.

Note that if p is odd, T is homogeneous of degree -1. So, if $T(x_1) = \lambda x_1$ for some $x_1 \in \text{int}(K_1)$ and $\lambda > 0$, then for $t = \sqrt{\lambda}$ we have that $T(tx_1) = t^{-1} T(x_1) = t^{-1} \lambda x_1 = tx_1$. □

Remark that the maps Λ_i are order-preserving and homogeneous of degree 1, and the maps R_i are order-reversing and homogeneous of degree -1. So, if p is even, T is order-preserving with respect to K_1 and homogeneous of degree 1. On the other hand, if p is odd, T is order-reversing with respect to K_1 and homogeneous of degree -1. In both cases we deduce from Corollaries 2.1.4 and 2.1.5 that T is non-expansive under Hilbert's metric.

As we are primarily concerned with ordinary DAD problems (A, α, β), we state the following consequence of Theorem 7.1.2.

Corollary 7.1.3 *Let* $R_1(x) = x^{-1}$ *for* $x \in \text{int}(\mathbb{R}^n_+)$, *and* $R_2(y) = y^{-1}$ *for* $y \in \text{int}(\mathbb{R}^m_+)$. *A* DAD *problem* (A, α, β) *has a solution if and only if the homogeneous order-preserving map* $T\colon \text{int}(\mathbb{R}^n_+) \to \text{int}(\mathbb{R}^n_+)$ *given by*

$$T(x) = (A^T M_\beta R_2 A M_\alpha R_1)(x) \quad \text{for } x \in \text{int}(\mathbb{R}^n_+) \tag{7.5}$$

has a fixed point in $\text{int}(\mathbb{R}^n_+)$.

Proof Recall that the DAD problem (A, α, β) corresponds to the generalized DAD problem (A, A^T, α, β). Moreover, as $\sum_i \beta_i = \sum_j \alpha_j$, the parameter λ has to equal 1. □

There is an important special case in which the map T in Theorem 7.1.2 has a unique, up to positive scalar multiples, eigenvector in $\text{int}(K_1)$.

Theorem 7.1.4 *If $(L_1, \dots, L_p, \alpha_1, \dots, \alpha_p)$ is a generalized DAD problem and there exists $1 \le j \le p$ such that*

$$\Delta(L_j) = \sup\{d_H(L_jx, L_jy) \colon x, y \in \text{int}(K_j)\} < \infty,$$

then T has a unique, up to positive scalar multiples, eigenvector in $\text{int}(K_1)$.

Proof For $x, y \in \text{int}(K_i)$ it is easy to verify that $d_H(R_ix, R_iy) = d_H(x, y)$ and $d_H(\alpha_i x, \alpha_i y) = d_H(x, y)$. As the linear map L_i maps $\text{int}(K_i)$ into $\text{int}(K_{i+1})$, we also know that $d_H(L_ix, L_iy) \le d_H(x, y)$ for all $x, y \in \text{int}(K_i)$. Moreover, it follows from the Birkhoff–Hopf Theorem A.4.1 that

$$d_H(L_jx, L_jy) \le \kappa(L_j)d_H(x, y)$$

for all $x, y \in \text{int}(K_j)$, where

$$\kappa(L_j) = \tanh\left(\frac{1}{4}\Delta(L_j)\right) < 1.$$

Consequently, $d_H(Tx, Ty) \le \kappa(L_j)d_H(x, y)$ for all $x, y \in \text{int}(K_1)$.

Let $S = \{x \in \text{int}(K_1) \colon \|x\| = 1\}$. We proved in Proposition 2.5.4 that (S, d_H) is a complete metric space if K_1 is a finite-dimensional solid closed cone. This result, however, holds more generally for normal cones in a Banach space, and, in particular, for the cone K_1 in $(C(Q_i), \|\cdot\|_\infty)$. The reader can find a proof in [158].

Let $g \colon S \to S$ be given by

$$g(x) = \frac{T(x)}{\|T(x)\|} \quad \text{for } x \in \text{int}(K_1).$$

So, for each $x, y \in S$,

$$d_H(g(x), g(y)) = d_H(T(x), T(y)) \le \kappa(L_j)d_H(x, y).$$

As $\kappa(L_j) < 1$, it follows from Banach's contraction Theorem 3.2.1 that g has a unique fixed point $u \in S$; so, $T(u) = \lambda u$ for some $\lambda > 0$.

If $v \in \text{int}(K_1)$ is an eigenvector of T, then $w = v/\|v\|$ is a fixed point of g, and hence $w = u$, which implies that $v = \|v\|u$. □

The basic approach to solving the original DAD problem (A, α, β) is to apply the perturbation method discussed in Section 6.2 to prove the existence of eigenvectors in the interior of $\mathbb{R}^n_+$ of the homogeneous order-preserving map $T \colon \text{int}(\mathbb{R}^n_+) \to \text{int}(\mathbb{R}^n_+)$ given in Corollary 7.1.3. Let U denote the $n \times m$ matrix with all entries equal to 1. For $\varepsilon > 0$, consider $A^T + \varepsilon U$, and note that $A^T + \varepsilon U$

defines a linear map $B_\varepsilon : \mathbb{R}^m \to \mathbb{R}^n$ which maps $\mathbb{R}^m_+$ into $\text{int}(\mathbb{R}^n_+)$. As all the entries of $A^T + \varepsilon U$ are positive, we know by Theorem A.6.2 that $\Delta(B_\varepsilon) < \infty$, and hence Theorem 7.1.4 implies that the homogeneous order-preserving map $T_\varepsilon : \text{int}(\mathbb{R}^n_+) \to \text{int}(\mathbb{R}^n_+)$ with

$$T_\varepsilon(x) = (B_\varepsilon M_\beta R_2 A M_\alpha R_1)(x) \quad \text{for } x \in \text{int}(\mathbb{R}^n_+) \tag{7.6}$$

has a unique, up to positive scalar multiples, eigenvector $u_\varepsilon \in \text{int}(\mathbb{R}^n_+)$. To summarize, we have the following lemma.

Lemma 7.1.5 *If (A, α, β) is a DAD problem, then for each $\varepsilon > 0$ the map $T_\varepsilon : \text{int}(\mathbb{R}^n_+) \to \text{int}(\mathbb{R}^n_+)$ defined by (7.6) has a unique eigenvector $u_\varepsilon \in \text{int}(\mathbb{R}^n_+)$ with $\|u_\varepsilon\|_\infty = 1$.*

In the next section we shall discuss a compatibility condition, which is necessary in order for the DAD problem (A, α, β) to have a solution, and which ensures that there exists a sequence $\varepsilon_k \to 0^+$ such that $u_{\varepsilon_k} \to u \in \text{int}(\mathbb{R}^n_+)$ and $T(u) = u$, where T is given by (7.5).

Of course the maps T_ε and T can be continuously extended to $\mathbb{R}^n_+$ by Theorem 5.1.5, but we shall not need that here. In the analysis of infinite-dimensional DAD problems, however, the corresponding map may fail to have a continuous extension to the boundary of the cone (see [163, p. 60]), which causes serious complications.

7.2 The compatibility condition

If I is a finite set, $|I|$ will denote the cardinality of I. Given an $m \times n$ matrix $A = (a_{ij})$ and $I \subseteq \{1, \ldots, m\}$ and $J \subseteq \{1, \ldots, n\}$, we shall write $A_{IJ} = 0$ if $a_{ij} = 0$ for all $(i, j) \in I \times J$, and we shall write $A_{IJ} \neq 0$ if there exists $(i, j) \in I \times J$ with $a_{ij} \neq 0$. We shall also use the notation $I^c = \{1, \ldots, m\} \setminus I$ and $J^c = \{1, \ldots, n\} \setminus J$ to denote the complements of I and J.

Let (A, α, β) be a DAD problem with solution $D = \text{diag}(d_1, \ldots, d_m)$ and $E = \text{diag}(e_1, \ldots, e_n)$. Thus, $d_1, \ldots, d_m > 0$ and $e_1, \ldots, e_n > 0$ satisfy equations (7.1) and (7.2). Assume that $I \subseteq \{1, \ldots, m\}$ and $J \subseteq \{1, \ldots, n\}$ are such that $A_{I^c J^c} = 0$. Writing $\hat{a}_{ij} = d_i a_{ij} e_j$ we see that

$$\sum_{j \in J} \alpha_j = \sum_{j \in J} \sum_{i \in I} \hat{a}_{ij} + \sum_{j \in J} \sum_{i \in I^c} \hat{a}_{ij}$$

and

$$\sum_{i \in I^c} \beta_i = \sum_{i \in I^c} \sum_{j \in J} \hat{a}_{ij} + \sum_{i \in I^c} \sum_{j \in J^c} \hat{a}_{ij} = \sum_{j \in J} \sum_{i \in I^c} \hat{a}_{ij},$$

so that

$$\sum_{i\in I^c}\beta_i \le \sum_{j\in J}\alpha_j. \tag{7.7}$$

Moreover, the inequality is strict if and only if $A_{IJ} \neq 0$. A similar argument shows that

$$\sum_{j\in J^c}\alpha_j \le \sum_{i\in I}\beta_i, \tag{7.8}$$

with equality if and only if $A_{IJ} = 0$. It should be noted, however, that (7.8) also follows from (7.7) and the fact that $\sum_i \beta_i = \sum_j \alpha_j$.

By using the convention that $\sum_{k\in\emptyset}\gamma_k = 0$ and $A_{I^cJ^c} = 0$ if I^c or J^c is empty, we can check that the inequalities (7.7) and (7.8) hold. Moreover, if I^c is empty, there is equality in (7.7) if and only if J is empty. Similarly, if J^c is empty there is equality in (7.8) if and only if I is empty. It follows that if $|I^c| = 1$, $|J^c| = n$, and $A_{I^cJ^c} = 0$, then (7.7) cannot hold, which shows that (7.7) implies that A has no row identically zero. In a similar way it can be shown that (7.8) implies that A has no zero columns.

In view of these observations we make the following definition.

Definition 7.2.1 A DAD problem (A, α, β) is said to satisfy the *compatibility condition* if for every $I \subseteq \{1, \ldots, m\}$ and $J \subseteq \{1, \ldots, n\}$ with $A_{I^cJ^c} = 0$ the inequality

$$\sum_{i\in I^c}\beta_i \le \sum_{j\in J}\alpha_j \tag{7.9}$$

holds, and the inequality is strict if and only if $A_{IJ} \neq 0$.

By the previous observations we know that DAD problems with a solution have to satisfy the compatibility condition.

Corollary 7.2.2 *If a DAD problem (A, α, β) has a solution, then it satisfies the compatibility condition.*

As we shall see later in this section the compatibility condition is also a sufficient condition for the DAD problem to have a solution. For the moment, however, it will be convenient to impose an additional condition.

Lemma 7.2.3 *If (A, α, β) is a DAD problem satisfying the compatibility condition and for all proper non-empty subsets I of $\{1, \ldots, m\}$ and J of $\{1, \ldots, n\}$ we have that $A_{I^cJ^c} = 0$ implies $A_{IJ} \neq 0$, then (A, α, β) has a solution.*

Proof Using the notation of Lemma 7.1.5 and Equation (7.6) we know that for each $\varepsilon > 0$ there exists $u_\varepsilon \in \mathbb{R}^n_+$, with $\|u_\varepsilon\|_\infty = \max_i (u_\varepsilon)_i = 1$, such that $T_\varepsilon(u_\varepsilon) = \lambda_\varepsilon u_\varepsilon$ for some $\lambda_\varepsilon > 0$. As

$$(AM_\alpha R_1(u_\varepsilon))_i = \sum_{j=1}^n a_{ij}\alpha_j (u_\varepsilon)_j^{-1} \geq \sum_{j=1}^n a_{ij}\alpha_j > 0,$$

for all $1 \leq i \leq m$, there exists a constant $0 < C < \infty$ such that

$$(R_2 AM_\alpha R_1(u_\varepsilon))_i \leq \Big(\sum_{j=1}^n a_{ij}\alpha_j\Big)^{-1} < C,$$

for all $\varepsilon > 0$ and $1 \leq i \leq m$.

Letting $z_\varepsilon = R_2 AM_\alpha R_1(u_\varepsilon)$ we find that $\|z_\varepsilon\|_\infty < C$ for all $\varepsilon > 0$. Thus, we can find a sequence $\varepsilon_k \to 0^+$ such that $z_{\varepsilon_k} \to z \in \mathbb{R}^m_+$ and $u_{\varepsilon_k} \to u \in \mathbb{R}^n_+$ with $\|u\|_\infty = 1$. Note that if for fixed $k \geq 1$ we choose j such that $(u_{\varepsilon_k})_j = 1$, then

$$\lambda_{\varepsilon_k} = \big((A^T + \varepsilon_k U)M_\beta z_{\varepsilon_k}\big)_j,$$

where U is the $m \times n$ matrix with all entries equal to 1. Thus, there exists a constant $C' > 0$ such that $0 < \lambda_{\varepsilon_k} < C'$ for all $k \geq 1$. By taking a further subsequence we may assume that $\lambda_{\varepsilon_k} \to \lambda$.

By applying Theorem 7.1.2 with $p = 2$, $L_1 = A$, $L_2 = A^T + \varepsilon_k U$, $\alpha_1 = \alpha$, and $\alpha_2 = \beta$, we find that there exist $d_1^k, \ldots, d_m^k > 0$, and $e_1^k, \ldots, e_n^k > 0$ such that

$$\sum_{j=1}^n d_i^k a_{ij} e_j^k = \beta_i \quad \text{for } 1 \leq i \leq m$$

and

$$\sum_{i=1}^m d_i^k (a_{ij} + \varepsilon_k) e_j^k = \lambda_{\varepsilon_k}\alpha_j \quad \text{for } 1 \leq j \leq n.$$

It follows that

$$\sum_{i=1}^m \sum_{j=1}^n d_i^k a_{ij} e_j^k = \sum_{i=1}^m \beta_i$$

and

$$\sum_{j=1}^n \sum_{i=1}^m d_i^k (a_{ij} + \varepsilon_k) e_j^k = \lambda_{\varepsilon_k} \sum_{j=1}^n \alpha_j.$$

As $\sum_i \beta_i = \sum_j \alpha_j$, we conclude that $\lambda_{\varepsilon_k} \geq 1$ for all $k \geq 1$, and hence $\lambda \geq 1$.

Given z and u as above, define $J = \{j : u_j > 0\}$. Clearly if $J = \{1, \ldots, n\}$, we can directly take limits for $k \to \infty$ in the equation $T_{\varepsilon_k}(u_{\varepsilon_k}) = \lambda_{\varepsilon_k} u_{\varepsilon_k}$ and obtain $T(u) = \lambda u$, where T is given by (7.5). In that case it follows from Theorem 7.1.2 and Corollary 7.1.3 that $\lambda = 1$ and the DAD problem (A, α, β) has a solution.

Now suppose that J is a proper subset of $\{1, \ldots, n\}$. We will show that this leads to a contradiction. Let $I = \{i : z_i = 0\}$, and note that if I is empty then $\lambda u = A^T M_\beta(z)$ is positive and hence $J = \{1, \ldots, n\}$, which is impossible. On the other hand, if $I = \{1, \ldots, m\}$, so $z = 0$, then $u = 0$, which contradicts the fact that $\|u\|_\infty = 1$. Thus, I is a proper non-empty subset of $\{1, \ldots, m\}$.

Taking limits in the equation $T_{\varepsilon_k}(u_{\varepsilon_k}) = \lambda_{\varepsilon_k} u_{\varepsilon_k}$ we see that

$$\lambda u_j = \sum_{i=1}^{m} a_{ij}\beta_i z_i = \sum_{i \in I^c} a_{ij}\beta_i z_i. \tag{7.10}$$

As $\beta_i z_i > 0$ for all $i \in I^c$, we conclude that $A_{I^c J^c} = 0$. By assumption $A_{IJ} \neq 0$, so that the compatibility condition implies that

$$\sum_{i \in I^c} \beta_i < \sum_{j \in J} \alpha_j. \tag{7.11}$$

For $j \in J$ it follows from (7.10) that

$$\lambda \alpha_j = \alpha_j u_j^{-1} \sum_{i \in I^c} a_{ij}\beta_i z_i,$$

which gives

$$\lambda \sum_{j \in J} \alpha_j = \sum_{j \in J} \sum_{i \in I^c} \alpha_j u_j^{-1} a_{ij}\beta_i z_i. \tag{7.12}$$

For $i \in I^c$, however, we know that as $A_{I^c J^c} = 0$,

$$\begin{aligned} z_i &= \lim_{k \to \infty} \Big(\sum_{j=1}^{n} a_{ij}\alpha_j (u_{\varepsilon_k})_j^{-1} \Big)^{-1} \\ &= \lim_{k \to \infty} \Big(\sum_{j \in J} a_{ij}\alpha_j (u_{\varepsilon_k})_j^{-1} \Big)^{-1} \\ &= \Big(\sum_{j \in J} a_{ij}\alpha_j u_j^{-1} \Big)^{-1}. \end{aligned}$$

It follows that

$$\sum_{i \in I^c} \beta_i z_i \Big(\sum_{j \in J} a_{ij}\alpha_j u_j^{-1} \Big) = \sum_{i \in I^c} \beta_i. \tag{7.13}$$

Combining (7.12) and (7.13) gives $\lambda \sum_{j \in J} \alpha_j = \sum_{i \in I^c} \beta_i$, which contradicts (7.11), as $\lambda \geq 1$. □

The additional assumption on A in Lemma 7.2.3 is redundant. To prove this we need the following auxiliary lemma.

Lemma 7.2.4 *If $B = (b_{ij})$ is an $m \times n$ nonnegative matrix with no row or column identically zero, then there exist partitions $\cup_{p=1}^r S_p = \{1, \ldots, m\}$ and $\cup_{p=1}^r T_p = \{1, \ldots, n\}$ such that*

(i) for each $1 \leq p \leq r$ and all proper non-empty subsets S of S_p and T of T_p we have that $B_{(S_p \setminus S)(T_p \setminus T)} = 0$ implies $B_{ST} \neq 0$, and
(ii) $b_{ij} = 0$ for all $(i, j) \notin \cup_{p=1}^r S_p \times T_p$.

Proof We argue by induction on $k = \max\{m, n\}$. If $k = 1$, then $B = (b_{11})$ with $b_{11} > 0$, and we can take $r = 1$ and $S_1 = T_1 = \{1\}$. Now assume that this assertion is true for all $m \times n$ nonnegative matrices B with no zero row or column and $k = \max\{m, n\}$. Let B be a nonnegative matrix with $k + 1 = \max\{m, n\}$ and no row or column identically zero. If B has the property that for all proper non-empty subsets $I \subseteq \{1, \ldots, m\}$ and $J \subseteq \{1, \ldots, n\}$ we have that $B_{I^c J^c} = 0$ implies $B_{IJ} \neq 0$, we can take $r = 1$, $S_1 = \{1, \ldots, m\}$, and $T_1 = \{1, \ldots, n\}$, and we are done.

So, suppose that there exist proper non-empty subsets $I \subseteq \{1, \ldots, m\}$ and $J \subseteq \{1, \ldots, n\}$ such that $B_{I^c J^c} = 0$ and $B_{IJ} = 0$. Clearly $I \cup I^c$ is a partition of $\{1, \ldots, m\}$ and $J \cup J^c$ is a partition of $\{1, \ldots, n\}$. Moreover, $b_{ij} = 0$ for all pairs $(i, j) \notin (I \times J^c) \cup (I^c \times J)$. As B has no row or column identically zero, B_{IJ^c} and $B_{I^c J}$ have no zero row or column. Thus we can apply the induction hypothesis to B_{IJ^c} and $B_{I^c J}$ to obtain the result. □

Using this lemma we now prove the following theorem.

Theorem 7.2.5 *A DAD problem (A, α, β) has a solution if and only if it satisfies the compatibility condition.*

Proof By Corollary 7.2.2 it suffices to show that if (A, α, β) satisfies the compatibility condition, then it has a solution. Let S_p and T_p, $1 \leq p \leq r$, be as in Lemma 7.2.4. Since $A_{S_p^c T_p} = 0$ and $A_{S_p T_p^c} = 0$, no row or column of $A_{S_p T_p}$ is identically zero, and by using the compatibility condition we find that

$$\sum_{i \in S_p} \beta_i = \sum_{j \in T_p} \alpha_j. \tag{7.14}$$

The main idea is to apply Lemma 7.2.3 to $(A_{S_p T_p}, \alpha_{T_p}, \beta_{S_p})$ for $1 \leq p \leq r$ and combine the solutions to create a solution to the original problem. Here

α_{T_p} denotes the restriction of α to the components $j \in T_p$, and β_{S_p} is the restriction of β to S_p.

Recall that S_p and T_p, $1 \le p \le r$, satisfy the assertions in Lemma 7.2.4. So, to apply Lemma 7.2.3 to $(A_{S_pT_p}, \alpha_{T_p}, \beta_{S_p})$, it remains to show that if $I_p \subseteq S_p$, $J_p \subseteq T_p$, and $A_{(S_p\setminus I_p)(T_p\setminus J_p)} = 0$, then

$$\sum_{i\in S_p\setminus I_p} \beta_i \le \sum_{j\in J_p} \alpha_j, \tag{7.15}$$

and the inequality is strict if and only if $A_{I_pJ_p} \ne 0$.

First note that if $I_p = S_p$, then the inequality (7.15) holds, and the equality holds if and only if J_p is empty. As $A_{S_pT_p}$ has no zero row or column, we know that $A_{S_pJ_p} = 0$ if and only if J_p is empty. Likewise, if $J_p = T_p$, then (7.15) holds with equality if and only if I_p is empty, which is equivalent to $A_{I_pT_p} = 0$. Also remark that, as $A_{S_pT_p}$ has no zero row or column, $I_p = S_p$, if J_p is empty. Similarly, if I_p is empty, then $J_p = T_p$.

Thus, it remains to consider the case where $I_p \subseteq S_p$ and $J_p \subseteq T_p$ are proper non-empty subsets. In that case we write $I = I_p \subseteq \{1, \ldots, m\}$ and $J = J_p \cup T_p^c \subseteq \{1, \ldots, n\}$. Clearly $I^c \times J^c = I_p^c \times (T_p \setminus J_p)$, $A_{I^cJ^c} = 0$, and $A_{IJ} \ne 0$, as $A_{(S_p\setminus I_p)(T_p\setminus J_p)} = 0$ and $A_{I_pJ_p} \ne 0$. It follows from the compatibility condition for (A, α, β) that

$$\sum_{j\in J} \alpha_j = \sum_{j\in J_p} \alpha_j + \sum_{j\in T_p^c} \alpha_j > \sum_{i\in I^c} \beta_i = \sum_{i\in S_p\setminus I_p} \beta_i + \sum_{i\in S_p^c} \beta_i.$$

As $\sum_i \beta_i = \sum_j \alpha_j$, it follows from (7.14) that

$$\sum_{i\in S_p^c} \beta_i = \sum_{j\in T_p^c} \alpha_j,$$

and hence $\sum_{j\in J_p} \alpha_j > \sum_{i\in S_p\setminus I_p} \beta_i$.

Thus, we can apply Lemma 7.2.3 to $(A_{S_pT_p}, \alpha_{T_p}, \beta_{S_p})$ for $1 \le p \le r$ and find $d_i > 0$, $i \in S_p$, and $e_j > 0$, $j \in T_p$, such that

$$\sum_{i\in S_p} d_i a_{ij} e_j = \alpha_j \quad \text{for all } j \in T_p$$

and

$$\sum_{j\in T_p} d_i a_{ij} e_j = \beta_i \quad \text{for all } i \in S_p.$$

Since $a_{ij} = 0$ for all $(i, j) \in (S_p \times T_p^c) \cup (S_p^c \times T_p)$, the above two equalities give, for $1 \le p \le r$,

$$\sum_{i=1}^{m} d_i a_{ij} e_j = \sum_{i \in S_p} d_i a_{ij} e_j = \alpha_j \quad \text{for all } j \in T_p$$

and

$$\sum_{j=1}^{n} d_i a_{ij} e_j = \sum_{j \in T_p} d_i a_{ij} e_j = \beta_i \quad \text{for all } i \in S_p.$$

As $\cup_{p=1}^r S_p = \{1, \ldots, m\}$ and $\cup_{p=1}^r T_p = \{1, \ldots, n\}$, we have obtained a solution to the original DAD problem (A, α, β). □

The compatibility condition does not depend on the exact values of the entries of A, but only its zero pattern. Recall that two $m \times n$ matrices A and B have the same zero pattern if for each i and j we have that $a_{ij} = 0$ if and only if $b_{ij} = 0$. So, we have the following simple consequence of Theorem 7.2.5.

Corollary 7.2.6 *If (A, α, β) and (B, α, β) are two DAD problems and A and B have the same zero pattern, i.e., $a_{ij} = 0$ if and only if $b_{ij} = 0$, then (A, α, β) has a solution if and only if (B, α, β) has a solution.*

Clearly, if DAE is a solution to the DAD problem (A, α, β), then for each $\lambda > 0$, $(\lambda D)A(\lambda^{-1}E)$ is also a solution. It is interesting to understand when the solution is unique, up to positive scalar multiples, or, equivalently, to know when T in Corollary 7.1.3 has a unique normalized fixed point in $\text{int}(\mathbb{R}^n_+)$. We already saw in Theorem 7.1.4 that if $\Delta(A) < \infty$ or $\Delta(A^T) < \infty$, then T has a unique normalized fixed point in $\text{int}(\mathbb{R}^n_+)$. In fact, in that case, it follows from the proof that T is a Lipschitz contraction under Hilbert's metric, which ensures that the fixed point of T in $\text{int}(\mathbb{R}^n_+)$ is unique, if it exists. An appropriate condition on A that makes T contractive is the so-called scrambling condition, which is discussed in Section B.7 of Appendix B. An $m \times n$ nonnegative matrix $A = (a_{ij})$, with no row identically zero, is called *scrambling* if for each $1 \le i_1 < i_2 \le m$ there exists $1 \le j \le n$ such that

$$a_{i_1 j} a_{i_2 j} > 0.$$

Theorem 7.2.7 *If (A, α, β) is a DAD problem such that A or A^T is scrambling, then either (A, α, β) has a unique, up to positive scalar multiples, solution, or it has no solution.*

Proof Recall that $T(x) = (A^T M_\beta R_2 A M_\alpha R_1)(x)$ for $x \in \text{int}(\mathbb{R}^n_+)$. We already saw that R_1 and M_α are isometries under d_H on $\text{int}(\mathbb{R}^n_+)$, and R_2 and M_β are isometries under d_H on $\text{int}(\mathbb{R}^m_+)$.

If A is scrambling, it follows from Corollary A.7.3 that $d_H(Ax, Ay) < d_H(x, y)$ for all $x, y \in \text{int}(\mathbb{R}^n_+)$ with $d_H(x, y) > 0$. Likewise A^T is contractive under d_H, if it is scrambling. So, in either case the map T is a contraction under Hilbert's metric on $\text{int}(\mathbb{R}^n_+)$. It follows that if $T(x) = x$ and $T(y) = y$ for some $x, y \in \text{int}(\mathbb{R}^n_+)$, then y is a positive multiple of x. It could, of course, also happen that T has no fixed point in $\text{int}(\mathbb{R}^n_+)$. □

7.3 Special *DAD* theorems

To determine whether a DAD problem (A, α, β) satisfies the compatibility condition can be computationally hard when the dimensions of A are large. In this section we discuss some alternative conditions that are easy to verify, but still ensure a solution for the DAD problem (A, α, β) in case A is square and $\alpha = \beta$. We begin with the following condition.

Definition 7.3.1 An $n \times n$ nonnegative matrix A is said to satisfy the *complements condition* if for each proper non-empty subset I of $\{1, \ldots, n\}$ with $A_{II^c} = 0$ we have that $A_{I^cI} = 0$.

Lemma 7.3.2 *If (A, α, α) satisfies the compatibility condition, then A satisfies the complements condition. Conversely, if (A, α, α) is a DAD problem where A satisfies the complements condition and $a_{ii} > 0$ for each $1 \leq i \leq n$, then (A, α, α) satisfies the compatibility condition.*

Proof If I is a proper non-empty subset of $\{1, \ldots, n\}$ such that $A_{II^c} = 0$, the compatibility condition for (A, α, α) implies that

$$\sum_{i \in I} \alpha_i \geq \sum_{i \in I} \alpha_i,$$

with strict inequality if $A_{I^cI} \neq 0$. So, in that case we must have that A satisfies the complements condition.

Now suppose that A in the DAD problem (A, α, α) satisfies the complements condition and A has a positive diagonal. Let $I, J \subseteq \{1, \ldots, n\}$ be such that $A_{I^cJ^c} = 0$. We have to show that $\sum_{j \in J} \alpha_j \geq \sum_{i \in I^c} \alpha_i$ and the inequality is strict if and only if $A_{IJ} \neq 0$.

Clearly $I^c \cap J^c = \emptyset$; otherwise, $i \in I^c \cap J^c$ and $a_{ii} > 0$, which is absurd. It follows that $J^c \subseteq I$ and $J \supseteq I^c$. Note that if $J \neq I^c$, then $I \cap J \neq \emptyset$, and hence $A_{IJ} \neq 0$. Also if $J \neq I^c$, then $\sum_{j \in J} \alpha_j > \sum_{i \in I^c} \alpha_i$, and we are done.

Now assume that $J = I^c$, so $\sum_{j \in J} \alpha_j = \sum_{i \in I^c} \alpha_i$. We must show that $A_{IJ} = 0$ in that case. If I is a proper non-empty subset of $\{1, \ldots, n\}$, then the complements condition implies that

$$0 = A_{I^cJ^c} = A_{I^cI} = A_{II^c} = A_{IJ}.$$

Finally note that if I is empty or $I = \{1, \ldots, n\}$, then $A_{IJ} = 0$, which completes the proof. □

Combining Lemma 7.3.2 with Theorem 7.2.5 directly gives the following result.

Theorem 7.3.3 *If (A, α, α) is a DAD problem and A has a positive diagonal, then (A, α, α) has a solution if and only if A satisfies the complements condition.*

Theorem 7.3.3 can be reformulated using the following notion. Given a proper non-empty subset $J = \{k_1, \ldots, k_m\}$ of $\{1, \ldots, n\}$ with $k_1 < k_2 < \ldots < k_m$ and a nonnegative matrix A, we say that A_{JJ} is *irreducible* if the matrix $(a_{k_ik_j})$ is irreducible. Furthermore we say that A_{JJ} has a *positive diagonal* if $a_{k_ik_i} > 0$ for all $1 \leq i \leq m$.

Theorem 7.3.4 *If (A, α, α) is a DAD problem and A is an $n \times n$ matrix with a positive diagonal, then (A, α, α) has a solution if and only if there exists a partition $\cup_{m=1}^{p} J_m$ of $\{1 \ldots, n\}$ such that*

(i) $A_{J_mJ_m}$ is irreducible for all $1 \leq m \leq p$, and
(ii) $a_{ij} = 0$ for all $(i, j) \notin \cup_{m=1}^{p} J_m \times J_m$.

Proof First suppose that there exists a partition $\cup_{m=1}^{p} J_m$ of $\{1, \ldots, n\}$ satisfying (i) and (ii). As $A_{J_mJ_m}$ is irreducible, there does not exist a non-empty subset L of J_m such that $A_{L(J_m \setminus L)} = 0$, so $A_{J_mJ_m}$ satisfies the complements condition. Furthermore, $A_{J_mJ_m}$ has a positive diagonal. It thus follows from Theorem 7.3.3 that there exist positive numbers d_i and e_j for $i, j \in J_m$ such that

$$\sum_{j \in J_m} d_i a_{ij} e_j = \alpha_i \quad \text{for all } i \in J_m,$$

and

$$\sum_{i \in J_m} d_i a_{ij} e_j = \alpha_j \quad \text{for all } j \in J_m.$$

Since $\cup_{m=1}^{p} J_m$ is a partition of $\{1, \ldots, n\}$, the sets J_m are pairwise disjoint, and we have obtained positive numbers d_i and e_j for all $i, j \in \{1, \ldots, n\}$. Moreover, as $a_{ij} = 0$ for all $(i, j) \notin \cup_{m=1}^{p} J_m \times J_m$, we see that

$$\sum_{j=1}^{n} d_i a_{ij} e_j = \sum_{j \in J_m} d_i a_{ij} e_j = \alpha_i \quad \text{for all } i \in J_m,$$

and

$$\sum_{i=1}^{n} d_i a_{ij} e_j = \sum_{i \in J_m} d_i a_{ij} e_j = \alpha_j \quad \text{for all } j \in J_m,$$

which shows that the DAD problem (A, α, α) has a solution.

Conversely, suppose that the DAD problem (A, α, α) has a solution and the $n \times n$ matrix A has a positive diagonal. To show that there exists a partition of $\{1, \dots, n\}$ satisfying (i) and (ii), we use induction on n. Obviously if A is a 1×1 matrix, the assertions hold.

Now suppose the assertions hold for all $k \times k$ nonnegative matrices B with a positive diagonal and $k < n$ and $n > 1$. As A has a positive diagonal and (A, α, α) has a solution, it follows from Theorem 7.3.3 that A satisfies the complements condition. If there does not exist a proper non-empty subset J of $\{1, \dots, n\}$ such that $A_{JJ^c} = 0$, A is irreducible, and we can take $p = 1$ and $J_1 = \{1, \dots, n\}$. If there exists a proper non-empty subset J of $\{1, \dots, n\}$ such that $A_{JJ^c} = 0$, then $A_{J^cJ} = 0$. Let $k = |J|$ and note that $1 \le k < n$. It follows that A_{JJ} is a $k \times k$ nonnegative matrix with a positive diagonal, and A_{JJ} satisfies the complements condition. Likewise, $A_{J^cJ^c}$ is a $(n - k) \times (n - k)$ nonnegative matrix, with a positive diagonal, satisfying the complements condition. Remark that, as $A_{JJ^c} = 0$ and $A_{J^cJ} = 0$, $(A_{JJ}, \alpha_J, \alpha_J)$ and $(A_{J^cJ^c}, \alpha_{J^c}, \alpha_{J^c})$ are DAD problems. So, by Theorem 7.3.3 we can apply the induction hypothesis to A_{JJ} and $A_{J^cJ^c}$ to obtain partitions $\cup_{m=1}^{r} J_m$ of J and $\cup_{m=r+1}^{p} J_m$ of J^c such that $A_{J_mJ_m}$ is irreducible for all $1 \le m \le p$, and $a_{ij} = 0$ for all

$$(i, j) \in \Big((J \times J) \setminus (\cup_{m=1}^{r} J_m \times J_m)\Big) \cup \Big((J^c \times J^c) \setminus (\cup_{m=r+1}^{p} J_m \times J_m)\Big).$$

From $A_{JJ^c} = 0$ and $A_{J^cJ} = 0$ it now follows that $a_{ij} = 0$ for all $(i, j) \notin \cup_{m=1}^{p} J_m \times J_m$. □

It may happen that the DAD problem (A, α, α) has a solution even if A does not have a positive diagonal. To illustrate this point we mention the following result.

Theorem 7.3.5 *If (A, α, α) is a DAD problem and there exist proper non-empty subsets S and T of $\{1, \dots, n\}$ such that $A_{ST} = 0$, $a_{ij} > 0$ for all $(i, j) \notin S \times T$, and*

$$\sum_{j \in T^c} \alpha_j > \sum_{i \in S} \alpha_i,$$

then (A, α, α) has a unique, up to positive scalar multiples, solution.

Proof We shall first prove that (A, α, α) satisfies the compatibility condition. Suppose that $I, J \subseteq \{1, \ldots, n\}$ and $A_{IJ} = 0$. Our hypotheses imply that $I \times J \subseteq S \times T$, and since $I^c \times J^c \supseteq S^c \times T^c$ and $A_{S^c T^c} \neq 0$, we must have that $A_{I^c J^c} \neq 0$. So, we need to show that

$$\sum_{j \in J^c} \alpha_j > \sum_{i \in I} \alpha_i.$$

As $J \subseteq T$ and $I \subseteq S$, $J^c \supseteq T^c$ and

$$\sum_{j \in J^c} \alpha_j \geq \sum_{j \in T^c} \alpha_j > \sum_{i \in S} \alpha_i \geq \sum_{i \in I} \alpha_i.$$

Thus, (A, α, α) satisfies the compatibility condition and hence it has a solution by Theorem 7.2.5. The uniqueness follows from Theorem 7.2.7, as the hypotheses imply that A is scrambling. □

7.4 Doubly stochastic matrices: the classic case

Let us now return to the *classic DAD problem* (A, α, β) where A is an $n \times n$ nonnegative matrix with no row or column identically zero, and $\alpha = \beta = \mathbb{1}$. In that case, elegant necessary and sufficient conditions were obtained independently by Sinkhorn and Knopp [206] and Brualdi, Parter, and Schneider [39]. We shall discuss these conditions here.

Given a permutation τ on $\{1, \ldots, n\}$ denote the associated $n \times n$ permutation matrix by P_τ; so, $P_\tau x = y$ where $y = (x_{\tau(1)}, \ldots, x_{\tau(n)})$ for all $x \in \mathbb{R}^n$. Note that P_τ is orthogonal, so that $P_\tau^{-1} = P_\tau^T$, and if A is an $n \times n$ matrix, then $C = P_\tau^T A P_\tau$ has entries

$$c_{ij} = a_{\tau(i)\tau(j)} \quad \text{for all } 1 \leq i, j \leq n.$$

More generally, one can verify that if σ and τ are two permutations on $\{1, \ldots, n\}$, then $B = P_\sigma A P_\tau$ has entries

$$b_{ij} = a_{\sigma^{-1}(i)\tau(j)} \quad \text{for all } 1 \leq i, j \leq n.$$

So, if A is doubly stochastic, then $P_\sigma A P_\tau$ is also doubly stochastic.

The following basic observation will be useful.

Lemma 7.4.1 *If* $(A, \mathbb{1}, \mathbb{1})$ *is a classic DAD problem and* P *and* Q *are* $n \times n$ *permutation matrices, then* $(A, \mathbb{1}, \mathbb{1})$ *has a solution if and only if* $(QAP, \mathbb{1}, \mathbb{1})$ *has a solution.*

Proof Suppose that $(A, \mathbb{1}, \mathbb{1})$ has a solution. Then there exist diagonal matrices D and E with positive diagonals such that $B = DAE$ is doubly stochastic. Let $D_1 = QDQ^T$ and $E_1 = P^T EP$, and note that D_1 and E_1 are diagonal matrices with positive diagonals. Moreover,

$$D_1 QAPE_1 = QBP.$$

Letting σ and τ be permutations on $\{1, \ldots, n\}$ so that $Q = Q_\sigma$ and $P = P_\tau$, we see that QBP has entries

$$b_{\sigma^{-1}(i)\tau(j)} \quad \text{for all } 1 \leq i, j \leq n.$$

It follows that QBP is doubly stochastic, as B is doubly stochastic

Conversely, if the classic DAD problem $(QAP, \mathbb{1}, \mathbb{1})$ has a solution, then $(Q^T(QAP)P^T, \mathbb{1}, \mathbb{1})$ has a solution by the previous argument. As $Q^{-1} = Q^T$ and $P^{-1} = P^T$, we conclude that $(A, \mathbb{1}, \mathbb{1})$ has a solution, and we are done. □

To proceed further in the analysis we shall need a combinatorial result concerning the permanent of nonnegative matrices known as the Frobenius–König theorem. Recall that the *permanent* of an $m \times n$ matrix $A = (a_{ij})$ with $m \leq n$ is defined by

$$\mathrm{Per}(A) = \sum_{\sigma \in S(m,n)} \left(\prod_{i=1}^{m} a_{i\sigma(i)} \right), \tag{7.16}$$

where $S(m, n)$ is the set of one-to-one maps $\sigma\colon \{1, \ldots, m\} \to \{1, \ldots, n\}$. Note that if A is nonnegative, then $\mathrm{Per}(A) \neq 0$ if and only if there exists one-to-one map $\sigma \in S(m, n)$ with

$$\prod_{i=1}^{m} a_{i\sigma(i)} \neq 0.$$

Theorem 7.4.2 (Frobenius–König) *If A is an $m \times n$ nonnegative matrix with $m \leq n$, then $\mathrm{Per}(A) = 0$ if and only if there exist non-empty sets $S \subseteq \{1, \ldots, m\}$ and $T \subseteq \{1, \ldots, n\}$ such that $A_{ST} = 0$ and $|S| + |T| = n + 1$.*

The reader is referred to [148, section 4.2] for a proof.

Lemma 7.4.3 *If $(A, \mathbb{1}, \mathbb{1})$ is a classic DAD problem satisfying the compatibility condition, then the following assertions hold:*

(i) If $S, T \subseteq \{1, \ldots, n\}$ are such that $A_{ST} = 0$, then $|S| + |T| \leq n$.
(ii) There exists a permutation τ on $\{1, \ldots, n\}$ such that $\prod_{i=1}^{n} a_{i\tau(i)} > 0$.

Proof The compatibility condition implies that

$$\sum_{i\in S} 1 = |S| \leq \sum_{j\in T^c} 1 = |T^c| = n - |T|,$$

so that $|S| + |T| \leq n$. Combining (i) and the Frobenius–König Theorem 7.4.2 we see that $\mathrm{Per}(A) \neq 0$, and hence there exists a permutation τ on $\{1, \dots, n\}$ such that $\prod_{i=1}^{n} a_{i\tau(i)} > 0$. □

Before we can state the necessary and sufficient conditions for the classic DAD problem to have a solution we need to introduce one more notion. If $B = (b_{ij})$ is an $n \times n$ matrix, we say that B is a *direct sum* of square matrices B_k, $1 \leq k \leq p$, if B is of the form

$$B = \begin{bmatrix} B_1 & 0 & \dots & 0 \\ 0 & B_2 & & 0 \\ \vdots & & \ddots & \vdots \\ 0 & 0 & \dots & B_p \end{bmatrix}.$$

In other words, there exist $1 \leq m_1 < m_2 < \dots < m_p = n$ such that if we let $J_1 = \{i : 1 \leq i \leq m_1\}$ and $J_k = \{i : \sum_{r=1}^{k-1} m_r < i \leq \sum_{r=1}^{k} m_r\}$ for $2 \leq k \leq p$, then $b_{ij} = 0$ for all $(i, j) \notin \cup_{k=1}^{p} J_k \times J_k$, and $B_k = B_{J_k J_k}$ for $1 \leq k \leq p$.

Theorem 7.4.4 *A classic DAD problem $(A, \mathbb{1}, \mathbb{1})$ has a solution if and only if there exist $n \times n$ permutation matrices P and Q such that the matrix $B = QAP$ has a positive diagonal and B is the direct sum of irreducible matrices B_k where $1 \leq k \leq p$.*

Proof First note that if there exist permutation matrices P and Q such that the matrix B satisfies the assertions of the theorem, it follows from Theorem 7.3.4 that the DAD problem $(QAP, \mathbb{1}, \mathbb{1})$ has a solution. Using Lemma 7.4.1 we see that $(A, \mathbb{1}, \mathbb{1})$ has a solution.

To prove the converse, assume that $(A, \mathbb{1}, \mathbb{1})$ has a solution. By Theorem 7.2.5 $(A, \mathbb{1}, \mathbb{1})$ satisfies the compatibility condition. Now using Lemma 7.4.3 we know there exists a permutation τ on $\{1, \dots, n\}$ such that AP_τ has a positive diagonal. Let $C = AP_\tau$ and note that it follows from Lemma 7.4.1 that $(C, \mathbb{1}, \mathbb{1})$ has a solution, as $(A, \mathbb{1}, \mathbb{1})$ has one. Thus, we can apply Theorem 7.3.4 to $(C, \mathbb{1}, \mathbb{1})$ and find a partition $\cup_{k=1}^{p} J_k$ of $\{1, \dots, n\}$ such that $C_{J_k J_k}$ is irreducible for all $1 \leq k \leq p$, and $c_{ij} = 0$ for all $(i, j) \notin \cup_{k=1}^{p} J_k \times J_k$.

To complete the proof define, for $1 \leq k \leq p$, $m_k = |J_k|$, and let

$$r_1 = m_1, \quad r_2 = m_1 + m_2, \quad \dots \quad , \quad r_p = \sum_{k=1}^{p} m_k.$$

Furthermore let $J_1' = \{1, \dots, r_1\}$ and $J_k' = \{r_{k-1} + 1, \dots, r_k\}$ for $2 \le k \le p$. Let σ be a permutation on $\{1, \dots, n\}$ such that $\sigma(J_k') = J_k$ for $1 \le k \le p$ and σ preserves the ordering on each J_k'. Define $B = P_\sigma^T C P_\sigma$. Clearly B has a positive diagonal, as C has one, and B is the direct sum of irreducible matrices $B_k = B_{J_k' J_k'}$ where $1 \le k \le p$. □

Theorem 7.4.4 is due to Sinkhorn and Knopp [206] and to Brualdi, Parter, and Schneider [39]. Usually it is stated in a slightly different, though equivalent, form. Recall (see [148]) that two $n \times n$ matrices are called *permutation equivalent* if there exist $n \times n$ permutation matrices P and Q such that $B = QAP$. An $n \times n$ matrix, with $n \ge 2$, is called *partly decomposable* if there exist non-empty subsets I and J of $\{1, \dots, n\}$ such that $|I| + |J| = n$ and $A_{IJ} = 0$. Equivalently, A is partly decomposable if and only if A is permutation equivalent to a matrix B of the form

$$B = \left[\begin{array}{c|c} X & Y \\ \hline 0 & Z \end{array}\right], \tag{7.17}$$

where X is $r \times r$ and Z is $(n - r) \times (n - r)$ for some $0 < r < n$. A 1×1 matrix $A = (a_{11})$ is called a *partly decomposable matrix* if $a_{11} = 0$. A matrix A is called *fully indecomposable* if it is not partly decomposable.

Obviously, if A and B are permutation equivalent, then A is fully indecomposable if and only if B is fully indecomposable. Moreover, if A is nonnegative and fully indecomposable, then A is irreducible, as there does not exist a permutation matrix P such that the matrix $B = P^T AP$ satisfies (7.17). The following result also appears in [148, theorem 4.4].

Lemma 7.4.5 *If A is a nonnegative $n \times n$ matrix, then A is fully indecomposable if and only if A is permutation equivalent to an irreducible matrix with a positive diagonal.*

Proof If A is fully indecomposable, it follows from the Frobenius–König Theorem 7.4.2 that there exists a permutation τ on $\{1, \dots, n\}$ such that $\prod_{i=1}^n a_{i\tau(i)} > 0$. Thus, AP_τ has a positive diagonal. Moreover, since AP_τ is fully indecomposable and nonnegative, AP_τ is irreducible.

Conversely, suppose there exist permutation matrices P and Q such that $B = QAP$ is irreducible and B has a positive diagonal. Note that if B is fully indecomposable, then A is fully indecomposable. Assume for the sake of contradiction that B is partly decomposable. Then there exist $I, J \subseteq \{1, \dots, n\}$ non-empty such that $|I| + |J| = n$ and $B_{IJ} = 0$. Remark that as B has a positive diagonal, $I \cap J$ is empty, and hence $J \subseteq I^c$. Using

$|I| + |J| = n$, we deduce that $J = I^c$. Thus, $B_{II^c} = 0$, which is impossible, as B is irreducible. □

Using Lemma 7.4.5, Theorem 7.4.4 can now be rephrased as follows.

Theorem 7.4.6 *A classic DAD problem* $(A, \mathbb{1}, \mathbb{1})$ *has a solution if and only if A is permutation equivalent to a matrix B with a positive diagonal and B is the direct sum of fully indecomposable matrices* B_k *where* $1 \leq k \leq p$.

7.5 Scaling to row stochastic matrices

In this section we will discuss the following problem.

Problem 7.5.1 *Given a nonnegative* $n \times n$ *matrix A, when does there exist a positive diagonal matrix E such that each row of EAE sums to 1?*

Obviously Problem 7.5.1 has a solution if and only if there exists x in $\mathrm{int}(\mathbb{R}^n_+)$ such that

$$\sum_{j=1}^{n} x_i a_{ij} x_j = 1 \quad \text{for all } 1 \leq i \leq n. \tag{7.18}$$

This equation corresponds to Equation (7.4) where $p = 1$. As before the fixed-point approach can be applied, but in this case we only find a partial solution to Problem 7.5.1.

Lemma 7.5.2 *Let* $T_{\mathrm{row}} \colon \mathrm{int}(\mathbb{R}^n_+) \to \mathrm{int}(\mathbb{R}^n_+)$ *be defined by*

$$T_{\mathrm{row}}(x) = R(Ax) \quad \text{for } x \in \mathrm{int}(\mathbb{R}^n_+),$$

where $R\colon y \mapsto y^{-1}$. *Then Problem 7.5.1 has a solution if and only if* T_{row} *has an eigenvector in* $\mathrm{int}(\mathbb{R}^n_+)$.

Proof If T_{row} has an eigenvector $u \in \mathrm{int}(\mathbb{R}^n_+)$ with $T_{\mathrm{row}}(u) = \lambda u$ for some $\lambda > 0$, then $v = \sqrt{\lambda} u$ satisfies

$$T_{\mathrm{row}}(v) = R(\sqrt{\lambda} A u) = \sqrt{\lambda^{-1}} R(Au) = \sqrt{\lambda} u = v.$$

So, $Av = v^{-1}$ and hence

$$\sum_{j=1}^{n} v_i a_{ij} v_j = 1 \quad \text{for all } 1 \leq i \leq n,$$

which shows that $E = \mathrm{diag}(v)$ is a solution of Problem 7.5.1 for A.

Conversely, if $E = \mathrm{diag}(x)$ with $x \in \mathrm{int}(\mathbb{R}^n_+)$ is a solution of Problem 7.5.1, then $Ax = x^{-1}$, so that $T_{\mathrm{row}}(x) = R(Ax) = x$. □

Using Lemma 7.5.2 the following partial result for Problem 7.5.1 can be proved.

Theorem 7.5.3 *If $A = (a_{ij})$ is a nonnegative $n \times n$ matrix with $a_{ii} > 0$ for $1 \leq i \leq n$, then there exists a positive diagonal matrix E such that each row of EAE sums to 1.*

Proof Let $\varepsilon > 0$ and define $A^\varepsilon = (a^\varepsilon_{ij})$ by

$$a^\varepsilon_{ij} = \left[\begin{array}{ll} a_{ij} & \text{if } a_{ij} > 0 \\ \varepsilon & \text{otherwise.} \end{array} \right.$$

It follows from the Birkhoff–Hopf Theorem A.4.1 and Theorem A.6.2 that $A^\varepsilon\colon \operatorname{int}(\mathbb{R}^n_+) \to \operatorname{int}(\mathbb{R}^n_+)$ is a Lipschitz contraction under Hilbert's metric. As R is an isometry under Hilbert's metric on $\operatorname{int}(\mathbb{R}^n_+)$, we find that the normalized map $\tau\colon \Delta^\circ_{n-1} \to \Delta^\circ_{n-1}$ given by

$$\tau(x) = \frac{R(A^\varepsilon x)}{\sum_{i=1}^n (R(A^\varepsilon x))_i} \quad \text{for } x \in \Delta^\circ_{n-1}$$

is a Lipschitz contraction on the open simplex Δ°_{n-1}. By Banach's contraction Theorem 3.2.1, τ has a unique fixed point in Δ°_{n-1}. Thus, $R \circ A^\varepsilon$ has a unique eigenvector $x^\varepsilon \in \operatorname{int}(\mathbb{R}^n_+)$ with

$$R(A^\varepsilon x^\varepsilon) = x^\varepsilon \tag{7.19}$$

by Lemma 7.5.2.

It follows from (7.19) that

$$x^\varepsilon_i = \Big(\sum_{j=1}^n a^\varepsilon_{ij} x^\varepsilon_j\Big)^{-1} \leq (a^\varepsilon_{ii} x^\varepsilon_i)^{-1},$$

so that

$$x^\varepsilon_i \leq \sqrt{a_{ii}^{-1}} \quad \text{for all } 1 \leq i \leq n. \tag{7.20}$$

It follows from (7.20), and the fact that $a_{ii} > 0$ for all i, that

$$\|x^\varepsilon\|_\infty \leq M_1 < \infty, \tag{7.21}$$

where $M_1 = \max_{1 \leq i \leq n} \sqrt{a_{ii}^{-1}}$.

Now assume that $0 < \varepsilon \leq 1$. Then

$$1 = \sum_{j=1}^n x^\varepsilon_i a^\varepsilon_{ij} x^\varepsilon_j \leq x^\varepsilon_i M_1 \Big(\max_{1 \leq i \leq n} \sum_{j=1}^n a^1_{ij}\Big) = x^\varepsilon_i M_2 \tag{7.22}$$

for some constant $M_2 > 0$, and hence

$$x_i^{\varepsilon} \geq M_2^{-1} \quad \text{for all } 1 \leq i \leq n. \tag{7.23}$$

It follows from (7.21) and (7.23) that there exist a sequence $(\varepsilon_k)_k$ of positive reals and $x \in \operatorname{int}(\mathbb{R}^n_+)$ such that $\varepsilon_k \to 0$ and

$$\lim_{k\to\infty} x^{\varepsilon_k} = x.$$

By taking limits in (7.22) we see that

$$1 = \sum_{j=1}^{n} x_i a_{ij} x_j \quad \text{for } 1 \leq j \leq n,$$

which shows that $E = \operatorname{diag}(x)$ is a solution of the scaling problem for A. □

Theorem 7.5.3 was proved by Brualdi, Parter, and Schneider [39, corollary 7.7] by a different method. It was also noted in [39, remark 8.3] that Problem 7.5.1 may have a solution even if $a_{ii} = 0$ for some indices i. Consider, for example, the 2×2 matrix

$$\begin{bmatrix} 0 & b \\ c & d \end{bmatrix},$$

where $b > 0$, $c > 0$, and $d > 0$. It is easy to show that there exists a positive diagonal matrix $E = \operatorname{diag}(x_1, x_2)$ such that each row of EAE sums to 1 if and only if $b > c$. In fact, if $b > c$, an explicit solution is given by

$$x_1 = \frac{1}{\sqrt{b}}\sqrt{\frac{d}{b-c}} \quad \text{and} \quad x_2 = \frac{1}{\sqrt{b}}\sqrt{\frac{b-c}{d}}.$$

It is curious that apparently no complete solution to Problem 7.5.1 is known for general square nonnegative matrices. We state without proof a result which indicates that such a solution is likely to be complicated.

Theorem 7.5.4 *If A is an $n \times n$ nonnegative scrambling matrix such that $a_{11} = 0$, $a_{ii} > 0$ for $2 \leq i \leq n$, and $a_{i1} > 0$ for $2 \leq i \leq n$, then there exists a positive diagonal matrix E such that each row of EAE sums to 1 if and only if*

$$\sum_{i=2}^{n} \frac{a_{1i}}{a_{i1}} > 1. \tag{7.24}$$

Moreover, if (7.24) holds, the solution E is unique.

Note that if in Theorem 7.5.4 one also has that $a_{1i} > 0$ for $2 \leq i \leq n$, then the matrix A is automatically scrambling. However, if $n > 2$, A may be scrambling even if $a_{1i} = 0$ for some indices i.

8

Dynamics of subhomogeneous maps

In this chapter we discuss the long-term behavior of the iterates of continuous order-preserving subhomogeneous maps on cones. By using results from Chapter 4, it will be shown that if the cone is polyhedral, then each norm-bounded orbit of a continuous order-preserving subhomogeneous maps converges to a periodic orbit. Moreover, there exists an a-priori upper bound for the periods of periodic points in terms of the number of facets of the polyhedral cone. In the case of the standard positive cone, a complete description of the set of possible periods of periodic points will be given. We will also discuss Denjoy–Wolff-type theorems for order-preserving homogeneous maps on general cones, which provide information about the behavior of orbits in the absence of an eigenvector in the interior of the cone.

8.1 Iterations on polyhedral cones

In this section we start the analysis of the iterative behavior of order-preserving subhomogeneous maps on polyhedral cones. Among other things it will be shown that each norm-bounded orbit converges to a periodic orbit.

Recall from Lemma 2.1.7 that every order-preserving subhomogeneous map $f: K \to K$ on a closed cone $K \subseteq V$ is non-expansive under Thompson's metric on each part of K, so

$$d_T(f(x), f(y)) \leq d_T(x, y) \quad \text{for all } x \sim_K y.$$

Moreover, if P is a part of a polyhedral cone, then (P, d_T) can be isometrically embedded into $(\mathbb{R}^m, \|\cdot\|_\infty)$ by Lemma 2.2.2, where $m = |I(P)|$ and $I(P) = \{i : \psi_i(x) > 0 \text{ for some } x \in P\}$. (Here $\psi_1, \ldots, \psi_N$ are facet-defining functionals of K.) Thus we can use the results from Chapter 4 to analyze the dynamics of order-preserving subhomogeneous maps on polyhedral cones. In particular, we have the following result.

Theorem 8.1.1 *Let P be a part of a polyhedral cone $K \subseteq V$ and let $m = |I(P)|$. If $f\colon P \to P$ is non-expansive under d_T and there exists $z \in P$ such that $\mathcal{O}(z)$ has a compact closure in P, then*

$$|\omega(x)| \leq \max_{1\leq k\leq m} 2^k \binom{m}{k}$$

for all $x \in P$.

Proof Let $\Psi\colon (P, d_T) \to (\mathrm{int}(\mathbb{R}^m_+), d_T)$ be the isometry from Lemma 2.2.2. Further let $L\colon (\mathrm{int}(\mathbb{R}^m_+), d_T) \to (\mathbb{R}^m, \|\cdot\|_\infty)$ be the coordinatewise log function, and recall that L is an isometry by Proposition 2.2.1. Put $X = L(\Psi(P)) \subseteq \mathbb{R}^m$ and let $g\colon X \to X$ be such that $L \circ \Psi \circ f = g \circ L \circ \Psi$. The map g is sup-norm non-expansive. Moreover, the orbit of $L(\Psi(z)) \in X$ under g has a compact closure in X. It therefore follows from Theorem 4.1.5 that

$$|\omega(x; f)| = |\omega(L(\Psi(x)); g)| \leq \max_{1\leq k\leq m} 2^k \binom{m}{k}$$

for all $x \in P$ □

In Theorem 8.1.1 a difficulty arises when $\mathcal{O}(z)$ is not contained in a single part, or if $\mathcal{O}(z)$ is contained in a part but its closure is not. In that case we cannot rely on non-expansiveness only. To proceed we need to make essential use of the order-preserving property of the map.

Recall that $\trianglelefteq$ is a partial ordering on the set of parts $\mathcal{P}(K)$ given by $P \trianglelefteq Q$ if there exists $x \in P$ and $y \in Q$ such that y dominates x.

Lemma 8.1.2 *If $K \subseteq V$ is a closed cone and $f\colon K \to K$ is non-expansive map under d_T on each part of K, then the following assertions hold:*

(i) f maps parts into parts, i.e., $f([x]) \subseteq [f(x)]$ for all $x \in K$.
(ii) The quotient map $\mathcal{F}\colon \mathcal{P}(K) \to \mathcal{P}(K)$ given by

$$\mathcal{F}(P) = [f(x)] \quad \text{for } x \in P \tag{8.1}$$

is well defined.
(iii) If, in addition, f is order-preserving, then $\mathcal{F}$ preserves the partial ordering $\trianglelefteq$ on $\mathcal{P}(K)$.

Proof To prove the first assertion we note that if $y \in f([x])$, then there exists $z \in [x]$ such that $f(z) = y$. Since $z \sim_K x$, we get that $d_T(f(x), y) = d_T(f(x), f(z)) \leq d_T(x, z) < \infty$. This implies that $f(x) \sim_K y$ and therefore

$f([x]) \subseteq [f(x)]$. We also see that the quotient map $\mathcal{F}: \mathcal{P}(K) \to \mathcal{P}(K)$ given by (8.1) is well defined.

Now suppose, in addition, that f is order-preserving. Assume that $P, Q \in \mathcal{P}(K)$ and $P \trianglelefteq Q$. There exist $x \in P$, $y \in Q$, and $\beta \geq 1$ such that $x \leq \beta y$. Since f is order-preserving and non-expansive under d_T, we know that f is subhomogeneous by Lemma 2.1.7. This implies that $\beta^{-1} f(x) \leq f(\beta^{-1} x) \leq f(y)$. Thus, $f(y)$ dominates $f(x)$, so that $[f(x)] \trianglelefteq [f(y)]$. From this we conclude that $\mathcal{F}(P) \trianglelefteq \mathcal{F}(Q)$. □

By using the quotient map $\mathcal{F}$ and the fact that each polyhedral cone has only finitely many parts, we now prove the following lemma.

Lemma 8.1.3 *If $K \subseteq V$ is a polyhedral cone and $f: K \to K$ is non-expansive under d_T on each part of K, then there exists $m \in \mathbb{N}$ such that $f^{2m}(x) \sim_K f^m(x)$ for all $x \in K$.*

Proof Let $\mathcal{F}: \mathcal{P}(K) \to \mathcal{P}(K)$ be the quotient map defined in (8.1). As K is polyhedral, $|\mathcal{P}(K)| \leq 2^N$, where N is the number of facets of K by Lemma 1.2.3. This implies that each $P \in \mathcal{P}(K)$ is eventually periodic under $\mathcal{F}$, i.e., there exist integers $r \geq 0$ and $p \geq 1$ such that $F^r(P) = F^{r+kp}(P)$ for all $k \in \mathbb{N}$. By the pigeonhole principle, we can take $r + p \leq 2^N$. Put $m = \mathrm{lcm}(1, \ldots, 2^N)$, so that $r \leq m$ and p divides m. Then $\mathcal{F}^m(P) = \mathcal{F}^{2m}(P)$ for all $P \in \mathcal{P}(K)$. By taking $P = [x]$ we get that $[f^m(x)] = \mathcal{F}^m(P) = \mathcal{F}^{2m}(P) = [f^{2m}(x)]$, so that $f^m(x) \sim_K f^{2m}(x)$. □

The following lemma shows that if the orbit $\mathcal{O}(x)$ of a continuous order-preserving subhomogeneous map is norm-bounded and contained in a part P of the polyhedral cone K, then there exists a part $Q \trianglelefteq P$ such that $\omega(x) \subseteq Q$.

Lemma 8.1.4 *If $f: K \to K$ is a continuous order-preserving subhomogeneous map on a polyhedral cone K and $\mathcal{O}(x)$ is a norm bounded orbit in a part P of K, then there exists $Q \trianglelefteq P$ such that $\omega(x) \subseteq Q$.*

Proof Suppose that $P \in \mathcal{P}(K)$ is such that $\mathcal{O}(x) \subseteq P$. If $P = \{0\}$, then $\mathcal{O}(x) = \{0\}$, and we are done. On the other hand, if $P \neq \{0\}$, then $I(P)$ is non-empty. We claim that there exists $c \geq 1$ such that $y \leq cx$ for all $y \in \mathcal{O}(x)$. Indeed, as $\mathcal{O}(x) \subseteq P$, we get that $I_y = I_x = I(P)$ for all $y \in \mathcal{O}(x)$. This implies that $\psi_i(y) > 0$, if and only if $i \in I(P)$, where $\psi_1, \ldots, \psi_N$ denote the facet-defining functionals of K. Put $c \geq 0$ equal to

$$c = \sup\{\psi_i(y)/\psi_i(x) \colon y \in \mathcal{O}(x) \text{ and } i \in I(P)\}.$$

Remark that $c < \infty$, as $\psi_i(x) > 0$ for all $i \in I(P)$ and $\mathcal{O}(x)$ is bounded. Moreover, $c \geq 1$, since $x \in \mathcal{O}(x)$ and $I(P) \neq \emptyset$. Now consider $y \in \mathcal{O}(x)$. By definition of c we know that $\psi_i(y - cx) \leq 0$ for all $i \in I(P)$. From this we deduce that $y \leq cx$, as $\psi_i(y) = \psi_i(x) = 0$ for all $i \notin I(P)$.

Using the claim, the proof is completed in the following manner. First note that $\{y \in K : y \leq cx\}$ is a closed set containing $\mathcal{O}(x)$, and hence it also contains $\omega(x)$. As f is order-preserving and subhomogeneous, we get that $f^k(y) \leq f^k(cx) \leq cf^k(x)$ for all $y \in \omega(x)$ and $k \geq 1$. As f is continuous and $\mathcal{O}(x)$ is bounded, f maps $\omega(x)$ onto itself by Lemma 3.1.2. Therefore $y \leq cf^k(x)$ for all $y \in \omega(x)$ and $k \geq 1$. As $\{z \in K : c^{-1}y \leq z\}$ is closed, we deduce that $y \leq cz$ for all $y, z \in \omega(x)$. This implies that $y \sim_K z$ for all $y, z \in \omega(x)$. Moreover, if we take $y \in \omega(x)$ and put $Q = [y]$, then Q satisfies $Q \trianglelefteq P = [x]$, as $y \leq cx$. □

To show that every norm-bounded orbit of a continuous order-preserving subhomogeneous map on a polyhedral cone converges to a periodic orbit, two more lemmas are needed.

Lemma 8.1.5 *Let P be a part of a polyhedral cone $K \subseteq V$ and $C \subseteq P$ be compact. If $f : C \to C$ is surjective and non-expansive under d_T, then each $x \in C$ is a periodic point of f.*

Proof We first note that as C is a compact subset of a part P of K, (C, d_T) is a compact metric space. Since f is non-expansive under d_T and maps C onto itself, it follows from Lemma 3.1.4 that f is an isometry. Let $x \in C$ and recall that $\omega(x)$ is finite by Theorem 8.1.1. As $\mathcal{O}(x) \subseteq C$, it has a compact closure, so that it follows from Lemma 3.1.3 that there exists a periodic point $\xi \in C$ with period $p = |\omega(x)|$ such that $\lim_{k\to\infty} f^{kp}(x) = \xi$ and $\omega(x) = \mathcal{O}(\xi)$. Since f is an isometry,

$$d_T(f^p(x), x) = d_T(f^{(k+1)p}(x), f^{kp}(x))$$

for all $k \geq 1$. Now remark that the right-hand side converges to 0 as $k \to \infty$, and hence $f^p(x) = x$. □

The following lemma is of a technical nature, but useful for our purposes.

Lemma 8.1.6 *If $f : K \to K$ is an order-preserving subhomogeneous map on a polyhedral cone $K \subseteq V$ and $x \in K$, then for each $y, \xi \in \omega(x)$ with ξ a periodic point of f, there exists an integer $\tau \geq 1$ such that $f^\tau(\xi) \leq y$. Moreover, if there exists a periodic point $\xi \in \omega(x)$, then $\mathcal{O}(\xi)$ is the only periodic orbit in $\omega(x)$.*

Proof Suppose that $\xi \in \omega(x)$ is a periodic point of f with period p. Then there exists a subsequence $(k_i)_i$ such that $f^{k_i}(x) \to \xi$ as $i \to \infty$. By taking a further subsequence we may assume that there exists $0 \leq \sigma < p$ such that

$k_i \equiv \sigma \bmod p$ for all $i \geq 1$. Let $0 < \lambda < 1$. As K is polyhedral, it satisfies condition G at ξ by Lemma 5.1.4. Therefore $\lambda\xi \leq f^{k_i}(x)$ for all i sufficiently large. Now suppose that $y \in \omega(x)$ and let $(m_i)_i$ be such that $f^{m_i}(x) \to y$ as $i \to \infty$. Again we may assume that there exists $0 \leq \tau < p$ such that $m_i \geq k_i$ and $m_i - k_i \equiv \tau \bmod p$ for all $i \geq 1$. Thus, for sufficiently large i we have that

$$f^{m_i}(x) = f^{m_i-k_i}(f^{k_i}(x)) \geq f^{m_i-k_i}(\lambda\xi) \geq \lambda f^{m_i-k_i}(\xi) = \lambda f^{\tau}(\xi).$$

By letting $i \to \infty$, we deduce that $\lambda f^{\tau}(\xi) \leq y$. Now let $\lambda \to 1^-$, to conclude that $f^{\tau}(\xi) \leq y$, which shows the first statement.

To prove the second statement we let $\xi, \eta \in \omega(x)$ be two periodic points of f with periods p and q, respectively. By the first assertion there exist $0 \leq \tau_1 < p$ and $0 \leq \tau_2 < q$ such that $f^{\tau_1}(\xi) \leq \eta$ and $f^{\tau_2}(\eta) \leq \xi$. Since f is order-preserving, we get that $f^{\tau_1+\tau_2}(\xi) \leq f^{\tau_2}(\eta) \leq \xi$. But $\mathcal{O}(\xi)$ is an anti-chain in $(K, \leq)$, and hence $f^{\tau_1+\tau_2}(\xi) = f^{\tau_2}(\eta) = \xi$. As ξ and η are periodic points of f, we conclude that $\mathcal{O}(\xi) = \mathcal{O}(\eta)$. □

Equipped with these two lemmas, we now prove the following theorem.

Theorem 8.1.7 *If $f: K \to K$ is a continuous order-preserving subhomogeneous map on a polyhedral cone $K \subseteq V$ and $x \in K$ has a norm-bounded orbit, then there exists a periodic point $\xi \in K$ of f with period p such that $\lim_{k\to\infty} f^{kp}(x) = \xi$.*

Proof We know that f is non-expansive under d_T on each part of K. Thus, it follows from Lemma 8.1.3 that there exists $m \in \mathbb{N}$ such that $f^m(x) \sim_K f^{2m}(x)$ for all $x \in K$. Put $g: K \to K$ equal to f^m, so $g(Q) \subseteq g^2(Q)$ for each $Q \in \mathcal{P}(K)$ and $\mathcal{O}(g(x); g) \subseteq P$ for some $P \in \mathcal{P}(K)$. Note that $\mathcal{O}(g(x); g)$ is bounded in norm, as it is a subset of $\mathcal{O}(x)$. Moreover, it is easy to verify that

$$\omega(x; f) = \omega(g(x); f) = \bigcup_{j=0}^{m-1} f^j(\omega(g(x); g)),$$

as f is continuous on K. Thus, by Lemma 3.1.3, it suffices to show that $\omega(g(x); g)$ is finite.

It follows from Lemma 8.1.2 that there exists a part $Q \trianglelefteq P$ such that $\omega(g(x); g) \subseteq Q$. Since $\omega(g(x); g)$ is a bounded closed subset of Q, it is compact with respect to d_T, and $g(\omega(g(x); g)) = \omega(g(x); g)$ by Lemma 3.1.2. Therefore it follows from Lemma 8.1.5 that every $y \in \omega(g(x); g)$ is a periodic point of g. We can now apply Lemma 8.1.6 to conclude that $\omega(g(x); g)$ is a periodic orbit, and hence it is finite. □

Thus every norm-bounded orbit of a continuous order-preserving subhomogeneous map on a polyhedral cone converges to a periodic orbit. In the next section the results from Chapter 4 will be used to derive an a-priori upper bound for the periods of periodic points in terms of the number of facets of the polyhedral cone.

8.2 Periodic orbits in polyhedral cones

The following result, which is a consequence of Theorem 4.3.10, provides an upper bound for the periods of periodic points in the interior of the standard positive cone.

Theorem 8.2.1 *If $f : \mathbb{R}^n_+ \to \mathbb{R}^n_+$ is an order-preserving subhomogeneous map and $x \in \mathrm{int}(\mathbb{R}^n_+)$ is a periodic point of f with period p, then*

$$p \leq \binom{n}{\lfloor n/2 \rfloor}.$$

Proof Let $x \in \mathrm{int}(\mathbb{R}^n_+)$ be a periodic point of $f : \mathbb{R}^n_+ \to \mathbb{R}^n_+$ with period p. As f is order-preserving and subhomogeneous, f maps parts into parts by Lemma 8.1.2. Thus, $[x] = \mathrm{int}(\mathbb{R}^n_+)$ is a periodic point of the quotient map $\mathcal{F}: \mathcal{P}(\mathbb{R}^n_+) \to \mathcal{P}(\mathbb{R}^n_+)$ given by (8.1). Recall that $\mathcal{F}$ preserves the partial ordering $\trianglelefteq$ and hence $\mathcal{O}([x]; \mathcal{F})$ is an anti-chain in $(\mathcal{P}(\mathbb{R}^n_+), \trianglelefteq)$. As $\mathrm{int}(\mathbb{R}^n_+)$ is a maximal element in $(\mathcal{P}(\mathbb{R}^n_+), \trianglelefteq)$, we find that $\mathrm{int}(\mathbb{R}^n_+)$ is a fixed point of $\mathcal{F}$ and hence $f(\mathrm{int}(\mathbb{R}^n_+)) \subseteq \mathrm{int}(\mathbb{R}^n_+)$.

Let $g : (\mathbb{R}^n, \|\cdot\|_\infty) \to (\mathbb{R}^n, \|\cdot\|_\infty)$ be the log-exp transform of f, so $g = L \circ f \circ E$. Remark that $g : \mathbb{R}^n \to \mathbb{R}^n$ is a sub-topical map. Moreover $L(x)$ is a periodic point of g with period p. It now follows from Theorem 4.3.10 that p does not exceed $\binom{n}{\lfloor n/2 \rfloor}$. □

By using the log-exp transform, we can turn any sub-topical map $g : \mathbb{R}^n \to \mathbb{R}^n$ into an order-preserving subhomogeneous map on $\mathrm{int}(\mathbb{R}^n_+)$. We also know from Theorem 4.3.10 that for each n there exists a sub-topical map on $\mathbb{R}^n$ that has a periodic point with period $\binom{n}{\lfloor n/2 \rfloor}$. So, we see that the upper bound in Theorem 8.2.1 is sharp.

It turns out that Theorem 8.2.1 can be generalized to polyhedral cones. To do this we need the following lemma.

Lemma 8.2.2 *Let $K \subseteq V$ be a closed cone and let P be a part of K. If $\mathcal{A}$ is an anti-chain in the partially ordered set $(P, \leq)$ and $f : P \to P$ is order-preserving and subhomogeneous, then $M(f(y)/f(x)) \leq M(y/x)$ for all $x, y \in \mathcal{A}$. Moreover, if $f(\mathcal{A}) \subseteq \mathcal{A}$ and each $x \in \mathcal{A}$ is a periodic point of f, then $M(f(y)/f(x)) = M(y/x)$ for all $x, y \in \mathcal{A}$.*

Proof Clearly the assertions are true if $x = y$. So suppose that $x, y \in \mathcal{A}$ and $x \neq y$. Let $\lambda = M(y/x)$ and note that $\lambda < \infty$, as $x, y \in P$. Furthermore, $y \leq \lambda x$ and $\lambda > 1$, since $\mathcal{A}$ is an anti-chain. From this it follows that $\lambda^{-1} f(y) \leq f(\lambda^{-1} y) \leq f(x)$, and hence $M(f(y)/f(x)) \leq \lambda = M(y/x)$.

To prove the second assertion, we remark that if $x, y \in \mathcal{A}$ are periodic points of f with periods p and q, respectively, then for $r = \mathrm{lcm}(p, q)$ we have that

$$M(y/x) = M(f^r(y)/f^r(x)) \leq M(f(y)/f(x)) \leq M(y/x).$$

This completes the proof. □

Proposition 8.2.3 *Let $K \subseteq V$ be a polyhedral cone and P be a part of K. If $f\colon K \to K$ is an order-preserving subhomogeneous map with $f(P) \subseteq P$, then the period of each periodic point $x \in P$ does not exceed $\binom{m}{\lfloor m/2 \rfloor}$, where $m = |I(P)|$.*

Proof Let $\psi_1, \ldots, \psi_N$ be the facet-defining functionals of K. By Lemma 1.2.3 there exist $\psi_{i_1}, \ldots, \psi_{i_m}$ such that

$$P = \{x \in K : \psi_{i_j}(x) > 0 \text{ for all } 1 \leq j \leq m \text{ and } \psi_i(x) = 0 \text{ otherwise}\},$$

as $m = |I(P)|$. Define $\Psi\colon (P, d_T) \to (\mathrm{int}(\mathbb{R}^m_+), d_T)$ by $\Psi_j(x) = \psi_{i_j}(x)$ for all $x \in P$ and $1 \leq j \leq m$. Then $M(\Psi(x)/\Psi(y)) = M(x/y)$ for all $x, y \in P$. Moreover, Ψ is injective. Indeed, $\Psi(x) = \Psi(y)$ implies $\Psi(x - y) = \Psi(y - x) = 0$. But we also know that $\psi_i(x - y) = \psi_i(y - x) = 0$ for all $i \notin I(P)$. Thus, $x - y \in K$ and $y - x \in K$, which shows that $x = y$.

Let $\mathcal{O}(\xi)$ be a periodic orbit of f in P with period p. It follows from Lemma 8.2.2 that

$$M(f^k(x)/f^k(y)) = M(x/y) \quad \text{for all } x, y \in \mathcal{O}(\xi) \text{ and } 0 \leq k < p. \tag{8.2}$$

Now consider the coordinatewise log function $L\colon \mathrm{int}(\mathbb{R}^m_+) \to \mathbb{R}^m$ and recall that

$$\mathrm{t}(L(\Psi(x)) - L(\Psi(y)) = \log M(\Psi(x)/\Psi(y)) = \log M(x/y) \tag{8.3}$$

for all $x, y \in P$. Put $\mathcal{A} = L(\Psi(\mathcal{O}(\xi)))$ and note that, as $\mathcal{O}(\xi)$ is an anti-chain in $(K, \leq_K)$, $\mathcal{A}$ is an anti-chain in $\mathbb{R}^m$ with respect to $\mathbb{R}^m_+$. For each $u, v \in \mathcal{O}(\xi)$ there exists an integer $k \geq 0$ such that $f^k(u) = v$. It therefore follows from (8.2) and (8.3) that $\mathcal{A}$ has a transitive cyclic group of top isometries. Using the fact that L and Ψ are injective, we conclude from Theorem 4.3.7 that

$$p = |\mathcal{O}(\xi)| = |\mathcal{A}| \leq \binom{m}{\lfloor m/2 \rfloor}.$$

□

We know that the estimate in Proposition 8.2.3 is sharp for $P = \mathrm{int}(\mathbb{R}^n_+)$.

Problem 8.2.4 *Does there exist for each solid polyhedral cone K, with N facets, an order-preserving subhomogeneous map $f: \mathrm{int}(K) \to \mathrm{int}(K)$ that has a periodic point with period $\binom{N}{\lfloor N/2 \rfloor}$?*

In general, periodic orbits of order-preserving subhomogeneous maps need not lie inside a single part of the cone. For instance, the map $f: \mathbb{R}^3_+ \to \mathbb{R}^3_+$ given by

$$f\begin{pmatrix} x_1 \\ x_2 \\ x_3 \end{pmatrix} = \begin{pmatrix} (3x_1 \wedge x_2) \vee (3x_2 \wedge x_3) \\ (3x_1 \wedge x_3) \vee (3x_3 \wedge x_2) \\ (3x_2 \wedge x_1) \vee (3x_3 \wedge x_1) \end{pmatrix} \quad \text{for } x \in \mathbb{R}^3_+$$

has $x = (1, 2, 0)$ as a period 6 point, whose orbit visits each facet of $\mathbb{R}^3_+$ twice. In fact, the reader can verify that if $x = (a, b, 0) \in \mathbb{R}^3_+$, where $a > 0$ and $b > 0$ are such that $3a \geq b$, $3b \geq a$, and $a \neq b$, then x is a periodic point of f with period 6. Furthermore, $\mathbb{1}$ is the unique normalized eigenvector of f in the interior of $\mathbb{R}^3_+$, and the map g given by

$$g(x) = \frac{f(x)}{(1/3)\sum_i f_i(x)} \quad \text{for } x \in \mathrm{int}(\mathbb{R}^3_+)$$

satisfies $\lim_{k\to\infty} g^k(x) = \mathbb{1}$ for all $x \in \mathrm{int}(\mathbb{R}^3_+)$.

For general periodic orbits, slightly more sophisticated arguments are needed to obtain a suitable upper bound for their periods. Recall that if $f: K \to K$ is order-preserving and subhomogeneous, then it maps parts into parts by Lemma 8.1.2. This places severe constraints on the periodic orbits. For instance, in Figure 8.1 two period 6 orbits in the boundary of $\mathbb{R}^3_+$ are depicted, but only the right-hand one may occur as a periodic orbit of an order-preserving subhomogeneous map.

In particular, we see that if $\mathcal{O}(\xi)$ is a periodic orbit with period p of an order-preserving subhomogeneous map $f: K \to K$, then $[\xi]$ is a periodic point of the quotient map $\mathcal{F}: \mathcal{P}(K) \to \mathcal{P}(K)$ given in (8.1), with period say $q_1 \geq 1$.

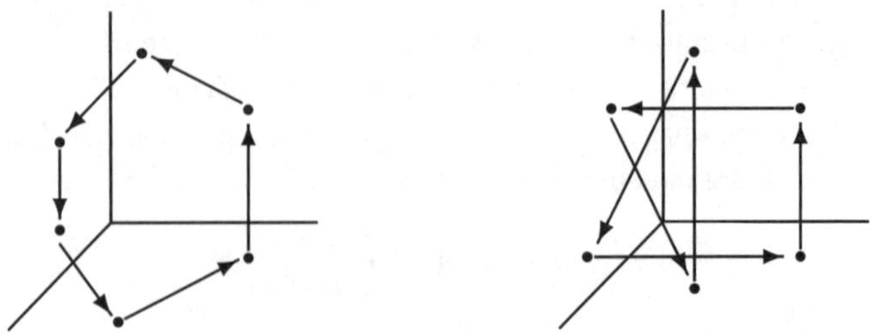

Figure 8.1 Periodic orbits with period 6 in $\partial\mathbb{R}^3_+$.

Moreover, $\mathcal{O}(\xi)$ visits each part in $\mathcal{O}([\xi]; \mathcal{F})$ exactly q_2 times for some integer $q_2 \geq 1$. Thus, the period p can be written as $p = q_1 q_2$. This observation leads to the following result.

Theorem 8.2.5 *Let $K \subseteq V$ be a polyhedral cone with N facets. If $f : K \to K$ is an order-preserving and subhomogeneous map and $x \in K$ is a periodic point of f with period p, then there exist integers $q_1, q_2 \geq 1$ such that $p = q_1 q_2$,*

$$1 \leq q_1 \leq \binom{N}{\max\{m, \lfloor N/2 \rfloor\}}, \quad \text{and} \quad 1 \leq q_2 \leq \binom{m}{\lfloor m/2 \rfloor},$$

where $m = \min\{|I_{f^k(x)}| : 0 \leq k < p\}$.

Proof Let $\mathcal{F}: \mathcal{P}(K) \to \mathcal{P}(K)$ be the quotient map defined in Lemma 8.1.2. Then $\mathcal{F}$ preserves the partial ordering $\trianglelefteq$ on $\mathcal{P}(K)$. The map $\pi : (\mathcal{P}(K), \trianglelefteq) \to (2^{\{1,\ldots,N\}}, \subseteq)$ given by $\pi(P) = I(P)$ is order-preserving and injective. Moreover, π^{-1} is also order-preserving on $\pi(\mathcal{P}(K))$. Let $G: \pi(\mathcal{P}(K)) \to \pi(\mathcal{P}(K))$ be defined by $G = \pi \circ \mathcal{F} \circ \pi^{-1}$. Then G preserves the partial ordering $\subseteq$ on $\pi(\mathcal{P}(K))$.

Suppose that $x \in K$ is a periodic point of f with period p, and write $m = \min\{|I_{f^k(x)}| : 0 \leq k < p\}$. Take $z \in \mathcal{O}(x)$ such that $|I_z| = m$ and put $Q = [z]$. We remark that $\mathcal{F}^k(Q) = [f^k(z)]$ for all $k \geq 1$, so that $\mathcal{F}^p(Q) = Q$. Let q_1 denote the period of Q under $\mathcal{F}$. Obviously q_1 divides p and $\pi(Q)$ is a periodic point of G with period q_1. Let $\mathcal{A} = \mathcal{O}(\pi(Q); G)$. As G is order-preserving, $\mathcal{A}$ is an anti-chain in $(2^{\{1,\ldots,N\}}, \subseteq)$.

We now use a LYM inequality to derive an upper bound for the cardinality of $\mathcal{A}$. As before, a maximal chain $\mathcal{C}$ in $(2^{\{1,\ldots,N\}}, \subseteq)$ is a sequence of $N + 1$ subsets $A_0 \subseteq A_1 \subseteq \ldots \subseteq A_N$ of $\{1, \ldots, N\}$ such that $|A_i| = i$ for all i. There are exactly $N!$ maximal chains in $(2^{\{1,\ldots,N\}}, \subseteq)$. If $A \subseteq \{1, \ldots, N\}$ and $|A| = s$, then there are exactly $s!(N - s)!$ maximal chains $\mathcal{C}$ that contain A. As $\mathcal{A}$ is an anti-chain, each maximal chain contains at most one element of $\mathcal{A}$. Since $m = \min\{|I_{f^k(x)}| : 0 \leq k < p\}$, we know that $|A| \geq m$ for all $A \in \mathcal{A}$. Indeed, $G^k(\pi(Q)) = \pi(\mathcal{F}^k(Q)) = \pi([f^k(z)]) = I_{f^k(z)}$ for all $k \geq 1$. We also see that $\mathcal{A} = \{I_{f^k(z)} : 0 \leq k < q_1\}$.

Now for $m \leq s \leq N$ let ν_s be the number of elements of $\mathcal{A}$ with cardinality s. As each maximal chain contains at most one element of $\mathcal{A}$ and each $A \in \mathcal{A}$ with $|A| = s$ is contained in $s!(N - s)!$ maximal chains, we find that

$$\sum_{s=m}^{N} \nu_s s!(N - s)! \leq N!,$$

so that

$$\sum_{s=m}^{N} \nu_s \binom{N}{s}^{-1} \leq 1.$$

Put $M(m) = \max_{m\leq s\leq N} \binom{N}{s}$ and note that $M(m) = \binom{N}{\lfloor N/2\rfloor}$ if $0 \leq m \leq \lfloor N/2\rfloor$, and $M(m) = \binom{N}{m}$ otherwise. From this we deduce that

$$q_1 = |\mathcal{A}| = \sum_{s=m}^{N} \nu_s \leq M(m) = \binom{N}{\max\{m, \lfloor N/2\rfloor\}}.$$

For q_1, z, and Q as above, we have that $f^{q_1}(Q) \subseteq [f^{q_1}(z)] = Q$. Now let $q_2 = p/q_1$ and remark that z is a periodic point of f^{q_1} with period q_2. Moreover, $\mathcal{O}(z; f^{q_1}) \subseteq Q$ and hence it follows from Proposition 8.2.3 that $1 \leq q_2 \leq \binom{m}{\lfloor m/2\rfloor}$, where $m = |I(Q)|$. □

As a corollary of Theorem 8.2.5 we obtain the following upper bound for the periods of periodic points.

Corollary 8.2.6 *If $f: K \to K$ is an order-preserving subhomogeneous map on a polyhedral cone $K \subseteq V$ with N facets, then the period of each periodic point of f does not exceed*

$$\beta_N = \max_{1\leq m\leq N} \binom{N}{\max\{m, \lfloor N/2\rfloor\}}\binom{m}{\lfloor m/2\rfloor} = \frac{N!}{\lfloor \frac{N}{3}\rfloor!\lfloor \frac{N+1}{3}\rfloor!\lfloor \frac{N+2}{3}\rfloor!}.$$

Proof By Theorem 8.2.5 it remains to be shown that the second equality holds. First we remark that for each $N \in \mathbb{N}$ we have that

$$\max_{1\leq m\leq N} \binom{N}{\max\{m, \lfloor N/2\rfloor\}}\binom{m}{\lfloor m/2\rfloor} = \max_{\lfloor N/2\rfloor\leq m\leq N} \binom{N}{m}\binom{m}{\lfloor m/2\rfloor}. \tag{8.4}$$

Furthermore,

$$\binom{N}{m}\binom{m}{\lfloor m/2\rfloor} = \frac{N!}{q!r!s!}, \tag{8.5}$$

where $q = N - m$, $r = \lfloor m/2\rfloor$, and $s = m - \lfloor m/2\rfloor$. This implies that

$$\max_{\lfloor N/2\rfloor\leq m\leq N} \binom{N}{m}\binom{m}{\lfloor m/2\rfloor} \leq \max_{q+r+s=N} \frac{N!}{q!r!s!}. \tag{8.6}$$

Let us consider the right-hand side of (8.6). Suppose that $0 \leq q^* \leq r^* \leq s^* \leq N$ are such that the maximum is attained. In that case $s^* \leq q^* + 1$. Otherwise $q^*!s^*! = q^*!(s^* - 1)!s^* > (q^* + 1)!(s^* - 1)!$, so that

$$\frac{N!}{q^*!r^*!s^*!} < \frac{N!}{(q^* + 1)!r^*!(s^* - 1)!},$$

which is a contradiction. As $q^* + r^* + s^* = N$ and $q^* \leq s^* \leq r^* \leq q^* + 1$, we have that $3q^* \leq N \leq 3q^* + 2$. This implies that $q^* = \lfloor N/3 \rfloor$. We also have that $r^* + s^* + q^* + 1 = N + 1$ and $r^* \leq s^* \leq q^* + 1 \leq r^* + 1$, so that $3r^* \leq N+1 \leq 3r^* + 2$ and therefore $r^* = \lfloor (N+1)/3 \rfloor$. In the same way one can show that $s^* = \lfloor (N+2)/3 \rfloor$. Thus, we find that

$$\max_{q+r+s=N} \frac{N!}{q!r!s!} = \frac{N!}{\lfloor \frac{N}{3} \rfloor! \lfloor \frac{N+1}{3} \rfloor! \lfloor \frac{N+2}{3} \rfloor!}. \tag{8.7}$$

Now put $m = \lfloor (N+1)/3 \rfloor + \lfloor (N+2)/3 \rfloor$ in (8.5). This gives $q = N - m = \lfloor N/3 \rfloor$, $r = \lfloor m/2 \rfloor = \lfloor (N+1)/3 \rfloor$, as $2\lfloor (N+1)/3 \rfloor \leq m \leq 2\lfloor (N+1)/3 \rfloor + 1$, and $s = m - \lfloor m/2 \rfloor = \lfloor (N+2)/3 \rfloor$. Thus, we find that

$$\max_{1 \leq m \leq N} \binom{N}{\max\{m, \lfloor N/2 \rfloor\}} \binom{m}{\lfloor m/2 \rfloor} \geq \frac{N!}{\lfloor \frac{N}{3} \rfloor! \lfloor \frac{N+1}{3} \rfloor! \lfloor \frac{N+2}{3} \rfloor!}.$$

Combining this inequality with (8.4), (8.6), and (8.7) gives the desired equality. □

Using Stirling's formula it is easy to verify that

$$\beta_N \sim \frac{3^{N+1}\sqrt{3}}{2\pi N}.$$

For a polyhedral cone K we denote by $\Gamma(K)$ the finite set consisting of those integers $p \geq 1$ for which there exists a continuous order-preserving subhomogeneous map $f : K \to K$ that has a periodic point with period p. If K has N facets, then by Theorem 8.2.5 we have the following inclusion:

$$\Gamma(K) \subseteq B(N), \tag{8.8}$$

where $B(N)$ consists of those integers $p \geq 1$ such that $p = q_1 q_2$, where the integers $q_1, q_2 \geq 1$ satisfy $1 \leq q_1 \leq \binom{N}{m}$ and $1 \leq q_2 \leq \binom{m}{\lfloor m/2 \rfloor}$ for some integer $1 \leq m \leq N$. In particular, we see that $\Gamma(\mathbb{R}^3_+) \subseteq \{1, 2, 3, 4, 6\}$; so, 5 is not in $\Gamma(\mathbb{R}^3_+)$. In view of the inclusion (8.8), it is natural to ask if we can completely determine $\Gamma(K)$ for a given polyhedral cone K. Particularly interesting is, of course, $\mathbb{R}^n_+$, for which there exists the following result.

Theorem 8.2.7 *For each $n \in \mathbb{N}$, $\Gamma(\mathbb{R}^n_+) = B(n)$.*

Proof By (8.8) it suffices to show that $B(n) \subseteq \Gamma(\mathbb{R}^n_+)$. So let $p \in B(n)$. Then there exist integers $q_1, q_2 \geq 1$ with $p = q_1 q_2$ such that $1 \leq q_1 \leq \binom{n}{m}$ and $1 \leq q_2 \leq \binom{m}{\lfloor m/2 \rfloor}$ for some integer $1 \leq m \leq N$. We first construct a continuous order-preserving homogeneous map $g : \mathbb{R}^m_+ \to \mathbb{R}^m_+$ that has a periodic orbit in $\mathrm{int}(\mathbb{R}^m_+)$ with period q_2.

Let $w^1, \ldots, w^{q_2}$ be distinct vectors in $\{1,2\}^m$ with exactly $\lfloor m/2 \rfloor$ coordinates equal to 2, and put $w^{q_2+1} = w^1$. (Note that there are $\binom{m}{\lfloor m/2 \rfloor}$ such vectors in $\{1,2\}^m$ and hence we can do this for all values of q_2.) It is straightforward to verify that $\{w^1, \ldots, w^{q_2}\}$ is a periodic orbit of the continuous order-preserving homogeneous map $g\colon \mathbb{R}^m_+ \to \mathbb{R}^m_+$ given by

$$g(z)_i = \bigvee_{k\colon w_i^{k+1}=2} \Big(\bigwedge_{j\colon w_j^k=2} z_j \Big)$$

for $1 \le i \le m$ and $z \in \mathbb{R}^m_+$. (In fact, the map g is nothing but the log-exp transform of the topical map defined in Proposition 4.3.1, where $X = \{L(w^1), \ldots, L(w^{q_2})\} \subseteq \mathbb{R}^m$ and h is a cyclic permutation of the elements of X.)

Subsequently we copy the q_2 points into q_1 distinct parts of the cone $\mathbb{R}^n_+$ by inserting zeros at appropriate coordinates, and construct a continuous order-preserving homogeneous map $f\colon \mathbb{R}^n_+ \to \mathbb{R}^n_+$ that cycles through the q_1q_2 points. Select q_1 distinct subsets $I_a \subseteq \{1, \ldots, n\}$, each of size m. We know from Lemma 1.2.3 that each I_a corresponds to a part $P_a \in \mathcal{P}(\mathbb{R}^n_+)$ by $P_a = \{x \in \mathbb{R}^n_+ \colon x_i > 0 \text{ if and only if } i \in I_a\}$. To formally insert the zeros it is convenient to introduce the following function. For $1 \le a \le q_1$ and $1 \le k \le m$ let $\eta(a,k)$ be the index of the k-th non-zero coordinate of the elements of P_a. For each $1 \le a \le q_1$ define a copying function $\eta_a\colon \mathbb{R}^m_+ \to \mathbb{R}^n_+$ by

$$\eta_a(z)_i = \left[\begin{array}{ll} z_k & \text{if } \eta(a,k) = i, \\ 0 & \text{otherwise.} \end{array}\right.$$

Note that by applying the functions η_a for all a, we obtain $p = q_1q_2$ distinct points in $\mathbb{R}^n_+$. We also use the following notation. For $1 \le a \le q_1$ and $1 \le b \le q_2$ let $y^{a,b} \in \{0,1,2\}^n$ be given by $y^{a,b} = \eta_a(w^b)$. Thus, $y^{a,b}$ is the copy of w^b in part P_a.

To complete the proof we construct a continuous order-preserving homogeneous map $f\colon \mathbb{R}^n_+ \to \mathbb{R}^n_+$ such that

$$f(y^{a,b}) = \left[\begin{array}{ll} y^{a+1,b} & \text{if } 1 \le a < q_1, \\ y^{1,b+1} & \text{if } a = q_1, \end{array}\right.$$

for $1 \le a \le q_1$ and $1 \le b \le q_2$. (Here the indices a and b are counted modulo q_1 and modulo q_2, respectively.) Clearly $y^{a,b}$ is a periodic point of f with period $p = q_1q_2$ in that case. To give a formula for the map f we need one more piece of notation. For $x \in \mathbb{R}^n_+$ we write $x_{|I_a}$ to denote the vector in $\mathbb{R}^m_+$ with coordinates $(x_{|I_a})_j = x_{\eta(a,j)}$. Now let $f\colon \mathbb{R}^n_+ \to \mathbb{R}^n_+$ be given by

Table 8.1 *The elements of* $\Gamma(\mathbb{R}^n_+)$ *for* $1 \le n \le 7$.

n	Elements of $\Gamma(\mathbb{R}^n_+)$
1	1
2	1, 2
3	1, 2, 3, 4, 6
4	1, 2, 3, 4, 5, 6, 8, 9, 10, 12
5	1, 2, 3, 4, 5, 6, 7, 8, 9, 10, 12, 14, 15, 16, 18, 20, 21, 24, 25, 27, 30
6	1, 2, 3, 4, 5, 6, 7, 8, 9, 10, 11, 12, 13, 14, 15, 16, 17, 18, 19, 20, 21, 22, 24, 25, 26, 27, 28, 30, 32, 33, 34, 35, 36, 38, 39, 40, 42, 44, 45, 48, 50, 51, 52, 54, 55, 56, 57, 60, 65, 66, 70, 72, 75, 78, 84, 90
7	1, 2, 3, 4, 5, 6, 7, 8, 9, 10, 11, 12, 13, 14, 15, 16, 17, 18, 19, 20, 21, 22 23, 24, 25, 26,27, 28, 29, 30, 31, 32, 33, 34, 35, 36, 38, 39, 40, 42, 44, 45, 46, 48, 49, 50, 51, 52, 54, 55, 56, 57, 58, 60, 62, 63, 64, 65, 66, 68, 69, 70, 72, 75, 76, 77, 78, 80, 81, 84, 85, 87, 88, 90, 91, 92, 93, 95, 96, 98, 99, 100, 102, 104, 105, 108, 110, 112, 114, 115, 116, 117, 119, 120, 124, 125, 126, 128, 130, 132, 133,135, 136, 138, 140, 144, 145, 147, 150, 152, 153, 155, 156, 160, 162, 165, 168, 170, 171, 174, 175, 180, 186, 189,190, 192, 198, 200, 204, 210

$$f(x)_i = \Big[\bigvee_{1\le k<q_1} \Big(\bigwedge_{j\in I_k} 2x_j\Big) \wedge \eta_{k+1}(x_{|I_k})_i \Big] \vee \Big[\Big(\bigwedge_{j\in I_{q_1}} 2x_j\Big) \wedge \eta_1(g(x_{|I_{q_1}}))_i \Big]$$

for $1 \le i \le n$ and $x \in \mathbb{R}^n_+$. Clearly f is a continuous order-preserving homogeneous map. We also remark that $\bigwedge_{j\in I_k} 2y_j^{a,b} = 0$ if $k \neq a$, and $\bigwedge_{j\in I_k} 2y^{a,b} = 2$ if $k = a$. Since $y^{a,b}_{|I_a} = w^b$, we find for $1 \le a < q_1$ that

$$f(y^{a,b})_i = \Big(\bigwedge_{j\in I_a} 2y_j^{a,b}\Big) \wedge \eta_{a+1}(w^b)_i = 2 \wedge y_i^{a+1,b} = y_i^{a+1,b}.$$

If $a = q_1$, we get that

$$f(y^{q-1,b})_i = \Big(\bigwedge_{j\in I_{q_1}} 2y_j^{q_1,b}\Big) \wedge \eta_1(g(w^b))_i = 2 \wedge \eta_1(w^{b+1})_i = y_i^{1,b+1},$$

which completes the proof. □

Knowing the equality $\Gamma(\mathbb{R}^n_+) = B(n)$, it is now easy to compute $\Gamma(\mathbb{R}^n_+)$; see Table 8.1. Furthermore, the proof of Theorem 8.2.7 provides for each $p \in \Gamma(\mathbb{R}^n_+)$ an explicit construction of a continuous homogeneous order-preserving map $f\colon \mathbb{R}^n_+ \to \mathbb{R}^n_+$, which has a periodic point with period p. For $n = 3$ and $p = 6$ we have already seen such an example on page 190. If $n = 3$ and $p = 4$ the construction the proof of Theorem 8.2.7 yields the map $f\colon \mathbb{R}^3_+ \to \mathbb{R}^3_+$ given by

$$f\begin{pmatrix} x_1 \\ x_2 \\ x_3 \end{pmatrix} = \begin{pmatrix} (2x_2 \wedge x_1) \vee (2x_1 \wedge x_3) \\ 2x_3 \wedge x_1 \\ 2x_1 \wedge x_2 \end{pmatrix} \quad \text{for } x \in \mathbb{R}^3_+,$$

which has $x = (1, 2, 0)$ as a period 4 point in $\partial\mathbb{R}^3_+$. For $n = 4$ and $p = 12$ there are two possible choices: (1) $q_1 = 4$ and $q_2 = 3$ and (2) $q_1 = 6$ and $q_2 = 2$. In case (1) we obtain the map $f\colon \mathbb{R}^4_+ \to \mathbb{R}^4_+$ given by

$$f\begin{pmatrix} x_1 \\ x_2 \\ x_3 \\ x_4 \end{pmatrix} = \begin{pmatrix} (2x_2 \wedge 2x_3 \wedge x_1) \vee (2x_2 \wedge 2x_4 \wedge x_1) \vee (2x_2 \wedge 2x_4 \wedge x_3) \\ (2x_1 \wedge 2x_3 \wedge x_2) \vee (2x_3 \wedge 2x_4 \wedge x_1) \vee (2x_2 \wedge 2x_3 \wedge x_4) \\ (2x_1 \wedge 2x_4 \wedge x_2) \vee (2x_1 \wedge 2x_4 \wedge x_3) \vee (2x_3 \wedge 2x_4 \wedge x_2) \\ (2x_1 \wedge 2x_2 \wedge x_3) \vee (2x_1 \wedge 2x_2 \wedge x_4) \vee (2x_1 \wedge 2x_3 \wedge x_4) \end{pmatrix}$$

for $x \in \mathbb{R}^4_+$, which has $x = (1, 1, 2, 0)$ as a period 12 point. Here we took $w^1 = (1, 1, 2)$, $w^2 = (1, 2, 1)$, $w^3 = (2, 1, 1)$, and the map $g\colon (z_1, z_2, z_3) \mapsto (z_2, z_3, z_1)$.

By using the fact that the standard positive cone is order isomorphic to any simplicial cone, we also know following.

Corollary 8.2.8 *If $K \subseteq V$ is a simplicial cone, then* $\Gamma(K) = B(n)$.

8.3 Denjoy–Wolff theorems for cone maps

In Section 3.4 a Denjoy–Wolff type theorem for fixed-point free non-expansive maps on proper metric spaces whose geometry resembles that of a hyperbolic space was discussed. The goal of the remaining two sections is to use the ideas from Section 3.4 to analyze the dynamics of homogeneous order-preserving maps on solid closed cones that have no eigenvector in the interior of the cone. We emphasize that throughout the remainder of this chapter we will use the following slightly different, but convenient, definition of ω-limit sets. Given a continuous map $f\colon X \to X$, with $X \subseteq V$ and $x \in X$, the *ω-limit set* of x is given by

$$\omega(x) = \bigcap_{m \geq 0} \mathrm{cl}(\{f^k(x)\colon k \geq m\}),$$

where the closure is with respect to the **norm topology on** V.

Suppose that $K \subseteq V$ is a solid closed cone and $f\colon \mathrm{int}(K) \to \mathrm{int}(K)$ is a continuous homogeneous order-preserving map which has no eigenvector in $\mathrm{int}(K)$. In that case we can take $\varphi \in \mathrm{int}(K^*)$ and consider the scaled map $g\colon \Sigma^\circ \to \Sigma^\circ$ given by

$$g(x) = \frac{f(x)}{\varphi(f(x))} \quad \text{for all } x \in \Sigma^\circ, \tag{8.9}$$

where $\Sigma^\circ = \{x \in \mathrm{int}(K) \colon \varphi(x) = 1\}$. So, g has no fixed point in Σ°, as f has no eigenvector in $\mathrm{int}(K)$, and g is non-expansive under Hilbert's metric by Lemma 2.1.6. It thus follows from Corollary 3.2.5 that $\Omega_f \cap \Sigma^\circ = \emptyset$, and hence the ω-limit set of each $x \in \Sigma^\circ$ under g is contained in the boundary of $\Sigma = \{x \in K \colon \varphi(x) = 1\}$. Recall that if the proper metric space (Σ°, d_H) satisfies Axioms I and II, then Theorem 3.4.6 implies that there exists a unique ray in ∂K that attracts all orbits of f.

As Hilbert's metric coincides with the cross-ratio metric κ on Σ° (see Theorem 2.1.2), we know from the results in Section 3.3 that (Σ°, d_H) satisfies Axiom I. But we also know from Example 3.4.3 that (Σ°, d_H) does not necessarily satisfy Axiom II. It was, however, shown by Beardon [18, 19] that if K is strictly convex, then (Σ°, d_H) also satisfies Axiom II. To prove this we follow Karlsson and Noskov [101].

Lemma 8.3.1 *Let $A \subseteq V$ be an open bounded convex set. If $(x_k)_k$ and $(y_k)_k$ are two convergent sequences in A with limits x and y in ∂A, respectively, and the straight-line segment $[x, y] \not\subseteq \partial A$, then the end-points x'_k and y'_k of the chord through x_k and y_k converge to x and y, respectively.*

Proof The situation is depicted in Figure 8.2 below. As $x \neq y$ and $x_k \to x$ and $y_k \to y$ as $k \to \infty$, each limit point of $(x'_k)_k$ and $(y'_k)_k$ must belong to the straight line $\ell_{x,y}$ through x and y. More to the point, each limit point of $(x'_k)_k$ belongs to the half-line ℓ_x^+ of $\ell_{x,z}$ emanating from x but not containing y. Similarly each limit point of $(y'_k)_k$ belongs to the half-line ℓ_y^+ emanating from y and not containing x. Remark that $[x, y] \cap \partial A = \{x, y\}$, as $[x, y] \not\subseteq \partial A$ and A is convex. Hence $\ell_x^+ \cap \partial A = \{x\}$ and $\ell_y^+ \cap \partial A = \{y\}$, which completes the proof. □

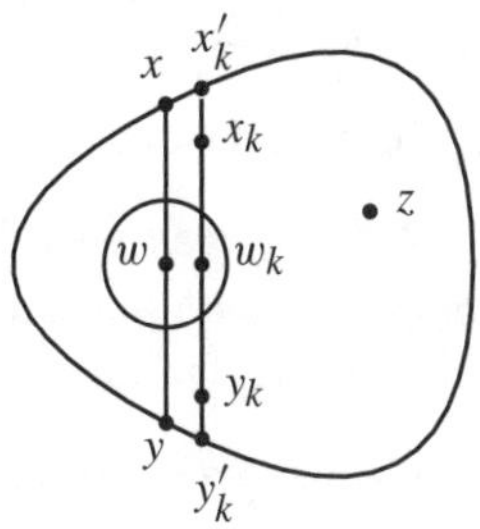

Figure 8.2 Proof of Proposition 8.3.2.

A key ingredient in our subsequent analysis will be the following proposition, which uses the so-called Gromov product. Given a metric space (X, d) and $x, y \in X$, recall that the *Gromov product* with respect to a reference point $z \in X$ is defined by

$$(x \mid y)_z = \frac{1}{2}\big(d(x, z) + d(y, z) - d(x, y)\big).$$

Proposition 8.3.2 *Let $A \subseteq V$ be an open bounded convex set equipped with the cross-ratio metric κ. If $(x_k)_k$ and $(y_k)_k$ are convergent sequences in A with limits x and y in ∂A, respectively, and the straight-line segment $[x, y] \nsubseteq \partial A$, then for each $z \in A$ there exists $C \geq 0$ such that*

$$\limsup_{k\to\infty} (x_k \mid y_k)_z \leq C.$$

Proof As $[x, y] \nsubseteq \partial A$ there exists a closed Euclidean ball $B_\varepsilon(w)$ with radius ε around the (Euclidean) mid-point w of the chord through x and y such that $B_\varepsilon(w) \subseteq A$. Recall that the norm topology coincides with the metric topology of κ on A by Corollary 2.5.6 and Theorem 2.1.2. Therefore $B_\varepsilon(w)$ is a compact subset of (A, κ). This implies that for each $z \in A$ there exists $C \geq 0$ such that

$$\sup_{u\in B_\varepsilon(w)} \kappa(u, z) \leq C.$$

Now let w_k denote the mid-point of the chord through x_k and y_k. Denote, as in Figure 8.2, the end-points of the chord through x_k and y_k by x_k' and y_k'.

By Lemma 8.3.1 we know that $x_k' \to x$ and $y_k' \to y$ as $k \to \infty$, since $[x, y] \nsubseteq \partial A$. Thus, $w_k \in B_\varepsilon(w)$ for all k sufficiently large. From Theorem 2.6.3 we know that each straight line is a geodesic in (A, κ); so,

$$\kappa(x_k, y_k) = \kappa(x_k, w_k) + \kappa(w_k, y_k)$$

for all k sufficiently large. This implies that

$$\begin{aligned}2(x_k \mid y_k)_z &= \kappa(x_k, z) + \kappa(y_k, z) - \kappa(x_k, y_k)\\ &= \kappa(x_k, z) + \kappa(y_k, z) - \kappa(x_k, w_k) - \kappa(w_k, y_k)\\ &\leq 2\kappa(w_k, z).\end{aligned}$$

Thus, $(x_k \mid y_k)_z \leq \kappa(w_k, z) \leq C$ for all sufficiently large k. □

Proposition 8.3.3 *Let $A \subseteq V$ be an open bounded convex set. If $(x_k)_k$ and $(y_k)_k$ are convergent sequences in A with limits x and y in ∂A, respectively, and the straight-line segment $[x, y] \nsubseteq \partial A$, then for each $z \in A$ we have that*

$$\liminf_{k\to\infty} \kappa(x_k, y_k) - \max\{\kappa(x_k, z), \kappa(y_k, z)\} = \infty.$$

In particular, (A, κ) satisfies Axiom II, if the closure of A is strictly convex.

Proof It follows from Proposition 8.3.2 that $\limsup_{k\to\infty}(x_k \mid y_k)_z \leq C$. Let $\varepsilon > 0$ and note that there exists $N \geq 1$ such that

$$2(x_k \mid y_k)_z \leq 2C + \varepsilon$$

for all $k \geq N$. This implies that

$$\begin{aligned}\max\{\kappa(x_k, z), \kappa(y_k, z)\} &= \kappa(x_k, z) + \kappa(y_k, z) - \min\{\kappa(x_k, z), \kappa(y_k, z)\}\\ &\leq \kappa(x_k, y_k) + 2C + \varepsilon - \min\{\kappa(x_k, z), \kappa(y_k, z)\}\end{aligned}$$

for all $k \geq N$. Hence

$$\kappa(x_k, y_k) - \max\{\kappa(x_k, z), \kappa(y_k, z)\} \geq \min\{\kappa(x_k, z), \kappa(y_k, z)\} - 2C - \varepsilon \to \infty,$$

as $k \to \infty$, which completes the proof. □

Now that we know that (A, κ) is a proper metric space satisfying Axioms I and II, if A is strictly convex, we can combine Corollary 3.2.5, Theorem 3.4.6, and Proposition 8.3.3 to directly obtain the following result by Beardon [19] and Karlsson [99].

Theorem 8.3.4 *Let $A \subseteq V$ be an open bounded strictly convex set. If $f\colon (A, \kappa) \to (A, \kappa)$ is a fixed-point free non-expansive map, then there exists $\zeta \in \partial A$ such that*

$$\lim_{k\to\infty} f^k(x) = \zeta \quad \textit{for all } x \in A,$$

and the convergence is uniform on compact subsets of A.

Theorem 8.3.4 has the following consequence for the dynamics of homogeneous order-preserving maps on strictly convex cones, which can be viewed as a Denjoy–Wolff type theorem in nonlinear Perron–Frobenius theory.

Theorem 8.3.5 *Let $K \subseteq V$ be a solid strictly convex closed cone and let $\varphi \in \mathrm{int}(K^*)$. If $f\colon \mathrm{int}(K) \to \mathrm{int}(K)$ is a homogeneous order-preserving map and f has no eigenvector in $\mathrm{int}(K)$, then there exists $\zeta \in \partial K$ such that*

$$\lim_{k\to\infty} \frac{f^k(x)}{\varphi(f^k(x))} = \zeta \quad \textit{for all } x \in \mathrm{int}(K).$$

Moreover, if f has a continuous extension to K such that $f(x) \neq 0$ for all $x \neq 0$ in K, then $\zeta \in \partial K$ is an eigenvector of f with eigenvalue $\varphi(f(\zeta)) > 0$.

Proof Let $\varphi \in \mathrm{int}(K^*)$ and define $\Sigma^\circ = \{x \in \mathrm{int}(K)\colon \varphi(x) = 1\}$. Also write Σ to denote the norm closure of Σ°. As $K \subseteq V$ is strictly convex, Σ° is a bounded strictly convex domain in an affine hyperplane in V. Equip Σ° with Hilbert's metric d_H and consider the scaled map $g\colon \Sigma^\circ \to \Sigma^\circ$ given by

$$g(x) = \frac{f(x)}{\varphi(f(x))} \quad \text{for } x \in \Sigma^\circ.$$

So, g is a fixed-point free non-expansive map on (Σ°, d_H), as f has no eigenvector in $\mathrm{int}(K)$. By Theorem 2.1.2, the metric d_H coincides with the cross-ratio metric on Σ°, and hence it follows from Theorem 8.3.4 that there exists $\zeta \in \partial\Sigma^\circ$ such that

$$\lim_{k\to\infty} \frac{f^k(x)}{\varphi(f^k(x))} = \lim_{k\to\infty} g^k(x) = \zeta \quad \text{for all } x \in \Sigma^\circ.$$

If f has a continuous extension to K such that $f(x) \neq 0$ for all $x \neq 0$ in K, then g has a continuous extension to Σ. In that case, $g(\zeta) = \zeta$, so that $f(\zeta) = \varphi(f(\zeta))\zeta$. Thus, ζ is an eigenvector of f in ∂K with eigenvalue $\varphi(f(\zeta)) > 0$. □

With regard to the previous theorem we mention the following open problem.

Problem 8.3.6 *If in Theorem 8.3.5 f has a continuous extension to K and $\varphi(f(\zeta)) > 0$, is $\varphi(f(\zeta)) = r_K(f)$?*

We shall see in the next section that, in general, the iterates of a fixed-point free non-expansive map $f\colon (A,\kappa) \to (A,\kappa)$ need not converge to a unique point in ∂A, if (A,κ) does not satisfy Axiom II. Karlsson and Nussbaum (see [168]), however, independently made the following conjecture.

Conjecture 8.3.7 (Karlsson and Nussbaum) *If $A \subseteq V$ is an open bounded convex set and $f\colon (A,\kappa) \to (A,\kappa)$ is a fixed-point free non-expansive map, then there exists a convex set $\Lambda \subseteq \partial A$ such that $\omega(x) \subseteq \Lambda$ for all $x \in A$.*

To prove this conjecture it suffices to show that there exists a point $z \in A$ such that the convex hull of its ω-limit set, denoted $\mathrm{co}(\omega(z))$, is contained in ∂A. Indeed, we have the following lemma.

Lemma 8.3.8 *Let $A \subseteq V$ be an open bounded convex set. If $f\colon (A,\kappa) \to (A,\kappa)$ is a fixed-point free non-expansive map such that $\mathrm{co}(\omega(z)) \subseteq \partial A$ for some $z \in A$, then there exists $\Lambda \subseteq \partial A$ convex such that $\omega(x) \subseteq \Lambda$ for all $x \in A$.*

Proof Let A^- denote the norm closure of A and let $A' = \{(1,a) \in \mathbb{R}\times V : a \in A^-\}$. Define a cone $K \subseteq \mathbb{R} \times V$ by

$$K = \{\lambda, u \in \mathbb{R} \times V : u \in A' \text{ and } \lambda \geq 0\}.$$

Clearly K is a solid closed cone in $\mathbb{R} \times V$, and $\varphi \in K^*$, given by $\varphi(x) = x_1$ for $x = (x_1, v) \in \mathbb{R} \times V$, is in the interior of K^*. Let $B = \{u \in K : \varphi(u) = 1\}$ and consider $r : A^- \to B$ given by $r(a) = (1, a)$ for $a \in A^-$. The restriction of r to A is an isometry from (A, κ) onto $(B \cap \mathrm{int}(K), d_H)$ by Theorem 2.1.2.

To prove the lemma we first show the following claim: if $(u_k)_k$ and $(v_k)_k$ are (norm) convergent sequences in $B \cap \mathrm{int}(K)$ with limits u and v in ∂K, respectively, and there exists $C \geq 0$ such that $d_H(u_k, v_k) \leq C$ for all $k \geq 1$, then $u \sim_K v$.

Indeed, as $u_k, v_k \in B \cap \mathrm{int}(K)$ and $d_H(u_k, v_k) \leq C$, there exist $0 < \alpha_k \leq \beta_k$ such that $\alpha_k u_k \leq_K v_k \leq_K \beta_k u_k$ and $\beta_k/\alpha_k \leq e^C$. By applying φ to the inequalities we get that $\alpha_k \leq 1 \leq \beta_k$, so that $\alpha_k \geq e^{-C}$ and $\beta_k \leq e^C$. By taking a subsequence we may assume that $\alpha_k \to \alpha \geq e^{-C}$ and $\beta_k \to \beta \leq e^C$, as $k \to \infty$. Since K is closed, this implies that $\alpha u \leq_K v \leq_K \beta u$, so that $d_H(u, v) \leq \log(\beta/\alpha) \leq 2C < \infty$. From this it follows that $u \sim_K v$.

Recall that, by Całka's Theorem 3.1.7, $\omega(x) \subseteq \partial A^-$ for all $x \in A$, as $\omega(z) \subseteq \partial A^-$. The proof of the lemma is now completed as follows. Suppose, for the sake of contradiction, that there exists $\zeta \in \mathrm{co}(\Omega_f)$ such that $\zeta \in A$. Then there exist $z^1, \ldots, z^m \in A$, $\zeta^1, \ldots, \zeta^m \in \partial A$, and $0 < \lambda_1, \ldots, \lambda_m < 1$ with $\sum_i \lambda_i = 1$ such that $\zeta = \sum_i \lambda_i \zeta^i$ and $\zeta^i \in \omega(z^i)$ for $1 \leq i \leq m$. Let $(k^i_j)_j$ be such that $f^{k^i_j}(z^i) \to \zeta^i$, as $j \to \infty$. By taking a further subsequence we may assume that $f^{k^i_j}(z) \to \xi^i$. As

$$d_H(r(f^{k^i_j}(z^i)), r(f^{k^i_j}(z))) = \kappa(f^{k^i_j}(z^i), f^{k^i_j}(z)) \leq \kappa(z^i, z),$$

it follows from the claim that $r(\zeta^i) \sim_K r(\xi^i)$ for $1 \leq i \leq m$. This implies that

$$r(\zeta) = r(\sum_i \lambda_i \zeta^i) \sim_K r(\sum_i \lambda_i \xi^i).$$

But $\sum_i \lambda_i \xi^i \in \partial A$, as $\mathrm{co}(\omega(z)) \subseteq \partial A$, and $\zeta = \sum_i \lambda_i \zeta^i \in A$, which is a contradiction. □

By using the convexity of the horoballs in (A, κ) and Theorem 3.3.5 the following partial result from [99] for Conjecture 8.3.7 can be proved.

Corollary 8.3.9 *Let $A \subseteq V$ be an open bounded convex set. If $f : (A, \kappa) \to (A, \kappa)$ is a fixed-point free non-expansive map such that*

$$\lim_{k \to \infty} \kappa(f^k(x), x)/k > 0,$$

then there exists $\Lambda \subseteq \partial A$ convex such that $\omega(x) \subseteq \Lambda$ for all $x \in A$.

Proof Let $x \in A$ and remark that by Theorem 3.3.5 there exists a horofunction $b_x : A \to \mathbb{R}$ such that

$$\lim_{k\to\infty} b_x(f^k(x)) = -\infty.$$

Let $H(-r)$ be the horoball with radius $-r$ of b_x. Clearly $\omega(x)$ is contained in $\bigcap_{r\geq 0}\mathrm{cl}(H(-r))$. Recall from Lemma 3.3.3 that $H(-r)$ is convex for each $r \geq 0$; so, $\bigcap_{r\geq 0}\mathrm{cl}(H(-r))$ is a convex subset of ∂A. Now let $y \in A$ and remark that

$$\begin{aligned} b_x(f^k(y)) &= \lim_{j\to\infty} \kappa(f^k(y), f^{k_j}(x)) - \kappa(f^{k_j}(x), x) \\ &\leq \lim_{j\to\infty} \kappa(f^k(x), f^{k_j}(x)) - \kappa(f^{k_j}(x), x) + \kappa(f^k(x), f^k(y)) \\ &\leq b_x(f^k(x)) + \kappa(x, y). \end{aligned}$$

Therefore $b_x(f^k(y)) \to -\infty$ as $k \to \infty$, and hence $\omega(y) \subseteq \bigcap_{r\geq 0}\mathrm{cl}(H(-r))$. The corollary now follows from Lemma 8.3.8. □

There are, however, many examples of fixed-point free non-expansive maps on (A, κ) for which $\lim_{k\to\infty} \kappa(f^k(x), x)/k = 0$. Consider for instance the map $f\colon \Delta_2^\circ \to \Delta_2^\circ$ given by

$$f(x_1, x_2) = \left(\frac{x_1 + x_2}{x_1 + 2x_2}, \frac{x_2}{x_1 + 2x_2}\right).$$

Other results concerning Conjecture 8.3.7 can be obtained by using Proposition 8.3.2. In particular, there exists the following result by Karlsson and Noskov [101, theorem 5.5].

Theorem 8.3.10 *Let $A \subseteq V$ be an open bounded convex set. If $f\colon (A, \kappa) \to (A, \kappa)$ is a fixed-point free non-expansive map, then there exists $\zeta \in \partial A$ such that for each $x \in A$ and $y \in \omega(x)$ we have that $[y, \zeta] \subseteq \partial A$.*

Proof Let $z \in A$ and $a_k = \kappa(f^k(z), z)$ for all $k \geq 1$. Note that $a_k \to \infty$ as $k \to \infty$ by Corollary 3.1.8. It follows from Lemma 3.3.4 that there exists a subsequence $(a_{k_i})_i$ such that $a_k < a_{k_i}$ for all $k < k_i$. By taking a further subsequence we may assume that $f^{k_i}(z) \to \zeta \in \partial A$ as $i \to \infty$.

Suppose that $y \in \omega(x)$. By definition there exists a subsequence $(m_i)_i$ such that $m_i < k_i$ for all i and $f^{m_i}(x) \to y$ as $i \to \infty$. Let us now consider the Gromov product $(f^{k_i}(z) \mid f^{m_i}(x))_z$. Remark that

$$\begin{aligned} 2(f^{k_i}(z) \mid f^{m_i}(x))_z &= a_{k_i} + \kappa(f^{m_i}(x), z) - \kappa(f^{k_i}(z), f^{m_i}(x)) \\ &\geq a_{k_i} + \kappa(f^{m_i}(x), x) - \kappa(x, z) \\ &\quad - \kappa(f^{k_i}(z), f^{m_i}(z)) - \kappa(f^{m_i}(z), f^{m_i}(x)) \\ &\geq a_{k_i} - a_{k_i - m_i} + \kappa(f^{m_i}(x), x) - 2\kappa(x, z) \\ &\geq \kappa(f^{m_i}(x), x) - 2\kappa(x, z), \end{aligned}$$

as f is non-expansive. The right-hand side goes to infinity as $i \to \infty$, and hence it follows from Proposition 8.3.2 that $[y, \zeta] \subseteq \partial A$. □

Further evidence supporting Conjecture 8.3.7 is the following result by Nussbaum [168], which complements Corollary 8.3.9.

Theorem 8.3.11 *Let $A \subseteq V$ be an open bounded convex set. If $f\colon (A,\kappa) \to (A,\kappa)$ is a fixed-point free non-expansive map such that*

$$\lim_{k\to\infty} \kappa(f^{k+1}(z), f^k(z)) = 0$$

for some $z \in A$, then there exists $\Lambda \subseteq \partial A$ convex such that $\omega(x) \subseteq \Lambda$ for all $x \in A$.

Proof By Lemma 8.3.8 it suffices to show that $\mathrm{co}(\omega(z)) \subseteq \partial A$. Suppose by way of contradiction that there exist $\zeta^1,\ldots,\zeta^m \in \omega(z)$ and $0 < \lambda_1,\ldots,\lambda_m < 1$ with $\sum_i \lambda_i = 1$ such that $\sum_i \lambda_i\zeta^i \in A$. As $\omega(z) \subseteq \partial A$, we may assume that $m \geq 2$ and is minimal. Let $(k_j^i)_j$ be such that $f^{k_j^i}(z) \to \zeta^i$, as $j \to \infty$. Put $\zeta = \zeta^1$ and $\eta = \sum_{i=2}^m \mu_i\zeta^i$, where $\mu_i = \lambda_i/(1-\lambda_1)$ for $2 \leq i \leq m$. Let $\eta^j = \sum_{i=2}^m \mu_i f^{k_j^i}(z)$ for all $j \geq 1$. Since m is minimal, we know that $\zeta, \eta \in \partial A$ and $\lambda_1\zeta + (1-\lambda_1)\eta \in A$. This implies that $\lambda\zeta + (1-\lambda)\eta \in A$ for all $0 < \lambda < 1$, as A is convex.

Consider the function

$$b_z(f^{k_j^1}(z))(x) = \kappa(x, f^{k_j^1}(z)) - \kappa(f^{k_j^1}(z), z).$$

By taking a further subsequence we may assume that the horofunction

$$b_z(x) = \lim_{j\to\infty} \kappa(x, f^{k_j^1}(z)) - \kappa(f^{k_j^1}(z), z)$$

exists for all $x \in A$ (see Proposition 3.3.2). Note that

$$\begin{aligned} b_z(f^k(z)) &= \lim_{j\to\infty} \kappa(f^k(z), f^{k_j^1}(z)) - \kappa(f^{k_j^1}(z), z) \\ &\leq \liminf_{j\to\infty} \kappa(f^k(z), f^{k_j^1+1}(z)) + \kappa(f^{k_j^1+1}(z), f^{k_j^1}(z)) - \kappa(f^{k_j^1}(z), z) \\ &\leq \liminf_{j\to\infty} \kappa(f^{k-1}(z), f^{k_j^1}(z)) - \kappa(f^{k_j^1}(z), z) + \kappa(f^{k_j^1+1}(z), f^{k_j^1}(z)) \\ &\leq b_z(f^{k-1}(z)), \end{aligned}$$

as $\lim_{k\to\infty} \kappa(f^{k+1}(z), f^k(z)) = 0$. From this it follows that $b_z(f^k(z)) \leq 0$ for all $k \geq 0$, since $b_z(z) = 0$.

Recall that the horoball of b_z with radius 0 is convex by Lemma 3.3.3, so that $b_z(\eta^j) \leq 0$ for all $j \geq 0$. Thus, there exists $k_j^1 \geq 1$ such that

$$\kappa(\eta^j, f^{k_n^1}(z)) - \kappa(f^{k_n^1}(z), z) \leq 1 \tag{8.10}$$

for all $k_n^1 \geq k_j^1$. By construction $\eta^j \to \eta \in \partial A$ and $f^{k_j^1}(z) \to \zeta \in \partial A$ as $j \to \infty$. Moreover, $[\zeta, \eta] \not\subseteq \partial A$, so that

$$\liminf_{j \to \infty} \kappa(\eta^j, f^{k_j^1}(z)) - \max\{\kappa(f^{k_j^1}(z), z), \kappa(\eta^j, z)\} = \infty$$

by Proposition 8.3.3. This, however, contradicts (8.10), and we are done. $\square$

8.4 A Denjoy–Wolff theorem for polyhedral cones

If $A \subseteq V$ is a bounded open convex set and A is not strictly convex, then (A, κ) does not satisfy Axiom II in general. In such metric spaces horoballs can meet the boundary in more than one point. For example, if A is the open two-dimensional simplex, then the horoballs can be as in Figure 8.3 (see [100]).

Therefore we cannot expect a result like Theorem 8.3.5 to hold for general solid closed cones in V. In fact, there exist simple examples that confirm this point. To see this, recall from Section 8.1 that every norm-bounded orbit of a continuous homogeneous order-preserving map $f\colon K \to K$ on a polyhedral cone K converges to a periodic orbit. Such periodic orbits may lie in the boundary, and may attract orbits from the interior of K. For example, the linear map $f\colon \mathbb{R}^3_+ \to \mathbb{R}^3_+$ given by

$$f\begin{pmatrix} x_1 \\ x_2 \\ x_3 \end{pmatrix} = \begin{pmatrix} 0 & 1 & 0 \\ 1 & 0 & 0 \\ 0 & 0 & 1/2 \end{pmatrix}\begin{pmatrix} x_1 \\ x_2 \\ x_3 \end{pmatrix}$$

for $x = (x_1, x_2, x_3) \in \mathbb{R}^3_+$. The ω-limit set of each $x \in \mathrm{int}(\mathbb{R}^3_+)$, with $x_1 \neq x_2$, is a periodic orbit of f in $\partial\mathbb{R}^3_+$. The scaled map $g\colon \Delta_2 \to \Delta_2$ given by

$$g(x) = \frac{f(x)}{\sum_i f_i(x)},$$

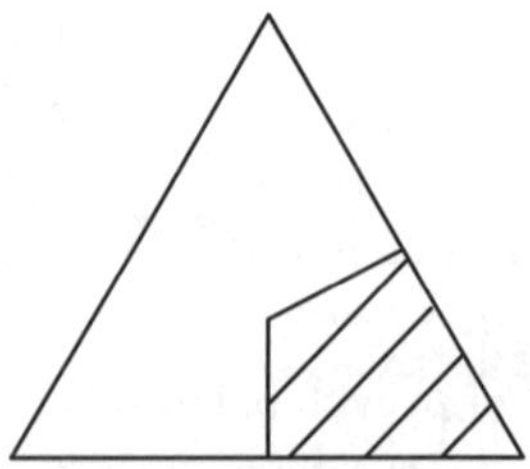

Figure 8.3 A horoball in Δ_2°.

for $x \in \Delta_2$, has the property that every orbit of x in Δ_2° converges to a periodic orbit in $\partial\Delta_2$ if $x_1 \neq x_2$. Obviously Theorem 8.3.5 fails in that case.

In this section we will discuss extensions of Theorem 8.3.5 to polyhedral cones. We begin with the following result.

Theorem 8.4.1 *Let $K \subseteq \mathbb{R}^n$ be a solid polyhedral cone, $\varphi \in \mathrm{int}(K^*)$, and $\Sigma^\circ = \{x \in \mathrm{int}(K) : \varphi(x) = 1\}$. Suppose that $f: K \to K$ is a continuous order-preserving homogeneous map with $f(\mathrm{int}(K)) \subseteq \mathrm{int}(K)$ and $r_K(f) = 1$, and let $g: \Sigma^\circ \to \Sigma^\circ$ be given by*

$$g(x) = \frac{f(x)}{\varphi(f(x))} \quad \text{for } x \in \Sigma^\circ.$$

If f has no eigenvector in $\mathrm{int}(K)$ and there exists $z \in \mathrm{int}(K)$ such that $\mathcal{O}(z; f)$ is bounded in norm, then there exists a part $P \subseteq \partial K$ of K such that $\omega(x; g) \subseteq P$ for all $x \in \mathrm{int}(K)$, and $\omega(x; g)$ is a finite set of size at most $\binom{N}{\lfloor N/2 \rfloor}$, where $N = |I(P)|$.

Proof Let $x \in \Sigma^\circ$ and note that there exists $\mu_x > 0$ such that $x \leq \mu_x z$. As f is order-preserving and homogeneous, we get that $f^k(x) \leq \mu_x f^k(z)$ for all $k \geq 0$. This implies that the orbit of x under f is bounded in norm, since $\mathcal{O}(z; f)$ is bounded in norm and K is a normal cone.

By Lemma 8.1.4 there exists a part $P_x \trianglelefteq \mathrm{int}(K)$ such that $\omega(x; f) \subseteq P_x$. By Theorem 8.1.7 and Proposition 8.2.3 it follows that $\omega(x; f)$ is a periodic orbit in P_x with period at most $\binom{N}{\lfloor N/2 \rfloor}$, where $N = |I(P_x)|$. We claim that there exists a part P of K such that $P = P_y$ for all $y \in \mathrm{int}(K)$. Indeed, if for some $v, w \in \mathrm{int}(K)$ we have that $P_v \neq P_w$, then there exists a subsequence $(k_i)_i$ such that $f^{k_i}(v) \to v^*$ and $f^{k_i}(w) \to w^*$ as $i \to \infty$, and $d_T(v^*, w^*) = \infty$. This, however, contradicts the inequality $d_T(v^*, w^*) = \lim_{i\to\infty} d_T(f^{k_i}(v), f^{k_i}(w)) \leq d_T(v, w) < \infty$.

As $r_K(f) = 1$, there exists an eigenvector $u \in \partial K$ of f with $u \neq 0$ and $f(u) = u$. For each $x \in \mathrm{int}(K)$ there exists $\delta > 0$ such that $\delta u \leq x$. This implies that $\varphi(f^k(x)) \geq \delta\varphi(u) > 0$ for all $k \geq 0$. Therefore $\omega(x; g)$ is a finite set of size at most $\binom{N}{\lfloor N/2 \rfloor}$, where $N = |I(P_x)|$. □

One might conjecture that the assertion in Theorem 8.4.1 also holds for unbounded orbits. The following example by Lins [131], however, shows that the ω-limit sets can be more complicated.

Example 8.4.2 To present the example we need a few preliminary remarks. Let $V = \{x \in \mathbb{R}^{n+1} : x_1 = 0\}$ and let $(x^k)_k$ be a sequence in V such that

$$x^{k-1} \leq x^k \quad \text{and} \quad x^{k+1} - x^k \leq x^k - x^{k-1} \quad \text{for all } k \geq 1, \tag{8.11}$$

where the partial ordering $\leq$ is induced by $\mathbb{R}^{n+1}_+$. We claim that there exists an order-preserving additively homogeneous map $g\colon \mathbb{R}^{n+1} \to \mathbb{R}^{n+1}$ such that $g(x^k) = x^{k+1}$ for all $k \geq 1$. Indeed, for each $1 \leq i \leq n+1$ there exists an increasing Lipschitz function $\gamma_i : \mathbb{R} \to \mathbb{R}$, with Lipschitz constant at most 1, such that $\gamma_i(x_i^k) = x_i^{k+1}$ for all $k \geq 1$. In fact, we can take γ_i to be piecewise linear. Now let $G\colon V \to V$ be defined by

$$G_i(x) = \gamma_i(x_i) \quad \text{for } 1 \leq i \leq n+1 \text{ and } x \in V.$$

Furthermore let $g\colon \mathbb{R}^{n+1} \to \mathbb{R}^{n+1}$ be given by

$$g(x) = G(x - x_1\mathbb{1}) + x_1\mathbb{1} \quad \text{for all } x \in \mathbb{R}^{n+1}.$$

It is straightforward to verify that g is additively homogeneous. To see that g is order-preserving with respect to $\mathbb{R}^{n+1}_+$, let $x \leq y$ in $\mathbb{R}^{n+1}$. It is convenient to distinguish two cases: $x_i - x_1 \leq y_i - y_1$ and $x_i - x_1 > y_i - y_1$. In the first case we have that $g_i(x) = \gamma_i(x_i - x_1) + x_1 \leq \gamma_i(y_i - y_1) + y_1 = g_i(y)$, since γ_i is increasing. On the other hand, if $x_i - x_1 > y_i - y_1$, then

$$0 \leq \gamma_i(x_i - x_1) - \gamma_i(y_i - y_1) \leq x_i - x_1 - y_i + y_1 \leq y_1 - x_1,$$

as γ_i is Lipschitz with constant at most 1. This implies that

$$g_i(x) = \gamma_i(x_i - x_1) + x_1 \leq \gamma_i(y_i - y_1) + y_1 = g_i(y).$$

Therefore g is order-preserving. Taking the log-exp transform of g yields a homogeneous order-preserving map $f\colon \mathrm{int}(\mathbb{R}^{n+1}_+) \to \mathrm{int}(\mathbb{R}^{n+1}_+)$; so, $f = E \circ g \circ L$.

The example will be a scaling of f, where $(x^k)_k$ is chosen appropriately. Let $h\colon \Delta_n^\circ \to \Delta_n^\circ$ be given by

$$h(x) = \frac{f(x)}{\sum_i f_i(x)} \quad \text{for } x \subset \Delta_n^\circ.$$

Let $y = \mathbb{1}/n \in \Delta_n^\circ$ and note that

$$h^k(y) = \frac{f^k(y)}{\sum_i f_i^k(y)} = \frac{f^k(\mathbb{1})}{\sum_i f_i^k(\mathbb{1})} = \frac{E(g^k(0))}{\sum_i E_i(g^k(0))}$$

for all $k \geq 1$, as f is homogeneous. Put $x^0 = 0$ and write $a^k = x^k - x^{k-1}$ to get

$$\begin{aligned} h^k(y) &= \frac{E(g^k(0))}{\sum_i E_i(g^k(0))} = \frac{E(x^k)}{\sum_i E_i(x^k)} = \frac{E(\sum_{j=1}^k a^j)}{\sum_i E_i(\sum_{j=1}^k a^j)} \\ &= \frac{\prod_{j=1}^k E(a^j)}{\sum_i \prod_{j=1}^k E_i(a^j)} \end{aligned}$$

for all $k \geq 1$, where the product is taken coordinatewise. For simplicity we write $b^j = E(a^j)$, so that

$$h^k(y) = \frac{\prod_{j=1}^k b^j}{\sum_i \prod_{j=1}^k b_i^j} \quad \text{for all } k \geq 1. \tag{8.12}$$

The properties of the sequence $(x^k)_k$ in (8.11) translate into the following properties for the sequence $(b^j)_j$:

$$b_1^j = 1,\ b^j \geq \mathbb{1},\ \text{and } b^{j+1} \leq b^j \quad \text{for all } j \geq 1.$$

The properties of the sequence $(b^j)_j$ allow us to construct maps $h\colon \Delta_n^\circ \to \Delta_n^\circ$ satisfying the following theorem due to Lins [131].

Theorem 8.4.3 *For each convex set $S \subseteq \partial\Delta_n^\circ$ there exist a map $h\colon \Delta_n^\circ \to \Delta_n^\circ$, which is non-expansive under Hilbert's metric, and a point $y \in \Delta_n^\circ$ such that $S \subseteq \omega(y; h)$.*

Proof Let $S \subseteq \partial\Delta_n^\circ$ be convex. Then there exist $1 \leq i \leq n+1$ such that $x_i = 0$ for all $x \in S$. By relabeling the coordinates we may assume that $i = 1$. We wish to chose a sequence $(b^j)_j$, with $b_1^j = 1$, $b_i^j > 1$ for $2 \leq i \leq n+1$, and $b^{j+1} \leq b^j$ for all $j \geq 1$, such that $(h^k(y))_k$ in (8.12) accumulates at a countable dense subset S' of S. Here $h\colon \Delta_n^\circ \to \Delta_n^\circ$ and $y = \mathbb{1}/n$ are as in Example 8.4.2.

Suppose that we have chosen vectors $b^1, \ldots, b^N$. Remark that if ξ and η are two points in $\mathrm{int}(\mathbb{R}_+^{n+1})$ such that $\xi_1 = \eta_1$ and $\xi < \eta$ for $2 \leq i \leq n-1$, then there exist $m \geq 1$ and vectors $b^{N+1}, \ldots, b^{N+m}$, with $b_1^{N+j} = 1, b_i^{N+j} > 1$ for $2 \leq i \leq n+1$, and $b^{N+j} \leq b^{N+j-1}$ for $1 \leq j \leq N$, such that the entrywise product of ξ with $\prod_{j=1}^m b^{N+j}$ is equal to η. Indeed, for sufficiently large m, $(\eta_i/\xi_i)^{1/m} \leq b_i^N$ for all i, so we can take

$$b_i^{N+j} = (\eta_i/\xi_i)^{1/m} \quad \text{for all } 1 \leq j \leq m \text{ and } 1 \leq i \leq n+1.$$

Now suppose that

$$\xi = h^N(y) = \frac{\prod_{j=1}^N b^j}{\sum_i \prod_{j=1}^N b_i^j} \in \Delta_n^\circ$$

and $s' \in S'$. Then there exists $\eta \geq \xi$, with $\eta_1 = \xi_1$ and $\eta_i > \xi_i$ for $2 \leq i \leq n-1$, such that $\eta/\sum_i \eta_i$ is arbitrarily close to s'. By taking $b^{N+1}, \ldots, b^{N+m}$ as in the previous paragraph we find that

$$h^{N+m}(y) = \frac{\prod_{j=1}^{N+m} b^j}{\sum_i \prod_{j=1}^{N+m} b_i^j} = \frac{\eta}{\sum_i \eta_i},$$

which is close to s'. By repeating this process we can secure that $(h^k(y))_k$ accumulates at a countable dense subset $S' = \{s^1, s^2, \ldots\}$ of S. Indeed, we can first get to within distance 1 of s^1, then to within distance 1/2 of s^2, then to within distance 1/2 of s^1, subsequently to within distance 1/3 of s^3, etc. As $\omega(y; h)$ is closed, $S = \mathrm{cl}(S') \subseteq \omega(y; h)$, and we are done. □

Lins [131, lemma 3.2] has proved that the ω-limit sets of the maps constructed in Theorem 8.4.3 are connected sets. We have also seen examples where each ω-limit set is finite. In this regard we mention the following open problem posed by Lins in [131].

Problem 8.4.4 *If $f\colon (\Delta_n^\circ, \kappa) \to (\Delta_n^\circ, \kappa)$ is a fixed-point free non-expansive map, is it true that each $\omega(x; f) \subseteq \partial\Delta_n^\circ$ is either finite or connected?*

Of course Theorem 8.4.3 does not contradict Conjecture 8.3.7. In fact, there exists the following result for polyhedral Hilbert's metric spaces by Lins [131].

Theorem 8.4.5 *If $A \subseteq \mathbb{R}^n$ is an open bounded polyhedron and $f\colon (A, \kappa) \to (A, \kappa)$ is a fixed-point free non-expansive map, then there exists a convex set $\Lambda \subseteq \partial A$ such that $\omega(x; f) \subseteq \Lambda$ for all $x \in A$.*

To prove this theorem we follow Lins [131] and make a digression into the analysis of horofunctions and unbounded orbits of fixed-point free non-expansive maps on finite-dimensional normed spaces. Subsequently we shall apply the isometric embedding of (A, κ) into $(\mathbb{R}^m, \|\cdot\|_\infty)$ for polyhedral domains A given in Proposition 2.2.3. We begin with a useful inequality.

Lemma 8.4.6 *Let $(X, \|\cdot\|)$ be a Banach space and $y \in X$ be such that $\|y\| = 1$. If $z \in X$ and there exist $0 < r < R$ such that $\|z\| \leq R$ and $\|z - ry\| > R - (3/4)r$, then $\|z - Ry\| > R/4$.*

Proof Let B_X denote the unit ball of X and let B_{X^*} denote the unit ball of the dual space of X^*, so $B_{X^*} = \{\varphi \in X^*\colon \varphi(x) \leq 1 \text{ for all } x \in B_X\}$. By the Hahn–Banach theorem there exists $\varphi \in B_{X^*}$ such that $\varphi(z - ry) = \|z - ry\| > R - (3/4)r$. This implies that $r\varphi(y) < \varphi(z) - R + (3/4)r$. As $\varphi(z) \leq \|z\| \leq R$, we get that $\varphi(ry) < (3/4)r$ and hence $\varphi(y) < 3/4$. This implies that $\varphi(z - Ry) = \varphi(z - ry) - \varphi(Ry - ry) > R - 3/4r - (R - r)3/4 = R/4$. □

This inequality is a key ingredient in the proof of the following result.

Theorem 8.4.7 *Let Y be a subset of a finite-dimensional normed space $(V, \|\cdot\|)$. If $f\colon Y \to Y$ is a non-expansive map and $x \in Y$ is such that $\lim_{k\to\infty} \|f^k(x)\| = \infty$, then there exists a subsequence $(f^{k_j}(x))_j$ that converges at infinity such that the horofunction*

$$b_0(z) = \lim_{j\to\infty} \|z - f^{k_j}(x)\| - \|f^{k_j}(x)\|$$

satisfies $\lim_{k\to\infty} b_0(f^k(x)) = -\infty$.

Proof As $\|f^k(x)\| \to \infty$ as $k \to \infty$, we can find a subsequence $(k_j)_j$ such that $\|f^{k_j}(x)\| > \|f^m(x)\|$ for all $m < k_j$ by Lemma 3.3.4. Moreover, it follows from Proposition 3.3.2 that we may assume, after taking a further subsequence, that the horofunction

$$b_0(z) = \lim_{j\to\infty} \|z - f^{k_j}(x)\| - \|f^{k_j}(x)\|$$

exists for all $z \in (V, \|\cdot\|)$. Since the unit ball in $(V, \|\cdot\|)$ is compact, there exists $y_0 \in V$ such that $\|y_0\| = 1$ and y_0 is an limit point of the sequence $(f^{k_j}(x)/\|f^{k_j}(x)\|)_j$. By taking a further subsequence we may assume that

$$\left\| f^{k_j}(x)/\|f^{k_j}(x)\| - y_0 \right\| \leq 2^{-j} \quad \text{for all } j \geq 1.$$

Write $y^j = \|f^{k_j}(x)\| y_0$ for each $j \geq 1$. So, we have that

$$\|f^{k_j}(x) - y^j\| \leq 2^{-j}\|f^{k_j}(x)\| \quad \text{for all } j \geq 1. \tag{8.13}$$

Now fix $m \in \mathbb{N}$ and remark that

$$\begin{aligned}\|f^{k_j-m}(x) - y^j\| &\leq \|f^{k_j-m}(x) - f^{k_j}(x)\| + \|f^{k_j}(x) - y^j\| \\ &\leq \sum_{l=1}^{m} \|f^{k_j-l}(x) - f^{k_j-l+1}(x)\| + 2^{-j}\|f^{k_j}(x)\| \\ &\leq m\|x - f(x)\| + 2^{-j}\|f^{k_j}(x)\|,\end{aligned}$$

as f is non-expansive. For j sufficiently large we get that

$$\|f^{k_j-m}(x) - y^j\| \leq \|f^{k_j}(x)\|/4.$$

Take $i \in \mathbb{N}$ fixed and remark that, as $\|f^{k_j-m}(x)\| < \|f^{k_j}(x)\|$, we can apply Lemma 8.4.6, where we interpret $R = \|f^{k_j}(x)\|$, $r = \|f^{k_i}(x)\|$, and $y = y_0$, to obtain

$$\|f^{k_j-m}(x) - y^i\| \leq \|f^{k_j}(x)\| - (3/4)\|f^{k_i}(x)\| \tag{8.14}$$

for all $j > i$ sufficiently large.

Let us now estimate $b_0(f^{k_i+m}(x))$. By definition

$$\begin{aligned}b_0(f^{k_i+m}(x)) &= \lim_{j\to\infty} \|f^{k_i+m}(x) - f^{k_j}(x)\| - \|f^{k_j}(x)\| \\ &\leq \liminf_{j\to\infty} \|f^{k_i}(x) - f^{k_j-m}(x)\| - \|f^{k_j}(x)\|,\end{aligned}$$

as f is non-expansive. It follows from (8.13) and (8.14) that

$$\|f^{k_i}(x) - f^{k_j-m}(x)\| \leq 2^{-i}\|f^{k_i}(x)\| + \|f^{k_j}(x)\| - (3/4)\|f^{k_i}(x)\|.$$

Thus, we find that

$$b_0(f^{k_i+m}(x)) \leq -3/4\|f^{k_i}(x)\| + 2^{-i}\|f^{k_i}(x)\| \to -\infty$$

as $i \to \infty$. This implies that $\lim_{k\to\infty} b_0(f^k(x)) = -\infty$. □

By applying the isometric embedding of (A, κ) into $(\mathbb{R}^m, \|\cdot\|_\infty)$ for polyhedral domains A and the previous theorem, Theorem 8.4.5 can be proved in the following manner.

Proof of Theorem 8.4.5 Let $A \subseteq \mathbb{R}^n$ be an open bounded polyhedron. We can view A as a subset of $\mathbb{R}^{n+1}$ by putting

$$A' = \{(1, a) \in \mathbb{R}^{n+1} : a \in A\}.$$

Let $K = \{\lambda x : \lambda \geq 0 \text{ and } x \in \mathrm{cl}(K)\}$ be the closed cone generated by $\mathrm{cl}(A')$ in $\mathbb{R}^{n+1}$. As A is an open polyhedron, K is a solid closed polyhedral cone. Remark that $\varphi \in K^*$, with $\varphi(x) = x_1$, is in $\mathrm{int}(K^*)$. Moreover,

$$A' = \{x \in \mathrm{int}(K) : \varphi(x) = 1\}$$

and hence (A', d_H) can be isometrically embedded into $(\mathbb{R}^m, \|\cdot\|_\infty)$ by Proposition 2.2.3. Since (A, κ) is isometric to (A', d_H), there exists an isometry $\Psi : (A, \kappa) \to (\mathbb{R}^m, \|\cdot\|_\infty)$. The map $g : \Psi(A) \to \Psi(A)$ given by $g = \Psi \circ f \circ \Psi^{-1}$ is sup-norm non-expansive. As f has no fixed point in A, we know that $\kappa(f^k(x), x) \to \infty$ as $k \to \infty$ for each $x \in A$ by Corollaries 3.1.8 and 3.2.5, so that $\lim_{k\to\infty} \|g^k(z) - z\|_\infty = \infty$ for each $z \in \Psi(A)$. By Theorem 8.4.7 there exists a horofunction

$$b_0(u) = \lim_{j\to\infty} \|u - g^{k_j}(z)\|_\infty - \|g^{k_j}(z)\|_\infty$$

such that $b_0(g^k(z)) \to -\infty$ as $k \to \infty$. Now pick $y \in A$. By using the isometry Ψ and Lemma 3.3.1(iii) it follows that the horofunction $b_y : A \to \mathbb{R}$ given by

$$b_y(x) = \lim_{k\to\infty} \kappa(x, f^{k_j}(x)) - \kappa(f^{k_j}(x), y)$$

exists and satisfies

$$b_y(x) = b_{\Psi(y)}(\Psi(x)) = b_0(\Psi(x)) - b_0(\Psi(y)).$$

This implies that $b_y(f^k(x)) = b_0(g^k(\Psi(x))) - b_0(\Psi(y))$ for all $k \geq 1$, and hence $\lim_{k\to\infty} b_y(f^k(x)) = -\infty$.

Recall that the horoball $H(-r) = \{u \in A : b_y(u) \leq -r\}$ is convex by Lemma 3.3.3. Let $\mathrm{cl}(H(-r))$ be the norm closure of $H(-r)$ and note that $\cap_{r\geq 0}\mathrm{cl}(H(-r))$ is a convex subset of ∂A. As $\omega(x; f) \subseteq H(-r)$ for all $r \geq 0$, we get that $\mathrm{co}(\omega(x; f)) \subseteq \cap_{r\geq 0}\mathrm{cl}(H(-r))$. From Lemma 8.3.8 we conclude that there exists a convex set $\Lambda \subseteq \partial A$ such that $\omega(z; f) \subseteq \Lambda$ for all $z \in A$. □

A careful look at the proof of Theorem 8.4.5 shows that if $f : A \to A$ is a fixed-point free non-expansive map on a polyhedral Hilbert's metric space, then there exists a horofunction $b : A \to \mathbb{R}$ such that $\lim_{k\to\infty} b(f^k(x)) = -\infty$ for all $x \in A$. For general Hilbert's metric spaces such a horofunction need not exist. In his thesis [130], Lins has given an example of a fixed-point free non-expansive map on (A, κ), where A is the open unit disc in $\mathbb{R}^2$, that illustrates this point.

9

Dynamics of integral-preserving maps

In a pioneering paper, Akcoglu and Krengel [2] studied the long-term behavior of discrete time diffusion processes on finite states. These processes are non-linear analogues of Markov chains and can be modeled by order-preserving integral-preserving maps on $\mathbb{R}^n_+$. Their work was further developed by Lemmens, Nussbaum, Scheutzow, and Verduyn Lunel [121, 122, 161, 169–171, 197, 198]. These studies provide detailed results on the iterative behavior of order-preserving integral-preserving maps and their periodic orbits. In this chapter we will discuss these results.

Recall (see Proposition 2.8.1) that order-preserving integral-preserving maps $f : \mathbb{R}^n_+ \to \mathbb{R}^n_+$ are non-expansive under the ℓ_1-norm and $f(0) = 0$. So, each orbit of f converges to a periodic orbit by Corollary 4.2.5 and Lemma 4.2.6. Moreover, there exists an a-priori upper bound for the largest possible period of a periodic point in terms of the dimension of the underlying space. Among other results we shall see that the periods of periodic points of order-preserving integral-preserving maps, $f : \mathbb{R}^n_+ \to \mathbb{R}^n_+$, are precisely the periods of so-called admissible arrays on n symbols. An admissible array is a combinatorial object involving n symbols, whose period is bounded by a number $\gamma(n)$ that only depends on the number of symbols. The important point is that if we fix the number of symbols, the set of possible periods of admissible arrays can be computed in finite time, and an explicit upper bound for $\gamma(n)$ can be given.

9.1 Lattice homomorphisms

Before we start we would like to emphasize that throughout this chapter the partial ordering will always be induced by $\mathbb{R}^n_+$ and the lattice operations $x \vee y$ and $x \wedge y$ on $\mathbb{R}^n$ correspond to $(x \vee y)_i = \max\{x_i, y_i\}$ and $(x \wedge y)_i = \min\{x_i, y_i\}$ for $1 \leq i \leq n$. Periodic orbits of order-preserving and integral-preserving maps on $\mathbb{R}^n_+$ are closely related to periodic orbits of lattice homomorphisms $g: V \to V$, where V is a finite lattice in $\mathbb{R}^n$, i.e., V is closed under the operations $\wedge$ and $\vee$. Before we explain this relation in detail it is convenient to introduce the following notation.

Definition 9.1.1 For a map $f: \mathbb{R}^n_+ \to \mathbb{R}^n_+$ we define the following conditions:

(C1) f is order-preserving and integral-preserving.
(C2) f is non-expansive under the ℓ_1-norm and $f(0) = 0$.
(C3) $f(\lambda \mathbb{1}) = \lambda \mathbb{1}$ for all $\lambda > 0$.

Using these conditions we subsequently define collections of maps $\mathcal{F}_i(n)$ for $i = 1, 2$, and 3 by

$$\mathcal{F}_1(n) = \{f: \mathbb{R}^n_+ \to \mathbb{R}^n_+ : f \text{ satisfies C1 and C3}\},$$

$$\mathcal{F}_2(n) = \{f: \mathbb{R}^n_+ \to \mathbb{R}^n_+ : f \text{ satisfies C1}\},$$

and

$$\mathcal{F}_3(n) = \{f: \mathbb{R}^n_+ \to \mathbb{R}^n_+ : f \text{ satisfies C2}\}.$$

Recall that if $f \in \mathcal{F}_1(n)$, then f is also sup-norm decreasing by Lemma 1.2.3. As each order-preserving integral-preserving map $f: \mathbb{R}^n_+ \to \mathbb{R}^n_+$ is non-expansive under the ℓ_1-norm by Proposition 2.8.1, we find for each $n \in \mathbb{N}$ that

$$\mathcal{F}_1(n) \subseteq \mathcal{F}_2(n) \subseteq \mathcal{F}_3(n). \tag{9.1}$$

It follows from Lemma 4.2.6 that there exists an a-priori upper bound for the periods of periodic points of maps in $\mathcal{F}_3(n)$. Therefore it is natural to define finite sets $P_i(n)$ for $i = 1, 2, 3$ as follows:

$$P_i(n) = \{p \in \mathbb{N}: \text{there exists } f \in \mathcal{F}_i(n) \text{ with a periodic point of period } p\}. \tag{9.2}$$

Thus, by (9.1) we get that

$$P_1(n) \subseteq P_2(n) \subseteq P_3(n) \quad \text{for each } n \in \mathbb{N}. \tag{9.3}$$

The main goal of this chapter is to analyze the finite sets $P_i(n)$. To do this we first need to establish a connection with periodic points of lattice homomorphisms.

For $x, y, z \in \mathbb{R}^n$ define

$$[x, y]_1 = \{z \in \mathbb{R}^n : \|x - y\|_1 = \|x - z\|_1 + \|z - y\|_1\}.$$

Given $a, b, c \in \mathbb{R}$, the *median* of a, b, and c, denoted $\mathrm{med}(a, b, c) \in \mathbb{R}$, is defined as follows: if a, b, and c are all distinct, then exactly one of the numbers lies between the other two. This number is called $\mathrm{med}(a, b, c)$ in this case. If the numbers are nondistinct, then at least two are equal, and $\mathrm{med}(a, b, c)$ is defined to be that number in that case. For example, $\mathrm{med}(0, -1, 1) = 0$ and $\mathrm{med}(1, 0, 1) = 1$. For $x, y, z \in \mathbb{R}^n$, we define $\mathrm{med}(x, y, z) \in \mathbb{R}^n$ to be a vector whose i-th coordinate is equal to $\mathrm{med}(x_i, y_i, z_i)$, e.g., if $x = (2, 1, 3)$, $y = (-1, 1, 3)$, and $z = (-2, 0, 3)$, then $\mathrm{med}(x, y, z) = (-1, 1, 3)$. A set $X \subseteq \mathbb{R}^n$ is called *median closed* if $\mathrm{med}(x, y, z) \in X$ for all $x, y, z \in X$. We observe that, for each $x, y, z \in \mathbb{R}^n$,

$$[x, y]_1 \cap [y, z]_1 \cap [x, z]_1 = \{\mathrm{med}(x, y, z)\}, \tag{9.4}$$

which has the following consequence.

Lemma 9.1.2 *Let $X \subseteq \mathbb{R}^n$ be median closed and $f \colon X \to \mathbb{R}^n$ be non-expansive under the ℓ_1-norm. If the restriction of f to $\{x, y, z\} \subseteq X$ is an ℓ_1-norm isometry, then*

$$f(\mathrm{med}(x, y, z)) = \mathrm{med}(f(x), f(y), f(z)).$$

Proof Let $m = \mathrm{med}(x, y, z)$ and remark that $\|f(x) - f(y)\|_1 = \|x - y\|_1 = \|x - m\|_1 + \|m - y\|_1$. As f is non-expansive, we deduce that

$$\|f(x) - f(y)\|_1 \geq \|f(x) - f(m)\|_1 + \|f(m) - f(y)\|_1 \geq \|f(x) - f(y)\|_1$$

and hence $f(m) \in [f(x), f(y)]_1$. In the same way we find that $f(m) \in [f(x), f(z)]_1$ and $f(m) \in [f(y), f(z)]_1$. Thus, $f(m) = \mathrm{med}(f(x), f(y), f(z))$ by (9.4). □

Now suppose that $x, y \in \mathbb{R}^n_+$ and $z = 0$. In that case $\mathrm{med}(x, y, z) = x \wedge y$. This observation and Lemma 9.1.2 provide a connection with so-called lower semi-lattice homomorphisms. A set $V \subseteq \mathbb{R}^n$ is called a *lower semi-lattice* if $x \wedge y \in V$ for all $x, y \in V$. Recall that V is a *lattice* if, in addition, $x \vee y$ is in V for each $x, y \in V$. For $S \subseteq \mathbb{R}^n$ the smallest lower semi-lattice containing S is called the *lower semi-lattice generated by* S and is denoted by V_S. A map $g \colon V \to V$, where V is a lower semi-lattice, is called a *lower semi-lattice homomorphism* if $g(x \wedge y) = g(x) \wedge g(y)$ for all $x, y \in V$. We remark that each lower semi-lattice homomorphism g is order-preserving, since $x \leq y$ implies $x = x \wedge y \leq y$, so that $g(x) = g(x \wedge y) = g(x) \wedge g(y) \leq g(y)$.

Proposition 9.1.3 *If $f \in \mathcal{F}_3(n)$ and $\xi \in \mathbb{R}^n_+$ is a periodic point of f with period p, then f restricted to the lower semi-lattice, $V_{\mathcal{O}}$, generated by $\mathcal{O} = \{f^k(\xi) \colon 0 \leq k < p\}$ is a lower semi-lattice homomorphism that maps $V_{\mathcal{O}}$ onto itself.*

Proof For $m \in \mathbb{N}$ define

$$U_m = \{\bigwedge_{i=1}^{k} y^i : 1 \leq k \leq m \text{ and } y^i \in \mathcal{O}\}.$$

We prove by induction on m that $f(\bigwedge_{i=1}^m y^i) = \bigwedge_{i=1}^m f(y^i)$, and f restricted to $U_m \cup \{0\}$ is an ℓ_1-norm isometry. Clearly the assertions are true for U_1. Suppose they hold for U_m. Let $\bigwedge_{i=1}^{m+1} y^i \in U_{m+1}$. Then y^{m+1} and $\bigwedge_{i=1}^m y^i$ are in U_m, so that f is an ℓ_1-norm isometry on $\{0, y^{m+1}, \bigwedge_{i=1}^m y^i\}$. Hence it follows from Lemma 9.1.2 and the induction hypothesis that

$$\begin{aligned} f(\bigwedge_{i=1}^{m+1} y^i) &= f(\mathrm{med}(y^{m+1}, \bigwedge_{i=1}^{m} y^i, 0)) \\ &= \mathrm{med}(f(y^{m+1}), f(\bigwedge_{i=1}^{m} y^i), 0) \\ &= \bigwedge_{i=1}^{m+1} f(y^i). \end{aligned}$$

This implies that $f^p(\bigwedge_{i=1}^{m+1} y^i) = \bigwedge_{i=1}^{m+1} f^p(y^i) = \bigwedge_{i=1}^{m+1} y^i$ and hence each element in U_{m+1} is a periodic point whose period divides p. From this we immediately deduce that f restricted to $U_{m+1} \cup \{0\}$ is an ℓ_1-norm isometry.

As $\mathcal{O}$ is finite, $V_\mathcal{O}$ is finite as well, and hence $V_\mathcal{O} = U_m$ for some $m \geq 1$. Thus, we find that f is a lower semi-lattice homomorphism that maps $V_\mathcal{O}$ onto itself. □

It turns out that if $f \in \mathcal{F}_2(n)$ and L is the lattice generated by a periodic orbit of f, then the restriction of f to L is a *lattice homomorphism*; so, $f(x \vee y) = f(x) \vee f(y)$ and $f(x \wedge y) = f(x) \wedge f(y)$ for all $x, y \in L$. To see this we first prove the following lemma.

Lemma 9.1.4 *If $f \in \mathcal{F}_2(n)$ and f has a periodic point $\xi \in \mathbb{R}^n_+$ with period p, then $\bigvee_{k=0}^{p-1} f^k(\xi)$ is a fixed point of f.*

Proof Let $z = \bigvee_{k=0}^{p-1} f^k(\xi)$. As f is order-preserving, $f(z) \geq f^k(\xi)$ for all $0 \leq k < p$ and hence $f(z)_i \geq \bigvee_{k=0}^{p-1} f^k(\xi)_i = z_i$ for all i. But $\sum_i f(z)_i = \sum_i z_i$, since f is integral-preserving, and therefore $f(z) = z$. □

Remark that if $x, y, z \in \mathbb{R}^n$ and $z \geq x \vee y$, then $\mathrm{med}(x, y, z) = x \vee y$. This observation and Lemma 9.1.4 together yield the next proposition. To state it we introduce the following notion. For $S \subseteq \mathbb{R}^n$, the smallest lattice in $\mathbb{R}^n$ containing S is called the *lattice generated by* S and is denoted by L_S.

Proposition 9.1.5 *If $f \in \mathcal{F}_2(n)$ and ξ is a periodic point of f with period p, then f restricted to the lattice, $L_{\mathcal{O}}$, generated by $\mathcal{O} = \{f^k(\xi) : 0 \le k < p\}$ is a lattice homomorphism that maps $L_{\mathcal{O}}$ onto itself.*

Proof Let $V_{\mathcal{O}}$ denote the lower semi-lattice generated by $\mathcal{O}$ and let $z = \bigvee_{k=0}^{p-1} f^k(\xi)$ be the fixed point of f from Lemma 9.1.4. For each $m \in \mathbb{N}$ define

$$L_m = \{\bigvee_{i=1}^{k} v^i : 1 \le k \le m \text{ and } v^i \in V_{\mathcal{O}}\}.$$

Clearly there exists $M \in \mathbb{N}$ such that $L_{\mathcal{O}} = L_M$, as $\mathcal{O}$ is finite. It therefore suffices to prove, by induction on m, that

$$f(\bigvee_{i=1}^{m} v^i) = \bigvee_{i=1}^{m} f(v^i) \quad \text{for all } \bigvee_{i=1}^{m} v^i \in L_m,$$

and f is an ℓ_1-norm isometry on $L_m \cup \{z\}$.

By Proposition 9.1.3 we know the assertions are true for $m = 1$. Now suppose that the assertion holds for each L_k with $1 \le k < m$. If $\bigvee_{i=1}^{m} v^i \in L_m$, then $\bigvee_{i=1}^{m-1} v^i \in L_{m-1}$ and $v^m \in L_{m-1}$. By the induction hypothesis, f is an ℓ_1-norm isometry on $\{\bigvee_{i=1}^{m-1} v^i, v^m, z\}$. Now remark that as $w \le z$ for all $w \in L_{\mathcal{O}}$,

$$\text{med}(\bigvee_{i=1}^{m-1} v^i, v^m, z) = (\bigvee_{i=1}^{m-1} v^i) \vee v^m = \bigvee_{i=1}^{m} v^i.$$

Therefore it follows from Lemma 9.1.2 and the induction hypothesis that

$$\begin{aligned} f(\bigvee_{i=1}^{m} v^i) &= f(\text{med}(\bigvee_{i=1}^{m-1} v^i, v^m, z)) \\ &= \text{med}(\bigvee_{i=1}^{m-1} f(v^i), f(v^m), z) \\ &= \bigvee_{i=1}^{m} f(v^i). \end{aligned}$$

This also implies that $f^p(\bigvee_{i=1}^{m}(v^i)) = \bigvee_{i=1}^{m} f^p(v^i) = \bigvee_{i=1}^{m} v^i$, as each $v^i \in V_{\mathcal{O}}$ satisfies $f^p(v^i) = v^i$ by Proposition 9.1.3. Thus, each element of L_m is a periodic point of f, and hence f is an ℓ_1-norm isometry on $L_m \cup \{z\}$. □

Define $Q_{\text{lat}}(n)$ (and $Q_{\text{sem}}(n)$), respectively, as the sets of possible periods of periodic points of (lower semi-)lattice homomorphisms $g : V \to V$, where

V is a (lower semi-)lattice in $\mathbb{R}^n$. We obtain the following inclusions from Propositions 9.1.3 and 9.1.5:

$$P_1(n) \subseteq P_2(n) \subseteq Q_{\text{lat}}(n) \subseteq Q_{\text{sem}}(n) \quad \text{and} \quad P_3(n) \subseteq Q_{\text{sem}}(n). \tag{9.5}$$

To analyze the set $Q_{\text{sem}}(n)$ we need to further study lower semi-lattice homomorphisms.

9.2 Periodic orbits of lower semi-lattice homomorphisms

Let $V \subseteq \mathbb{R}^n$ be a finite lower semi-lattice in $\mathbb{R}^n$. Two points u and v in V are called *comparable* if either $u \leq v$ or $v \leq u$. If $A \subseteq V$ and there exists $b \in V$ such that $a \leq b$ for all $a \in A$, then A is said to be *bounded from above in V* and b is called an *upper bound* of A in V. Likewise A is said to be *bounded from below in V* if there exists $b \in V$ such that $b \leq a$ for all $a \in A$ and b is called a *lower bound* of A in V. If A is bounded from above in V, there exists a unique upper bound β of A in V such that $\beta \leq b$ for all upper bounds b of A in V. The element β is called the *supremum* of A in V and is denoted by $\sup_V(A)$. As V is a finite lower semi-lattice in $\mathbb{R}^n$ any set $A \subseteq V$ is bounded from below in V and there exists a unique lower bound α of A in V such that $b \leq \alpha$ for each lower bound b of A in V. In fact, $\alpha = \wedge_{a \in A} a$. We call α the *infimum* of A in V and we write $\inf_V(A)$. If A is bounded above in V, then it is easy to see that $\sup_V(A) = \inf_V(B)$, where $B = \{b \in V : b \text{ is an upper bound of } A \text{ in } V\}$.

Lemma 9.2.1 *Let $V \subseteq \mathbb{R}^n$ be a finite lower semi-lattice and let $g: V \to V$ be a one-to-one lower semi-lattice homomorphism. If $A \subseteq V$ is bounded from above in V, then $g(\sup_V(A)) = \sup_V(g(A))$.*

Proof As V is finite and $g: V \to V$ is one-to-one, there exists an integer $p \geq 1$ such that $g^{p-1} = g^{-1}$. This implies that g^{-1} is order-preserving, since g is order-preserving. From this we deduce that

$$\begin{aligned} g(\sup_V(A)) &= \bigwedge\{g(b) : b \text{ is an upper bound of } A \text{ in } V\} \\ &= \bigwedge\{g(b) \colon g(b) \text{ is an upper bound of } g(A) \text{ in } V\} \\ &= \sup_V(g(A)), \end{aligned}$$

and we are done. □

If $V \subseteq \mathbb{R}^n$ is a finite lower semi-lattice, then for $x \in V$ with $x \neq \inf_V(V)$, we define the *height of x in V* by

$$h_V(x) = \max\{k \geq 1\colon \text{there exist } y^0, \ldots, y^k \in V \text{ with } y^k = x \\ \text{and } y^i < y^{i+1} \text{ for } 0 \leq i < k\}.$$

We define $h_V(x) = 0$ if $x = \inf_V(V)$, and we note that $h_V(x) > 0$ for all $x \in V$ with $x \neq \inf_V(V)$. For $x \in V$ we write $S_x = \{v \in V\colon v < x\}$ and we call x *irreducible in* V if either $S_x = \emptyset$ or

$$\sup_V(S_x) < x. \tag{9.6}$$

Irreducible elements play a key role in the analysis of lower semi-lattice homomorphisms. The following basic result from lattice theory will be particularly useful.

Lemma 9.2.2 *If $V \subseteq \mathbb{R}^n$ is a finite lower semi-lattice, then for each $x \in V$ we have that*

$$x = \sup_V\{v \in V\colon v \leq x \text{ and } v \text{ irreducible in } V\}. \tag{9.7}$$

Proof To prove the lemma we first show the following claim: if for each irreducible v in V with $v \leq x$ we have that $v \leq y$, then $x \leq y$.

To prove the claim we suppose that $x \not\leq y$. Consider the set $A = \{v \in V\colon v \leq x \text{ and } v \not\leq y\}$. Clearly A is non-empty, as $x \in A$. Since V is finite, A has a minimal element $w \in A$, i.e., there exists no $u \in A$ with $u \leq w$ and $u \neq w$. To prove the claim it suffices to show that w is irreducible in V. Consider $\sup_V(S_w)$. For each $v \in S_w$ we have that $v \leq x$ and $v \leq y$, as w is a minimal element of A. This implies that $\sup_V(S_w) \leq y$. As $w \not\leq y$, we get that $w \neq \sup_V(S_w)$ and hence w is irreducible in V.

Using the claim it is now easy to prove (9.7). Indeed, let $y \in V$ be an upper bound of $\{v \in V\colon v \leq x \text{ and } v \text{ irreducible in } V\}$. Then for each irreducible $v \in V$ with $v \leq x$ we have that $v \leq y$. Thus, by the claim, $x \leq y$ and hence (9.7) follows. □

If x is an irreducible element in a finite lower semi-lattice $V \subseteq \mathbb{R}^n$ and $S_x \neq \emptyset$, then we define $I_V(x)$ by

$$I_V(x) = \{i\colon \sup_V(S_x)_i < x_i\}.$$

If S_x is empty, so $x = \inf_V(V)$, then we put $I_V(x) = \{1, \ldots, n\}$. The irreducible elements turn out to be very useful in the study of periodic points of lower semi-lattice homomorphisms, as we shall see now.

Lemma 9.2.3 *If $V \subseteq \mathbb{R}^n$ is a finite lower semi-lattice and $f\colon V \to V$ is a one-to-one lower semi-lattice homomorphism, then the following assertions are true:*

(i) If $y \in V$, then $h_V(y) = h_V(f^k(y))$ for all $k \geq 1$.
(ii) If y is irreducible in V, then $f^k(y)$ is irreducible in V for all $k \geq 1$.
(iii) If y and y' are irreducible elements in V, and y and y' are not comparable, then

$$I_V(y) \cap I_V(y') = \emptyset.$$

(iv) If y is an irreducible element in V and y is a periodic point of f with period p, then $1 \leq p \leq n$.

Proof To prove the first statement we remark that, as V is finite and f is a one-to-one lower semi-lattice homomorphism, there exists $q \geq 1$ such that $f^q(v) = v$ for all $v \in V$. This implies that f^{-1} and f are both order-preserving and hence $h_V(y) = h_V(f^k(y))$ for all $k \geq 1$.

If y is irreducible in V and $h_V(y) = 0$, then $h_V(f^k(y)) = 0$ for all $k \geq 1$, so that $f^k(y) = y = \inf_V(V)$ and the assertion follows. Now if $h_V(y) > 0$, then S_y and $S_{f^k(y)}$ are non-empty. Since f and f^{-1} are order-preserving, we find that $f^k(S_y) = S_{f^k(y)}$. It therefore follows from Lemma 9.2.1 that

$$\sup_V(S_{f^k(y)}) = f^k(\sup_V(S_y)) < f^k(y),$$

as y is irreducible. Therefore $f^k(y)$ is irreducible in V.

To show the third assertion we suppose that y and y' are two non-comparable irreducible elements in V and $i \in I_V(y) \cap I_V(y')$. As y and y' are not comparable, $y \wedge y' < y$ and $y \wedge y' < y'$. Thus $(y \wedge y')_i < y_i$ and $(y \wedge y')_i < y'_i$, as y and y' are irreducible in V, which is impossible.

To verify the last assertion we remark that, by (i),

$$h_V(y) = h_V(f(y)) = \ldots = h_V(f^{p-1}(y)),$$

so that $y, f(y), \ldots, f^{p-1}(y)$ are noncomparable irreducible elements by (ii). It now follows from (iii) that $I_V(f^k(y))$ are pairwise disjoint non-empty subsets of $\{1, \ldots, n\}$ for $0 \leq k < p$. This implies that $p \leq n$. □

Lemma 9.2.3 has the following immediate consequence for the periods of maps $f \in \mathcal{F}_3(n)$. Recall that the *period* of $f : \mathbb{R}^n_+ \to \mathbb{R}^n_+$ in $\mathcal{F}_3(n)$ is the least integer $p \geq 1$ such that $\lim_{k\to\infty} f^{kp}(x)$ exists for all $x \in \mathbb{R}^n_+$, and $\mathrm{lcm}(S)$ denotes the least common multiple of elements in S.

Corollary 9.2.4 *If $f \in \mathcal{F}_3(n)$, then f is periodic and its period divides* $\mathrm{lcm}(1, 2, \ldots, n)$.

Proof By Corollary 4.2.5 and Lemma 4.2.6 it suffices to show that the period of each periodic point of f divides $\mathrm{lcm}(1, 2, \ldots, n)$. Let $x \in \mathbb{R}^n_+$ be a periodic point of $f \in \mathcal{F}_3(n)$ with period p. Let $\mathcal{O}(x)$ denote the orbit of x under f

and let V be the lower semi-lattice generated by $\mathcal{O}(x)$. If g is the restriction of f to V, it follows from Proposition 9.1.3 that $g\colon V \to V$ is a lower semi-lattice homomorphism that maps V onto itself. By Lemma 9.2.2 we know that $x = \sup_v(S)$, where $S = \{v \in V : v \leq x \text{ and } v \text{ irreducible in } V\}$. It follows from Lemma 9.2.3(iv) that each $v \in S$ has period p_v, where $1 \leq p_v \leq n$. Now let $q = \operatorname{lcm}(\{p_v : v \in S\})$ and remark that

$$g^q(x) = g^q(\sup_V(S)) = \sup_V(g^q(S)) = \sup_V(S) = x$$

by Lemma 9.2.1; so, p divides q. As q divides $\operatorname{lcm}(1, 2, \ldots, n)$, we are done. □

It turns out that the upper bound in Corollary 9.2.4 is sharp. In fact, we will see that there exists a map $f \in \mathcal{F}_1(n)$ that has periodic points with period p for all $p \in P_1(n)$. As $\{1, 2, \ldots, n\} \subseteq P_1(n)$, it follows that $\operatorname{lcm}(1, 2, \ldots, n)$ is a tight upper bound for the possible periods of maps in $\mathcal{F}_3(n)$. The construction of this map is somewhat laborious, but, as it will be useful to us later, we discuss it in detail here.

We write $\mathbb{R}^n_- = \{x \in \mathbb{R}^n : x \leq 0\}$ and denote $\mathbb{D}^n = \mathbb{R}^n_+ \cup \mathbb{R}^n_-$. Let $S\colon \mathbb{R}^2 \to \mathbb{R}^2$ be given by

$$S(x_1, x_2) = ((x_1 \vee 0) + (x_2 \wedge 0), (x_1 \wedge 0) + (x_2 \vee 0)) \quad \text{for all } (x_1, x_2) \in \mathbb{R}^2.$$

We claim that S^2 is an order-preserving integral-preserving sup-norm decreasing retraction onto $\mathbb{D}^2$. Indeed, it is easy to verify that S is order-preserving and integral-preserving. Moreover, $S(\lambda \mathbb{1}) = \lambda \mathbb{1}$ for all $\lambda \in \mathbb{R}$ and hence S is also sup-norm decreasing. The map S^2 also has these properties. To see that S^2 is a retraction onto $\mathbb{D}^2$ we remark that if $x_1 \geq 0$ and $x_2 \leq 0$, then

$$S^2(x_1, x_2) = \begin{cases} (x_1 + x_2, 0) & \text{if } x_1 + x_2 \geq 0 \\ (0, x_1 + x_2) & \text{if } x_1 + x_2 \leq 0. \end{cases}$$

Similarly, if $x_1 \leq 0$ and $x_2 \geq 0$, then

$$S^2(x_1, x_2) = \begin{cases} (x_1 + x_2, 0) & \text{if } x_1 + x_2 \leq 0 \\ (0, x_1 + x_2) & \text{if } x_1 + x_2 \geq 0. \end{cases}$$

From these observations it is easy to deduce that S^2 is a retraction onto $\mathbb{D}^2$. Next we prove an auxiliary lemma involving the map S.

Lemma 9.2.5 *There exists a retraction $\varphi\colon \mathbb{R}^n \to \mathbb{R}^n$ onto $\mathbb{D}^n$ which is order-preserving, integral-preserving, and sup-norm decreasing.*

Proof For $1 \leq i, j \leq n$ with $i \neq j$ let $S_{ij} : \mathbb{R}^n \to \mathbb{R}^n$ be given by

$$S_{ij}(x)_k = \begin{cases} x_k & \text{if } i \neq k \text{ and } j \neq k \\ (x_i \vee 0) + (x_j \wedge 0) & \text{if } k = i \\ (x_i \wedge 0) + (x_j \vee 0) & \text{if } k = j. \end{cases}$$

Define $R_{ij}(x) = S_{ij}^2(x)$ for all $x \in \mathbb{R}^n$. As in the paragraph preceding this lemma, it can be shown that

(1) R_{ij} is order-preserving, integral-preserving, and sup-norm decreasing.
(2) If $k \neq i$ and $k \neq j$, and we define $y = R_{ij}(x)$, then $y_k = x_k$.
(3) R_{ij} is a retraction onto $D_{ij} = \{x \in \mathbb{R}^n : x_i x_j \geq 0\}$.
(4) If $x_i \leq 0$ and $x_j \geq 0$, then either $y_i \leq 0$ and $y_j = 0$, or $y_i = 0$ and $y_j \geq 0$. The exactly analogous statement holds if $x_i \geq 0$ and $x_j \leq 0$.

Using these properties of the maps R_{ij} we complete the proof by induction on n. We already know that the assertion is true for $n = 2$. Assume that the lemma is true for all $m \leq n - 1$, where $n \geq 3$. Let $\varphi' : \mathbb{R}^{n-1} \to \mathbb{R}^{n-1}$ be an order-preserving integral-preserving sup-norm decreasing retraction onto $\mathbb{D}^{n-1}$. Define $\psi : \mathbb{R}^n \to \mathbb{R}^n$ by

$$\psi(x_1, x) = (x_1, \varphi'(x)) \quad \text{for } x_1 \in \mathbb{R} \text{ and } x \in \mathbb{R}^{n-1}.$$

Clearly ψ is the identity on $\mathbb{D}^n$ and ψ is order-preserving, integral-preserving, and sup-norm decreasing.

By using the map ψ we subsequently define the retraction $\varphi : \mathbb{R}^n \to \mathbb{R}^n$ by $\varphi = R_{12} \circ R_{13} \circ \cdots \circ R_{1n} \circ \psi$. By construction φ is order-preserving, integral-preserving, and sup-norm decreasing. Moreover φ is the identity on $\mathbb{D}^n$, and hence we only need to show that φ maps $\mathbb{R}^n$ onto $\mathbb{D}^n$.

Put $y = \psi(x)$ and remark that by the induction hypothesis either $y_i \geq 0$ for all $2 \leq i \leq n$, or $y_i \leq 0$ for all $2 \leq i \leq n$. Suppose that $y_i \geq 0$ for all $2 \leq i \leq n$. (The proof for the other case is similar to this one.) If $y_1 \geq 0$, we are done. So assume that $y_1 < 0$. If $y_1 + y_n \geq 0$, then a simple calculation shows that $R_{1n}(y) = (0, y_2, y_3, \ldots, y_1 + y_n) \in \mathbb{D}^n$. This implies that $\varphi(x) \in \mathbb{D}^n$ in that case. Now assume that $y_1 + y_n < 0$. In that case $R_{1n}(y) = (y_1 + y_n, y_2, \ldots, y_{n-1}, 0)$. If we subsequently apply the map $R_{1(n-1)}$, we find that $R_{1(n-1)}(R_{1n}(y)) = (0, y_2, y_3, \ldots, y_{n-2}, y_1 + y_{n-1} + y_n, 0)$ if $y_1 + y_{n-1} + y_n \geq 0$, and $R_{1(n-1)}(R_{1n}(y)) = (y_1 + y_{n-1} + y_n, , y_2, \ldots, y_{n-2}, 0, 0)$ if $y_1 + y_{n-1} + y_n < 0$. Thus, if $y_1 + y_{n-1} + y_n \geq 0$, we see that $\varphi(x) \in \mathbb{D}^n$, and we are done. On the other hand, if we assume that $y_1 + y_{n-1} + y_n < 0$, we can repeat the argument to conclude that either $R_{1j} \circ \cdots \circ R_{1n}(y) \in \mathbb{D}^n$ for some $j \geq 3$, in which case $\varphi(x) \in \mathbb{D}^n$, or $R_{13} \circ \cdots \circ R_{1n}(y) \notin \mathbb{D}^n$ and

$$R_{13}\circ\cdots\circ R_{1n}(y) = (y_1 + \sum_{k=3}^{n} y_k, y_2, 0, \ldots, 0),$$

where $y_1 + \sum_{k=3}^{n} y_k < 0$ and $y_2 \geq 0$. Let $z = (z_1, z_2, 0, \ldots, 0) = R_{13}\circ\cdots\circ R_{1n}(y)$. A simple computation shows that

$$R_{12}(z) = \begin{cases} (0, z_1 + z_2, 0, \ldots, 0) & \text{if } z_1 + z_2 \geq 0 \\ (z_1 + z_2, 0, 0, \ldots, 0) & \text{if } z_1 + z_2 < 0. \end{cases}$$

Thus, $\varphi(x) = R_{12}(z) \in \mathbb{D}^n$, and we are done. □

The retraction in the previous lemma is the main ingredient in the proof of the following proposition.

Proposition 9.2.6 *If* $0 = a_0 < a_1 < \ldots < a_k$ *in* $\mathbb{R}$ *and*

$$B_j = \{x \in \mathbb{R}^n_+ : a_j\mathbb{1} \leq x \leq a_{j+1}\mathbb{1}\} \quad \textit{for } 0 \leq j < k$$

and

$$B_k = \{x \in \mathbb{R}^n_+ : a_k\mathbf{1} \leq x\},$$

then there exists a retraction $r \in \mathcal{F}_1(n)$ *onto* $B = \cup_{j=0}^{k} B_j$.

Proof The proof goes by induction on k. First assume that $k = 1$. Let $\varphi\colon \mathbb{R}^n \to \mathbb{R}^n$ be the retraction from Lemma 9.2.5 and define $r\colon \mathbb{R}^n_+ \to \mathbb{R}^n_+$ by

$$r(x) = \varphi(x - a_1\mathbb{1}) + a_1\mathbb{1} \quad \text{for all } x \in \mathbb{R}^n_+.$$

For each $x \in \mathbb{R}^n_+$ we have that $r(x) \geq r(0) = \varphi(-a_1\mathbb{1}) + a_1\mathbb{1} = 0$, so that $r(x) \in \mathbb{R}^n_+$. It is easy to verify that $r \in \mathcal{F}_1(n)$. As $\varphi(x - a_1\mathbb{1}) \in \mathbb{D}^n$, we find that $r(x) \in B_0$ or $r(x) \in B_1$. But also if $x \in B_0 \cup B_1$, then $x - a_1\mathbb{1} \in \mathbb{D}^n$ and hence $\varphi(x - a_1\mathbb{1}) = x - a_1\mathbb{1}$, so that $r(x) = x$. Thus, r is a retraction onto $B = B_0 \cup B_1$.

Now let $0 = a_0 < a_1 < \ldots < a_k < a_{k+1}$ in $\mathbb{R}$ and let $B'_j = B_j$ for $0 \leq j < k$, where B_j is as in the statement of the proposition. Furthermore let $B'_k = \{x \in \mathbb{R}^n_+ : a_k\mathbb{1} \leq x \leq a_{k+1}\mathbb{1}\}$ and $B'_{k+1} = \{x \in \mathbb{R}^n_+ : a_{k+1}\mathbb{1} \leq x\}$. By induction there exists a retraction $\sigma \in \mathcal{F}_1(n)$ onto $B = \cup_{j=0}^{k} B_j$, where $B_j = B'_j$ for $0 \leq j < k$ and $B_k = \{x \in \mathbb{R}^n_+ : a_k\mathbb{1} \leq x\}$. Using the map $\varphi\colon \mathbb{R}^n \to \mathbb{R}^n$ from Lemma 9.2.5 we first define $r'\colon \mathbb{R}^n_+ \to \mathbb{R}^n_+$ by

$$r'(x) = \varphi(x - a_{k+1}\mathbb{1}) + a_{k+1}\mathbb{1} \quad \text{for all } x \in \mathbb{R}^n_+.$$

Subsequently let $r\colon \mathbb{R}^n_+ \to \mathbb{R}^n_+$ be given by $r = \sigma \circ r'$. As $r', \sigma \in \mathcal{F}_1(n)$, we have that $r \in \mathcal{F}_1(n)$. To see that r maps $\mathbb{R}^n_+$ onto $B' = \cup_{j=0}^{k+1} B'_j$, we let $x \in \mathbb{R}^n_+$ and put $\xi = r'(x)$. Then $a_{k+1}\mathbb{1} \leq \xi$ or $\xi \leq a_{k+1}\mathbb{1}$. In the first case $\xi \in B_k$, so that $\sigma(\xi) = \xi$ and hence $r(x) \in B'$. On the other hand if

$\xi \leq a_{k+1}\mathbb{1}$, we know that $\sigma(\xi) \in B$ and $\sigma(\xi) \leq \sigma(a_{k+1}\mathbb{1}) = a_{k+1}\mathbb{1}$ and therefore $\sigma(x) \in \cup_{j=0}^{k} B'_j \subset B'$. We leave it to the reader to verify that r is a retraction. □

The linear maps induced by the $n \times n$ permutation matrices are in $\mathcal{F}_1(n)$. So, the set

$$\begin{aligned} L(n) &= \{p \colon p \text{ is the order of a permutation on } n \text{ letters}\} \\ &= \{\mathrm{lcm}(p_1, \ldots, p_k) \colon p_1, \ldots, p_k \in \mathbb{N} \text{ and } \textstyle\sum_{j=1}^{k} p_j \leq n\} \end{aligned}$$

is contained in $P_1(n)$ for all n.

Theorem 9.2.7 *For $n \in \mathbb{N}$ there exists a map $f \in \mathcal{F}_1(n)$ such that for each $p \in P_1(n)$ the map f has a periodic point with period p. In particular, f has period* lcm $(1, 2, \ldots, n)$.

Proof As $P_1(n) = \{p_1, \ldots, p_m\}$ is finite, there exist maps $f_1, \ldots, f_m \in \mathcal{F}_1(n)$ such that f_j has a periodic point x^j with period p_j for $1 \leq j \leq m$. For each x^j we select $\alpha_j > 0$ such that $x^j < \alpha_j \mathbb{1}$. Define $a_k = \sum_{j=1}^{k} \alpha_j$ for $1 \leq k < m$ and put $a_0 = 0$. We know from Proposition 9.2.6 that there exists a retraction $r \in \mathcal{F}_1(n)$ onto $B = \cup_{j=0}^{m-1} B_j$, where $B_j = \{x \in \mathbb{R}^n_+ \colon a_j\mathbb{1} \leq x \leq a_{j+1}\mathbb{1}\}$ for $1 \leq j < m-1$ and $B_{m-1} = \{x \in \mathbb{R}^n_+ \colon a_{m-1}\mathbb{1} \leq x\}$.

Now consider $f \in \mathcal{F}_1(n)$ given by

$$f(x) = f_j(r(x) - a_{j-1}\mathbb{1}) + a_{j-1}\mathbb{1} \quad \text{for } r(x) \in B_{j-1}.$$

For $j \neq k$ we have that $B_j \cap B_k$ is either empty or consists of a single vector $\lambda\mathbb{1}$ with $\lambda > 0$. As each $f_j \in \mathcal{F}_1(n)$, we see that f is well defined. Moreover, if we let $\xi^j = x^j + a_{j-1}\mathbb{1}$, then $a_{j-1}\mathbb{1} \leq \xi^j \leq a_j\mathbb{1}$, so that $\xi \in B_{j-1}$. Therefore ξ^j is a periodic point of f with period p_j. We leave it to the reader to check that $f \in \mathcal{F}_1(n)$.

The map f has period lcm $(p_1, \ldots, p_m)$. As $L(n) \subseteq P_1(n)$, it follows that lcm $(1, 2, \ldots, n)$ divides the period of f, and hence we conclude from Corollary 9.2.4 that f has period lcm $(1, 2, \ldots, n)$. □

This shows that lcm $(1, 2, \ldots, n)$ is a sharp upper bound for the possible periods of maps in $\mathcal{F}_1(n)$. This upper bound is, however, far from optimal for the individual periods of the periodic points of maps in $\mathcal{F}_1(n)$. To gain further insight into the set of possible periods of the periodic points, we need to examine the periodic points of lower semi-lattice homomorphisms more closely. It will be convenient to introduce the following technical definition.

Definition 9.2.8 Suppose that W is a lower semi-lattice in $\mathbb{R}^n$ and that $g \colon W \to W$ is a lower semi-lattice homomorphism which has a periodic point

$\xi \in W$ with period p. Let V be the lower semi-lattice generated by the orbit of ξ under g and denote the restriction of g to V by f. A finite sequence $(y^i)_{i=1}^m \subseteq V$ is called a *complete sequence for* ξ if it satisfies

(i) $y^i \leq \xi$ for $1 \leq i \leq m$,
(ii) y^i is irreducible in V for $1 \leq i \leq m$,
(iii) $p = \operatorname{lcm}(p_1, \ldots, p_m)$, where $1 \leq p_i \leq n$ is the period of y^i under f,
(iv) $h_V(y^i) \leq h_V(y^{i+1})$ for $1 \leq i < m$,
(v) y^i and y^j are not comparable for all $i \neq j$, and
(vi) $\mathcal{O}(y^i; f) \cap \mathcal{O}(y^j; f) = \emptyset$ for all $i \neq j$.

The following result shows that each periodic point of a lower semi-lattice homomorphism has a complete sequence.

Proposition 9.2.9 *If W is a lower semi-lattice in $\mathbb{R}^n$ and $g\colon W \to W$ is a lower semi-lattice homomorphism which has a periodic point $\xi \in W$, then there exists a complete sequence for ξ.*

Proof Let V be the lower semi-lattice generated by the orbit of ξ under g and let f denote the restriction of g to V. By Lemma 9.2.2 we have that

$$\xi = \sup_V\{v \in V : v \leq \xi \text{ and } v \text{ irreducible in } V\}. \tag{9.8}$$

Now let U be a minimal subset of $\{v \in V : v \leq \xi$ and v irreducible in $V\}$ such that $\xi = \sup_V(U)$. Suppose that $|U| = m$ and write $U = \{z^1, \ldots, z^m\}$. By relabeling we may assume that $h_V(z^i) \leq h_V(z^{i+1})$ for all $1 \leq i < m$. As m is minimal and $\xi = \sup_V(U)$, we know that for each $1 \leq k \leq m$ there exists $1 \leq i \leq n$ such that $z_i^k > \max\{z_i^j : 1 \leq j \leq m \text{ and } j \neq k\}$. Therefore $z^1, z^2, \ldots, z^m$ are pairwise incomparable. From Lemma 9.2.3 it follows that the period, p_k, of z^k under f is at most n. Moreover, each p_k divides p, as $f^p(v) = v$ for all $v \in V$. Thus, if we define $q = \operatorname{lcm}(p_1, p_2, \ldots, p_m)$, then q divides p. On the other hand,

$$f^q(\xi) = f^q(\sup_V(\{z^1, \ldots, z^m\})) = \sup_V(\{z^1, \ldots, z^m\}) = \xi$$

by Lemma 9.2.1. Therefore p divides q, so that $p = q$.

Thus $(z^k)_{k=1}^m$ is a sequence in V that satisfies properties (i)–(v) in Definition 9.2.8. To make it satisfy the last property we remove elements in the following way. Let $y^1 = z^1$. If $y^j = z^{\sigma(j)}$ for $1 \leq j \leq k$, where $\sigma(j) < \sigma(j+1)$ for $1 \leq j < k$, then we put $y^{k+1} = z^{\sigma(k+1)}$, where $\sigma(k+1)$ is the first index with $\sigma(k) < \sigma(k+1) \leq m$ such that the orbit $\mathcal{O}(z^{\sigma(k+1)}; f)$ is disjoint from all orbits $\mathcal{O}(z^{\sigma(j)}; f)$ with $1 \leq j \leq k$. By construction the finite sequence $(y^i)_{i=1}^r$ satisfies properties (i), (ii), and (iv)–(vi) of Definition 9.2.8. It also satisfies the third property. Indeed, $p_i = p_j$ whenever the orbits of z^i and z^j are not disjoint, so that

$$p = \mathrm{lcm}(p_1, \ldots, p_m) = \mathrm{lcm}(\{q_i : q_i \text{ period of } y^i\}).$$

Thus, $(y^i)_{i=1}^r$ is a complete sequence for ξ. □

Using the complete sequences we can associate a semi-infinite integer matrix (a_{ij}), where $1 \leq i \leq m$ and $j \in \mathbb{Z}$, with a periodic point of a lower semi-lattice homomorphism in the following way. Let $\xi \in W$ be a periodic point of a lower semi-lattice homomorphism $g: W \to W$ with period p. Let $V \subseteq W$ be the finite lower semi-lattice generated by the orbit of ξ under g and denote the restriction of g to V by f. Thus, $f: V \to V$ is a one-to-one lower semi-lattice homomorphism on V. By Proposition 9.2.9 there exists a complete sequence $(y^i)_{i=1}^m$ for ξ. Let p_i denote the period of y^i under f. It follows from Definition 9.2.8(ii) and Lemma 9.2.3(ii) that $f^j(y^i)$ is irreducible in V for all $1 \leq i \leq m$ and $j \in \mathbb{Z}$. Therefore, $I_V(f^j(y^i))$ is non-empty, and hence we can select for each $1 \leq i \leq m$ and $0 \leq j < p_i$ an integer $a_{ij} \in I_V(f^j(y^i))$. For general $j \in \mathbb{Z}$ we define a_{ij} by

$$a_{ij} = a_{ik} \quad \text{where } 0 \leq k < p_i \text{ and } j \equiv k \bmod p_i.$$

We note that $a_{ij} \in I_V(f^j(y^i))$ for all $1 \leq i \leq m$ and $j \in \mathbb{Z}$, because y^i has period p_i under f. The resulting semi-infinite matrix (a_{ij}) is called an *array of* ξ. It turns out that these arrays satisfy certain arithmetical and combinatorial constraints. To analyze these constraints we introduce in the next section the notion of an admissible array on n symbols.

9.3 Periodic points and admissible arrays

An admissible array on n symbols is a semi-infinite matrix in which the entries are taken from a finite symbol set $\Sigma = \{1, 2, \ldots, n\}$ and the rows have to satisfy certain arithmetical and combinatorial conditions. The precise definition is the following.

Definition 9.3.1 Let $(L, <)$ be a finite totally ordered set and let $\Sigma = \{1, 2, \ldots, n\}$. For each $\lambda \in L$ let $\vartheta_\lambda : \mathbb{Z} \to \Sigma$ be a map. The semi-infinite matrix $\vartheta = (\vartheta_\lambda(j))$, where $\lambda \in L$, $j \in \mathbb{Z}$, and the rows are ordered according to their height in $(L, <)$, is called an *admissible array on n symbols* if ϑ satisfies the following:

(i) For each $\lambda \in L$ there exists an integer $1 \leq p_\lambda \leq n$ such that $\vartheta_\lambda : \mathbb{Z} \to \Sigma$ is periodic with period p_λ; so, $\vartheta_\lambda(s + p_\lambda) = \vartheta_\lambda(s)$ for all $s \in \mathbb{Z}$ and $\vartheta_\lambda(s) \neq \vartheta_\lambda(t)$ for all $1 \leq s < t \leq p_\lambda$.

(ii) If $\lambda_1 < \lambda_2 < \ldots < \lambda_{r+1}$ is a sequence of distinct elements in L and

$$\vartheta_{\lambda_i}(s_i) = \vartheta_{\lambda_{i+1}}(t_i) \quad \text{for } 1 \leq i \leq r,$$

then

$$\sum_{i=1}^{r}(t_i - s_i) \not\equiv 0 \bmod \rho, \quad \text{where } \rho = \gcd(\{p_{\lambda_i} : 1 \leq i \leq r+1\}).$$

Here $\gcd(S)$ denotes the greatest common divisor of the elements of S. The *period* of an admissible array is defined to be the $\mathrm{lcm}(\{p_\lambda : \lambda \in L\})$. Furthermore, for $n \in \mathbb{N}$ we let

$$Q(n) = \{p \in \mathbb{N} : p \text{ is the period of an admissible array on } n \text{ symbols}\}.$$

We shall see that any array of a periodic point of a lower semi-lattice homomorphism $g : W \to W$, where $W \subseteq \mathbb{R}^n$, is an admissible array on n symbols. Before explaining this relation in detail, we make some basic observations about admissible arrays. First, we remark that we can always take $L = \{1, 2, \ldots, m\}$ where the ordering is the usual ordering. The advantage of the more general definition, however, is that we can easily talk about sub-arrays. In particular, if $L' \subseteq L$ and the ordering on L' coincides with the ordering on L, then $\vartheta' = (\vartheta_\lambda(j))$ where $\lambda \in L'$ is also an admissible array on n symbols. Such an array is called a *sub-array of* ϑ.

It is easy to see how the first property of an admissible array fits with the arrays of periodic points of lower semi-lattice homomorphisms. Indeed, if (a_{ij}) is an array of a periodic point of a lower semi-lattice homomorphism, then $a_i(j) \mapsto a_{ij}$ is periodic map from $\mathbb{Z}$ to $\{1, 2, \ldots, n\}$ with period p_i, and $1 \leq p_i \leq n$ by Definition 9.2.8(ii) and Lemma 9.2.3(iv). Moreover, if $a_{ij} = a_{ik}$ for some $1 \leq j < k \leq p_i$, then by construction $I_V(f^j(y^i)) \cap I_V(f^k(y^i))$ is non-empty, so that $f^j(y^i)$ and $f^k(y^i)$ are comparable by Lemma 9.2.3, which is impossible, as the orbit of y^i is an anti-chain. Thus, $a_{ij} \neq a_{ik}$ for all $1 \leq j < k \leq p_i$. This shows that the array (a_{ij}) satisfies the first property in Definition 9.3.1.

It turns out that the second property in Definition 9.3.1 also holds. To see this we need to think about the information that is encoded in the array (a_{ij}) of the periodic point of the lower semi-lattice homomorphism. Recall that the entry a_{ij} corresponds to the element $f^j(y^i)$ for $1 \leq i \leq m$ and $j \in \mathbb{Z}$, and that both f and f^{-1} are order-preserving on the lower semi-lattice generated by the periodic orbit of ξ. From the array (a_{ij}) we can read off which elements in $\{f^j(y^i): 1 \le i \le m \text{ and } j \in \mathbb{Z}\}$ are comparable. For example, suppose that there exist $i_1 < i_2 < i_3$ such that $a_{i_1 s_1} = a_{i_2 t_1}$ and $a_{i_2 s_2} = a_{i_3 t_2}$ for some $s_1, t_1, s_2, t_2 \in \mathbb{Z}$. Then we know that $f^{s_1}(y^{i_1})$ and $f^{t_1}(y^{i_2})$ are comparable and $f^{s_2}(y^{i_2})$ and $f^{t_2}(y^{i_3})$ are comparable by Lemma 9.2.3. Moreover it follows from properties (iv) and (vi) in Definition 9.2.8 that

$$f^{s_1}(y^{i_1}) \le f^{t_1}(y^{i_2}) \quad \text{and} \quad f^{s_2}(y^{i_2}) \le f^{t_2}(y^{i_3}),$$

as $i_1 < i_2 < i_3$. Since f and f^{-1} are order-preserving, we know that

$$f^{s_1+k}(y^{i_1}) \le f^{i_2+k}(y^{i_2}) \quad \text{and} \quad f^{s_2+m}(y^{i_2}) \le f^{t_2+m}(y^{i_3})$$

for all $k, m \in \mathbb{Z}$. In particular,

$$y^{i_1} \le f^{t_1-s_1}(y^{i_2}) \quad \text{and} \quad y^{i_2} \le f^{t_2-s_2}(y^{i_3})$$

and hencc

$$y^{i_1} \le f^{(t_1-s_1)+(t_2-s_2)}(y^{i_3}).$$

By construction, the elements of the complete sequencc $(y^i)_{i=1}^m$ are not comparable, and hence $(t_1-s_1)+(t_2-s_2) \not\equiv 0 \bmod p_{i_3}$. A pictorial representation of the situation is given in Figure 9.1.

We can generalize this idea slightly to deduce that

$$(t_1 - s_1) + (t_2 - s_2) \not\equiv 0 \bmod \rho,$$

where $\rho = \gcd(p_1, p_2, p_3)$, as follows. We know there exist $A_1, A_2, A_3 \in \mathbb{Z}$ such that

$$A_1 p_1 + A_2 p_2 + A_3 p_3 = \rho.$$

So, if $(t_1 - s_1) + (t_2 - s_2) \equiv 0 \bmod \rho$, then there exist $B_1, B_2, B_3 \in \mathbb{Z}$ such that

$$(t_1 - s_1) + (t_2 - s_2) - (B_1 p_1 + B_2 p_2 + B_3 p_3) = 0.$$

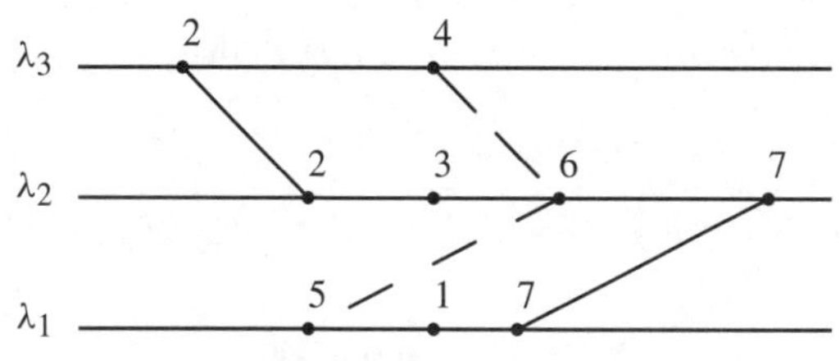

Figure 9.1 Admissible array.

Using the periodicity of y^1, y^2, and y^3, we find that

$$f^{s_1}(y^{i_1}) = f^{s_1+B_1p_1}(y^{i_1}) \leq f^{t_1}(y^{i_2}) \text{ and } f^{s_2}(y^{i_2}) = f^{s_2+B_2p_2}(y^{i_2}) \leq f^{t_2}(y^{i_3}).$$

This implies that

$$y^{i_1} \leq f^{(t_1-s_1)-B_1p_1}(y^{i_2}) \quad \text{and} \quad y^{i_2} \leq f^{(t_2-s_2)-B_2p_2}(y^{i_3}),$$

so that $y^{i_1} \leq f^{\mu}(y^{i_3})$, where $\mu = (t_1-s_2)+(t_2-s_2)-B_1p_1-B_2p_2$. From the fact that $f^{-B_3p_3}(y^{i_3}) = y^{i_3}$ we now deduce that $y^{i_1} \leq f^{\mu-B_3p_3}(y^{i_3}) = y^{i_3}$, which is impossible.

Generalizing these observations further leads to the following result.

Proposition 9.3.2 *Let W be a lower semi-lattice in $\mathbb{R}^n$ and let $g\colon W \to W$ be a lower semi-lattice homomorphism which has a periodic point $\xi \in W$ with period p. Let (a_{ij}), where $1 \leq i \leq m$ and $j \in \mathbb{Z}$, be an array of ξ. Put $\Sigma = \{1, 2, \ldots, n\}$ and let $L = \{1, \ldots, m\}$ be equipped with the usual ordering. If we define $\vartheta = (\vartheta_\lambda\colon \mathbb{Z} \to \Sigma)$ by*

$$\vartheta_\lambda(j) = a_{\lambda,j} \quad \textit{for all } \lambda \in L \textit{ and } j \in \mathbb{Z}, \tag{9.9}$$

then ϑ is an admissible array on n symbols with period p.

Proof Let $V \subseteq W$ be the finite lower semi-lattice generated by the orbit of ξ, and let f denote the restriction of g to V. Assume that (a_{ij}), where $1 \leq i \leq m$ and $j \in \mathbb{Z}$, is an array of ξ and suppose that $(y^i)_{i=1}^m$ is a complete sequence for ξ that generates this array. Also let p_i denote the period of y^i under f for $1 \leq i \leq m$. Put $\Sigma = \{1, \ldots, n\}$ and equip $L = \{1, \ldots, m\}$ with the usual ordering. Define $\vartheta = (\vartheta_\lambda(j))$ by (9.9).

By construction $a_{\lambda,j} = a_{\lambda,j+p_\lambda}$ for all $j \in \mathbb{Z}$ and $\lambda \in L$. Moreover, if $a_{\lambda,j} = a_{\lambda,k}$, then $I_V(f^j(y^\lambda)) \cap I_V(f^k(y^\lambda))$ is non-empty and hence Lemma 9.2.3 implies that $f^j(y^\lambda)$ and $f^k(y^\lambda)$ are comparable. But, as f is order-preserving and y^λ is a periodic point of f, the orbit of y^λ is an anti-chain and hence $f^j(y^\lambda) = f^k(y^\lambda)$. This implies that $j = k + mp_\lambda$ for some $m \in \mathbb{Z}$, which completes the proof of the first property in Definition 9.3.1.

To prove the second property suppose that there exist $s_1, \ldots, s_r, t_1, \ldots, t_r$ in $\mathbb{Z}$ and distinct $\lambda_1 < \lambda_2 < \ldots < \lambda_{r+1}$ in L such that

$$\vartheta_{\lambda_i}(s_i) = \vartheta_{\lambda_{i+1}}(t_i) \quad \text{for } 1 \leq i \leq r.$$

Let $\rho = \gcd(\{p_{\lambda_i} : 1 \leq i \leq r+1\})$. Suppose, by way of contradiction, that

$$\sum_{i=1}^{r}(t_i - s_i) \equiv 0 \bmod \rho.$$

Then there exist $B_1, \dots, B_{r+1} \in \mathbb{Z}$ such that

$$\sum_{i=1}^{r}(t_i - s_i) - \sum_{i=1}^{r+1} B_i p_{\lambda_i} = 0. \tag{9.10}$$

As $\vartheta_{\lambda_i}(s_i) = \vartheta_{\lambda_{i+1}}(t_i)$, we know that $f^{s_i}(y^{\lambda_i})$ and $f^{t_i}(y^{\lambda_{i+1}})$ are comparable by Lemma 9.2.3. Since $\lambda_i < \lambda_{i+1}$, we conclude from Definition 9.2.8 that

$$f^{s_i}(y^{\lambda_i}) < f^{t_i}(y^{\lambda_{i+1}}) \quad \text{for } 1 \le i \le r. \tag{9.11}$$

As f is order-preserving and y^{λ_i} has period p_{λ_i}, we deduce from (9.11) that

$$y^{\lambda_i} < f^{(t_i - s_i) - B_i p_{\lambda_i}}(y^{\lambda_{i+1}}) \quad \text{for } 1 \le i \le r. \tag{9.12}$$

Applying (9.12) iteratively gives

$$y^{\lambda_1} < f^{\nu}(y^{\lambda_{r+1}}), \quad \text{where } \nu = \sum_{i=1}^{r}(t_i - s_i) - \sum_{i=1}^{r} B_i p_{\lambda_i}.$$

Now put $\mu = -B_{r+1} p_{\lambda_{r+1}}$ and remark that, as $f^{\mu}(y^{\lambda_{r+1}}) = y^{\lambda_{r+1}}$, it follows from (9.10) and (9.12) that

$$y^{\lambda_1} < f^{\nu+\mu}(y^{\lambda_{r+1}}) = y^{\lambda_{r+1}}.$$

But this contradicts property (v) in Definition 9.2.8, and we are done. □

This proposition yields the following inclusions.

Theorem 9.3.3 *For each $n \in \mathbb{N}$ we have that*

$$P_1(n) \subseteq P_2(n) \subseteq Q_{\text{lat}}(n) \subseteq Q(n) \quad \textit{and} \quad P_3(n) \subseteq Q_{\text{sem}}(n) \subseteq Q(n).$$

Proof By (9.5) it remains to be shown that $Q_{\text{sem}} \subseteq Q(n)$, as $Q_{\text{lat}}(n) \subseteq Q_{\text{sem}}(n)$. If $p \in Q_{\text{sem}}(n)$, then there exist a lower semi-lattice homomorphism $g \colon W \to W$, where W is a lower semi-lattice in $\mathbb{R}^n$, and a periodic point $\xi \in W$ of g with period p. Let (a_{ij}) be an array of ξ and define ϑ as in Proposition 9.3.2. Then ϑ is an admissible array on n symbols with period p, and hence $p \in Q(n)$. □

Although the definition of an admissible array is somewhat complicated, the set $Q(n)$ can be computed in finite time. Of course this becomes increasingly hard as n increases. We shall discuss this issue in the next section, but first we prove the following striking equalities:

$$P_2(n) = P_3(n) = Q_{\text{lat}}(n) = Q_{\text{sem}} = Q(n) \quad \text{for all } n \in \mathbb{N}.$$

We note that these equalities do not involve the set $P_1(n)$. In fact, at present a complete combinatorial description of $P_1(n)$ is unknown.

To show that $P_2(n) = Q(n)$ we need to construct for each $p \in Q(n)$ an order-preserving integral-preserving map $f\colon \mathbb{R}^n_+ \to \mathbb{R}^n_+$ and a periodic point of f with period p. It turns out that we can associate with each admissible array ϑ a sand-shift map $f_\vartheta \in \mathcal{F}_2(n)$ that has a periodic point with the same period as ϑ.

Recall from Section 1.6 that a sand-shift map is defined in the following manner. Let $C_1, \ldots, C_n$ be containers each containing a finite amount of sand, x_i, and let $x = (x_1, \ldots, x_n) \in \mathbb{R}^n_+$. Assume that with each container C_i an infinite sequence of buckets $b_{i1}, b_{i2}, \ldots$ is associated. Let v_{ij} denote the volume of bucket b_{ij} and assume that $\sum_{m=1}^{\infty} v_{im} = \infty$ for all i. Now for each container C_i pour the contents of C_i into bucket b_{i1} until either b_{i1} is full or C_i is empty. If b_{i1} is full the remaining sand in C_i is poured into b_{i2} until either b_{i2} is full or C_i is empty. If b_{i2} is full the process is repeated until C_i is empty. The amount of sand in bucket b_{ik} is given by

$$M_{ik}(x) = \min\left(v_{ik}, \max\{x_i - \sum_{m=1}^{k-1} v_{im}, 0\}\right). \tag{9.13}$$

To get the sand back into the containers we use a fixed rule $\gamma\colon \{1, \ldots, n\} \times \mathbb{N} \to \{1, \ldots, n\}$ by pouring the contents of bucket b_{ik} into container $C_{\gamma(i,k)}$ for $1 \leq i \leq n$ and $k \in \mathbb{N}$. The new distribution of sand in the containers is given by $y = (y_1, \ldots, y_n) \in \mathbb{R}^n_+$, where

$$y_j = \sum_{\gamma(i,k)=j} M_{ik}(x) \quad \text{for } 1 \leq j \leq n.$$

The map $f_\gamma\colon \mathbb{R}^n_+ \to \mathbb{R}^n_+$ is called a *sand-shift map with rule* γ.

For our purposes we can assume that each bucket has volume 1. So, (9.13) reduces to

$$M_{ik}(x) = \min\left(1, \max\{x_i - (k-1), 0\}\right) \quad \text{for all } 1 \leq i \leq n \text{ and } k \in \mathbb{N}. \tag{9.14}$$

It turns out that given an admissible array ϑ we can specify a rule γ_ϑ, so that the associated sand-shift map has a periodic point whose period coincides with the period of ϑ.

To formulate the rule γ_ϑ, it is convenient to first collect some notation. Let $\vartheta = (\vartheta_\lambda\colon \mathbb{Z} \to \Sigma \mid \lambda \in L)$ be an admissible array on n symbols with period p. We write $R(\vartheta_\lambda) = \{\vartheta_\lambda(q)\colon q \in \mathbb{Z}\}$ to denote the range of ϑ_λ. For each symbol $a \in \Sigma$ we let $\Lambda_a = \{\lambda \in L\colon a \in R(\vartheta_\lambda)\}$ and we put $\kappa(a) = |\Lambda_a|$. If $\kappa(a) > 0$, we label the elements of Λ_a by $\lambda_1(a) < \lambda_2(a) < \ldots < \lambda_{\kappa(a)}(a)$. Using this notation we now define a rule γ_ϑ as follows.

Definition 9.3.4 Given an admissible array $\vartheta = (\vartheta_\lambda : \mathbb{Z} \to \Sigma \mid \lambda \in L)$ on n symbols with period p, the rule $\gamma_\vartheta : \{1, \ldots, n\} \times \mathbb{N} \to \{1, \ldots, n\}$ is defined as follows:

(a) If $1 \le i \le n$ and $1 \le k \le \kappa(i)$, then

$$\gamma_\vartheta(i, k) = \vartheta_{\lambda_k(i)}(s + 1), \text{ where } s \text{ is such that } i = \vartheta_{\lambda_k(i)}(s).$$

(b) If $1 \le i \le n$ and $k > \kappa(i)$, then

$$\gamma_\vartheta(i, k) = i.$$

Moreover, we denote by $f_\vartheta : \mathbb{R}^n_+ \to \mathbb{R}^n_+$ the sand-shift map with rule γ_ϑ and where each bucket has volume 1.

We remark that the rule γ_ϑ is well defined, as the maps ϑ_λ are periodic with period p_λ and $\vartheta_\lambda(s) \ne \vartheta_\lambda(t)$ whenever $0 < |s - t| < p_\lambda$.

The idea is to show that the sand-shift map $f_\vartheta : \mathbb{R}^n_+ \to \mathbb{R}^n_+$ has a periodic point with period p, if ϑ has period p. To construct this periodic point it is useful to introduce the following auxiliary numbers.

Definition 9.3.5 Given an admissible array $\vartheta = (\vartheta_\lambda : \mathbb{Z} \to \Sigma \mid \lambda \in L)$ on n symbols, we define, for each $q \in \mathbb{Z}$, $a \in \Sigma$, and $\lambda \in L$, numbers $\xi^q_{a,\lambda}$ by

1. $\xi^q_{a,\lambda} = 1/2$ if $\vartheta_\lambda(q) = a$,
2. $\xi^q_{a,\lambda} = 1$ if there exist distinct $\lambda_1 < \lambda_2 < \ldots < \lambda_{r+1}$ in L with $\lambda = \lambda_1$ and there exists $\delta \in \mathbb{Z}$ such that $a = \vartheta_\lambda(q - \delta)$,

$$\vartheta_{\lambda_i}(s_i) = \vartheta_{\lambda_{i+1}}(t_i) \quad \text{for } 1 \le i \le r,$$

and

$$\sum_{i=1}^{r} (t_i - s_i) \equiv \delta \bmod \rho, \quad \text{where } \rho = \gcd(\{p_{\lambda_i} : 1 \le i \le r + 1\}),$$

3. $\xi^q_{a,\lambda} = 0$ otherwise.

For example, in Figure 9.1, $\xi^0_{4,\lambda_3} = \xi^0_{3,\lambda_2} = \xi^0_{1,\lambda_1} = 1/2$ and $\xi^0_{6,\lambda_2} = \xi^0_{5,\lambda_1} = 1$. For each $q \in \mathbb{Z}$ we subsequently define $\xi^q \in \mathbb{R}^n_+$ by

$$\xi^q_i = \sum_{\lambda \in L} \xi^q_{i,\lambda} \quad \text{for all } 1 \le i \le n. \tag{9.15}$$

We shall see that ξ^q is a periodic point of the sand-shift map f_ϑ and has the same period as the admissible array ϑ. To prove this we first collect some basic properties of the numbers $\xi^q_{a,\lambda}$.

Lemma 9.3.6 *If* $\vartheta = (\vartheta_\lambda : \mathbb{Z} \to \Sigma \mid \lambda \in L)$ *is an admissible array on* n *symbols and the numbers* $\xi^q_{a,\lambda}$ *are defined as in Definition 9.3.5, then the following assertions hold:*

(i) The numbers $\xi^q_{a,\lambda}$ *are well defined.*
(ii) If $a \notin R(\vartheta_\lambda)$*, then* $\xi^q_{a,\lambda} = 0$.
(iii) If $\xi^q_{a,\lambda} > 0$*, then* $\xi^q_{a,\lambda'} = 1$ *for all* $\lambda' < \lambda$ *with* $a \in R(\vartheta_{\lambda'})$.

Proof The second assertion follows immediately from the definition of $\xi^q_{a,\lambda}$. To prove (i) we argue by contradiction. Assume that simultaneously $\xi^q_{a,\lambda} = 1/2$ and $\xi^q_{a,\lambda} = 1$. Then there exist distinct $\lambda_1 < \lambda_2 < \ldots < \lambda_{r+1}$ in L with $\lambda_1 = \lambda$ and there exists $\delta \in \mathbb{Z}$ such that $\vartheta_\lambda(q - \delta) = a$,

$$\vartheta_{\lambda_i}(s_i) = \vartheta_{\lambda_{i+1}}(t_i) \quad \text{for } 1 \le i \le r,$$

and

$$\sum_{i=1}^{r}(t_i - s_i) \equiv \delta \bmod \rho, \quad \text{where } \rho = \gcd(\{p_{\lambda_i} : 1 \le i \le r+1\}).$$

But also $\vartheta_\lambda(q - \delta) = \vartheta_\lambda(q) = a$, as $\xi^q_{a,\lambda} = 1/2$. Since ϑ_λ is periodic with period p_λ and $\vartheta_\lambda(s) \neq \vartheta_\lambda(t)$ for all $0 < |t - s| < p_\lambda$, we deduce that $\delta \equiv 0 \bmod p_\lambda$. This, however, implies that

$$\sum_{i=1}^{r}(t_i - s_i) \equiv 0 \bmod \rho,$$

which contradicts the second property in Definition 9.3.1.

To prove the last assertion we assume that $\xi^q_{a,\lambda} > 0$ and let $\lambda' < \lambda$ be such that $a \in R(\vartheta_{\lambda'})$. There are two cases to consider: $\xi^q_{a,\lambda} = 1/2$ and $\xi^q_{a,\lambda} = 1$. In the first case $\vartheta_\lambda(q) = a$. Since $a \in R(\vartheta_{\lambda'})$ there exists $k \in \mathbb{Z}$ such that $\vartheta_{\lambda'}(k) = a$. Put $\delta = q - k$ and remark that $\vartheta_{\lambda'}(k) = \vartheta_\lambda(q)$ and $a = \vartheta_{\lambda'}(q - \delta)$. Moreover, $q - k \equiv \delta \bmod \rho$, where $\rho = \gcd(p_{\lambda'}, p_\lambda)$ and hence $\xi^q_{a,\lambda'} = 1$. On the other hand, if $\xi^q_{a,\lambda} = 1$, then there exist distinct $\lambda_1 < \lambda_2 < \ldots < \lambda_{r+1}$ in L with $\lambda_1 = \lambda$ and there exists $\delta \in \mathbb{Z}$ such that $\vartheta_\lambda(q - \delta) = a$,

$$\vartheta_{\lambda_i}(s_i) = \vartheta_{\lambda_{i+1}}(t_i) \quad \text{for } 1 \le i \le r,$$

and

$$\sum_{i=1}^{r}(t_i - s_i) \equiv \delta \bmod \rho, \quad \text{where } \rho = \gcd(\{p_{\lambda_i} : 1 \le i \le r+1\}).$$

As $a \in R(\vartheta_{\lambda'})$, there exists $k \in \mathbb{Z}$ such that $\vartheta_{\lambda'}(k) = a$. Put $\lambda_0 = \lambda'$, $s_0 = k$, $t_0 = q - \delta$, $\delta' = q - k$, and $\rho' = \gcd(\rho, p_{\lambda_0})$. Clearly

$$\vartheta_{\lambda_i}(s_i) = \vartheta_{\lambda_{i+1}}(t_i) \quad \text{for } 0 \le i \le r$$

and $a = \vartheta_{\lambda_0}(q - \delta')$. Moreover, there exists $m \in \mathbb{Z}$ such that

$$\sum_{i=0}^{r} (t_i - s_i) = t_0 - s_o + \delta + m\rho = q - \delta - k + \delta + m\rho = q - k + m\rho.$$

Since ρ' divides ρ and $\delta' = q - k$, we find that

$$\sum_{i=0}^{r} (t_i - s_i) \equiv \delta' \bmod \rho'$$

and hence $\xi^q_{a,\lambda'} = 1$. □

We now use Lemma 9.3.6 to prove that ξ^q is a periodic point of the sand-shift map $f_\vartheta : \mathbb{R}^n_+ \to \mathbb{R}^n_+$ with the same period as ϑ.

Proposition 9.3.7 *Let $\vartheta = (\vartheta_\lambda : \mathbb{Z} \to \Sigma \mid \lambda \in L)$ be an admissible array on n symbols with period p and let $f_\vartheta . \mathbb{R}^n_+ \to \mathbb{R}^n_+$ be the sand-shift map with rule γ_ϑ given by Definition 9.3.4. If $\xi^q \in \mathbb{R}^n_+$ is given by (9.15), then $f_\vartheta(\xi^q) = \xi^{q+1}$ and ξ^q is a periodic point of f_ϑ with period p.*

Proof The proof of this proposition is based on the following two claims.
Claim 1.

$$\xi^q_{\vartheta_\lambda(s),\lambda} = \xi^{q+t}_{\vartheta_\lambda(s+t),\lambda} \quad \text{for all } q, s, t \in \mathbb{Z} \text{ and } \lambda \in L.$$

Claim 2.

$$f_\vartheta(\xi^q)_j = \sum_{\substack{(i,\lambda):\\ \vartheta_\lambda(s)=i,\ \vartheta_\lambda(s+1)=j}} \xi^q_{i,\lambda} \qquad \text{for all } q \in \mathbb{Z} \text{ and } 1 \le j \le n.$$

If we assume these claims for the moment, the proof of the proposition can be completed in the following manner. It follows from the claims and Lemma 9.3.6(ii) that

$$f_\vartheta(\xi^q)_j = \sum_{\substack{(i,\lambda):\\ \vartheta_\lambda(s)=i,\ \vartheta_\lambda(s+1)=j}} \xi^q_{i,\lambda} = \sum_{\substack{(i,\lambda):\\ \vartheta_\lambda(s)=i,\ \vartheta_\lambda(s+1)=j}} \xi^{q+1}_{\vartheta_\lambda(s+1),\lambda} = \sum_{\lambda \in L:\ j \in R(\vartheta_\lambda)} \xi^{q+1}_{j,\lambda} = \xi^{q+1}_j$$

for all $q \in \mathbb{Z}$ and $1 \le j \le n$. Thus, $f(\xi^q) = \xi^{q+1}$ for each $q \in \mathbb{Z}$. To verify that ξ^q is a periodic point of f_ϑ we first remark that if $i = \vartheta_\lambda(s)$, then

$$\xi^q_{i,\lambda} = \xi^q_{\vartheta_\lambda(s),\lambda} = \xi^{q+p_\lambda}_{\vartheta_\lambda(s+p_\lambda),\lambda} = \xi^{q+p_\lambda}_{i,\lambda}$$

by Claim 1. On the other hand if $i \notin R(\vartheta_\lambda)$, then $\xi^q_{i,\lambda} = \xi^{q+p_\lambda}_{i,\lambda} = 0$ by Lemma 9.3.6(ii). This implies that $\xi^{q+p} = \xi^q$ for all $q \in \mathbb{Z}$ and $p = \mathrm{lcm}(\{p_\lambda : \lambda \in L\})$, and therefore $f^p(\xi^q) = \xi^q$.

Let us now fix $q \in \mathbb{Z}$. To show that p is the period of ξ^q under f_ϑ, let $\mu \geq 1$ be the smallest integer such that $f^\mu(\xi^q) = \xi^q$. We need to prove that p_λ divides μ for all $\lambda \in L$. So, let $\lambda \in L$ and suppose that $\vartheta_\lambda(q) = j$. It follows from Lemma 9.3.6(iii) that λ is the only element in L such that $\xi^q_{j,\lambda} = 1/2$. Therefore $\xi^q_j = \sum_{\lambda \in L} \xi^q_{j,\lambda}$ is not an integer. This implies that $\xi^{q+\mu}_j$ is not an integer, as $\xi^{q+\mu}_j = f^\mu(\xi^q)_j = \xi^q_j$. Therefore there exists a unique $\lambda' \in L$ such that $\xi^{q+\mu}_{j,\lambda'} = 1/2$ and $\vartheta_{\lambda'}(q+\mu) = j$. If $\lambda' < \lambda$, then it follows from Lemma 9.3.6(iii) that $\xi^{q+\mu}_j < \xi^q_j$. Likewise, if $\lambda < \lambda'$, then $\xi^q_j < \xi^{q+\mu}_j$. Thus, $\lambda = \lambda'$, as $\xi^q_j = \xi^{q+\mu}_j$. From this equality it follows that $\vartheta_\lambda(q+\mu) = j = \vartheta_\lambda(q)$ and hence p_λ divides μ by Definition 9.3.1(i). Thus, ξ^q has period p under f_ϑ.

To complete the proof it remains to show the claims. To prove the first claim we distinguish two cases: $\xi^q_{\vartheta_\lambda(s),\lambda} = 1/2$ and $\xi^q_{\vartheta_\lambda(s),\lambda} = 1$. Note that if we can prove the equality in these two cases, the equality will also hold whenever $\xi^q_{\vartheta_\lambda(s),\lambda} = 0$. If $\xi^q_{\vartheta_\lambda(s),\lambda} = 1/2$, then $\vartheta_\lambda(s) = \vartheta_\lambda(q)$ and hence $\vartheta_\lambda(s+t) = \vartheta_\lambda(q+t)$, so that $\xi^{q+t}_{\vartheta_\lambda(q+t),\lambda} = 1/2 = \xi^q_{\vartheta_\lambda(s),\lambda}$. On the hand if $\xi^q_{\vartheta_\lambda(s),\lambda} = 1$, then there exist distinct $\lambda_1 < \lambda_2 < \ldots < \lambda_{r+1}$ in L with $\lambda_1 = \lambda$ and $\delta \in \mathbb{Z}$ such that $\vartheta_\lambda(s) = \vartheta_\lambda(q-\delta)$,

$$\vartheta_{\lambda_i}(s_i) = \vartheta_{\lambda_{i+1}}(t_i) \quad \text{for } 1 \leq i \leq r,$$

and

$$\sum_{i=1}^{r}(t_i - s_i) \equiv \delta \bmod \rho, \qquad \text{where } \rho = \gcd(\{p_{\lambda_i} : 1 \leq i \leq r+1\}).$$

As $\vartheta_\lambda(s) = \vartheta_\lambda(q-\delta)$, we know that $\vartheta_\lambda(s+t) = \vartheta_\lambda(q+t-\delta)$, so that $\xi^{q+t}_{\vartheta_\lambda(s+t),\lambda} = 1 = \xi^q_{\vartheta_\lambda(s),\lambda}$.

To obtain the second claim we remark that

$$f_\vartheta(\xi^q)_j = \sum_{\gamma_\vartheta(i,k)=j} M_{ik}(\xi^q) \quad \text{for } 1 \leq j \leq n,$$

where $M_{ik}(\xi^q) = \min\{1, \max\{\sum_{\lambda \in L} \xi^q_{i,\lambda} - (k-1), 0\}\}$, as the buckets have volume 1. It follows from Lemma 9.3.6 that

$$M_{ik}(\xi^q) = \begin{cases} \xi^q_{i,\lambda_k(i)} & \text{if } k \leq \kappa(i) \\ 0 & \text{otherwise.} \end{cases}$$

Now by using Definition 9.3.4 we deduce that

$$f_\vartheta(\xi^q)_j = \sum_{(i,k):\ \gamma_\vartheta(i,k)=j} M_{ik}(\xi^q) = \sum_{(i,k):\ \gamma_\vartheta(i,k)=j} \xi^q_{i,\lambda_k(i)} = \sum_{\substack{(i,\lambda):\\ \vartheta_\lambda(s)=i,\ \vartheta_\lambda(s+1)=j}} \xi^q_{i,\lambda},$$

which completes the proof. □

Combining this proposition with Theorem 9.3.3 yields the following equalities.

Theorem 9.3.8 *For each $n \in \mathbb{N}$ we have that*

$$P_2(n) = P_3(n) = Q_{\text{lat}}(n) = Q_{\text{sem}}(n) = Q(n).$$

Proof By Theorem 9.3.3 and the fact that $P_2(n) \subseteq P_3(n)$, it suffices to show that $Q(n) \subseteq P_2(n)$. So, suppose $p \in Q(n)$. Then there exists an admissible array on n symbols with period p. Let $f_\vartheta : \mathbb{R}^n_+ \to \mathbb{R}^n_+$ be the sand-shift map defined in Definition 9.3.4 and let ξ^0 be defined by (9.15). It follows from Proposition 9.3.7 that ξ^0 is a periodic point of f_ϑ with period p. As $f_\vartheta \in \mathcal{F}_2(n)$, we conclude that $p \in P_2(n)$. □

We mentioned earlier that this theorem leaves open the following problem.

Problem 9.3.9 *Does there exist a complete characterization of the set $P_1(n)$ in terms of arithmetical and combinatorial constraints?*

It is known [171] that $P_1(n) = Q(n)$ for all $n \leq 50$, which suggests that $P_1(n)$ may also equal $Q(n)$ for all $n \in \mathbb{N}$. The main obstruction to proving the equality for all n is that there appears to be no obvious class of maps in $P_1(n)$ that relates in a natural way to admissible arrays.

Given the construction of the sand-shift $f_\vartheta : \mathbb{R}^n_+ \to \mathbb{R}^n_+$ and the periodic point $\xi^0 \in \mathbb{R}^n_+$ it is natural to ask how the admissible array ϑ relates to the admissible array induced by an array of ξ^0. In fact, one may wonder if they are equal. It turns out that this is the case if the admissible array satisfies a simple additional condition. We shall explain this issue in detail in the remainder of the section, but it is not essential for the sequel.

An admissible array $\vartheta = (\vartheta_\lambda : \mathbb{Z} \to \Sigma \mid \lambda \in L)$ is said to be *minimal* if, for every proper subset L' of L, we have that

$$\operatorname{lcm}(\{p_\lambda : \lambda \in L'\}) < \operatorname{lcm}(\{p_\lambda : \lambda \in L\}).$$

We recall that each sub-array, $\vartheta' = (\vartheta_\lambda : \mathbb{Z} \to \Sigma \mid \lambda \in L')$, is also an admissible array on n symbols and hence

$$Q(n) = \{p \in \mathbb{N} : p \text{ is the period of a minimal admissible array on } n \text{ symbols}\}.$$

Under the additional assumption of minimality of the admissible array ϑ we shall see that the array of the periodic point ξ^0 of the sand-shift map f_ϑ coincides with ϑ. Before starting the proof of this result we remark that if $\vartheta = (\vartheta_\lambda : \mathbb{Z} \to \Sigma \mid \lambda \in L)$ is a minimal admissible array on n symbols, then for each $\lambda \in L$ there exists a prime power $m_\lambda^{\alpha_\lambda}$ such that $m_\lambda^{\alpha_\lambda}$ divides p_λ, but $m_\lambda^{\alpha_\lambda}$ does not divide $p_{\lambda'}$ for all $\lambda' \neq \lambda$. Using this observation we prove the following auxiliary lemma.

Lemma 9.3.10 *Let $\vartheta = (\vartheta_\lambda : \mathbb{Z} \to \Sigma \mid \lambda \in L)$ be a minimal admissible array on n symbols and let*

$$q_\lambda = \operatorname{lcm}(\{p_{\lambda'} : \lambda' \neq \lambda \text{ and } \lambda' \in L\} \cup \{p_\lambda / m_\lambda\}).$$

For each $a \in \Sigma$ and $\lambda \in L$ let

$$S_{a,\lambda} = \{m \in \mathbb{Z} : \xi_{a,\lambda}^m = 1\},$$

where $\xi_{a,\lambda}^m$ is defined as in Definition 9.3.5. Then we have that

$$S_{a,\lambda} = S_{a,\lambda} + rq_\lambda \quad \text{for all } r \in \mathbb{Z}.$$

Proof If $a \notin R(\vartheta_\lambda)$, then $\xi_{a,\lambda}^m = 0$ for all $m \in \mathbb{Z}$, so that $S_{a,\lambda} = \emptyset$. In this case clearly $S_{a,\lambda} = S_{a,\lambda} + rq_\lambda$ for all $r \in \mathbb{Z}$.

Now assume that $a \in R(\vartheta_\lambda)$. Then there exists $k \in \mathbb{Z}$ such that $\vartheta_\lambda(k) = a$. If $m \in S_{a,\lambda}$, then $\xi_{a,\lambda}^m = 1$, and hence there exist distinct $\lambda_1 < \ldots < \lambda_{r+1}$ in L, with $\lambda_1 = \lambda$, and $\delta \in \mathbb{Z}$ such that $a = \vartheta_\lambda(m - \delta)$,

$$\vartheta_{\lambda_i}(s_i) = \vartheta_{\lambda_{i+1}}(t_i) \quad \text{for } 1 \leq i \leq r,$$

and $\sum_{i=1}^r (t_i - s_i) \equiv \delta \bmod \rho$, where $\rho = \gcd(\{p_{\lambda_i} : 1 \leq i \leq r+1\})$. We remark that $\rho \mid q_\lambda$. For each $r \in \mathbb{Z}$, there exist $A_1, \ldots, A_{r+1} \in \mathbb{Z}$ such that

$$r\rho = A_1 p_{\lambda_1} + \cdots + A_{r+1} p_{\lambda_{r+1}}.$$

We shall now show that $\xi_{a,\lambda}^{m+r\rho} = 1$. For $1 \leq i \leq r$ let $s_i' = s_i - A_i p_{\lambda_i}$ and remark that $\vartheta_{\lambda_i}(s_i') = \vartheta_{\lambda_i}(s_i) = \vartheta_{\lambda_{i+1}}(t_i)$ for all $1 \leq i \leq r$. Moreover, if $\delta' = r\rho + \delta$, then $\vartheta_\lambda(m + r\rho - \delta') = \vartheta_\lambda(m - \delta) = a$ and

$$\sum_{i=1}^r (t_i - s_i') = \sum_{i=1}^r (t_i - s_i) + \sum_{i=1}^r A_i p_{\lambda_i} = \delta + r\rho - A_{r+1} p_{\lambda_{r+1}} \equiv \delta' \bmod \rho.$$

This implies that $\xi_{a,\lambda}^{m+r\rho} = 1$ and hence $m + r\rho \in S_{a,\lambda}$ for all $r \in \mathbb{Z}$. As ρ divides q_λ, we get that $m + rq_\lambda \in S_{a,\lambda}$ and hence $S_{a,\lambda} + rq_\lambda \subseteq S_{a,\lambda}$ for all $r \in \mathbb{Z}$.

To obtain the other inclusion we remark that if $m \in S_{a,\lambda}$ and $r \in \mathbb{Z}$, then $m - rq_\lambda \in S_{a,\lambda}$, so that $m = (m - rq_\lambda) + rq_\lambda \in S_{a,\lambda} + rq_\lambda$ and therefore $S_{a,\lambda} \subseteq S_{a,\lambda} + rq_\lambda$ for all $r \in \mathbb{Z}$. □

We can now prove the following theorem.

Theorem 9.3.11 *Let $\vartheta = (\vartheta_\lambda \colon \mathbb{Z} \to \Sigma \mid \lambda \in L)$ be a minimal admissible array on n symbols with period p and let $f_\vartheta \colon \mathbb{R}^n_+ \to \mathbb{R}^n_+$ be the sand-shift map associated with ϑ. Suppose that $\xi^m \in \mathbb{R}^n_+$ is defined by $\xi^m_i = \sum_{\lambda \in L} \xi^m_{i,\lambda}$, where $\xi^m_{i,\lambda}$ is given by Definition 9.3.5 for each $1 \le i \le n$ and $m \in \mathbb{Z}$. If we define*

$$S_\lambda = \{m \in \mathbb{Z} \colon \xi^m_{a,\lambda} \ge \xi^0_{a,\lambda} \text{ and } a = \vartheta_\lambda(0)\}$$

and

$$z^\lambda = \bigwedge_{m \in S_\lambda} \xi^m,$$

then the following assertions are true:

(i) *z^λ is an irreducible element in the lower semi-lattice V generated by $\{\xi^m \colon 0 \le m < p\}$.*
(ii) *z^λ is a periodic point of f_ϑ with period p_λ.*
(iii) *$I_V(f^k_\vartheta(z^\lambda)) = \{\vartheta_\lambda(k)\}$ for all $k \in \mathbb{Z}$.*
(iv) *z^λ and $z^{\lambda'}$ are not comparable for each $\lambda \ne \lambda'$ in L.*
(v) *$h_V(z^\lambda) < h_V(z^{\lambda'})$ whenever $\lambda < \lambda'$ and $R(\vartheta_\lambda) \cap R(\vartheta_{\lambda'}) \ne \emptyset$.*
(vi) *$\xi^0 = \bigvee_{\lambda \in L} z^\lambda$.*

Proof We first prove the second assertion. Let $S_{a,\lambda}$ be defined as in Lemma 9.3.10. If $m \in S_\lambda$, then $m + rp_\lambda \in S_\lambda$ by Claim 1 in the proof of Proposition 9.3.7, so that $S_\lambda = S_\lambda + rp_\lambda$ for all $r \in \mathbb{Z}$. Since $a = \vartheta_\lambda(0)$, $\xi^0_{a,\lambda} = 1/2$ and hence $\xi^m_{a,\lambda} \ge \xi^0_{a,\lambda}$ if and only if either $\xi^m_{a,\lambda} = 1/2$ or $\xi^m_{a,\lambda} = 1$. Remark that $\xi^m_{a,\lambda} = 1/2$ if and only if $\vartheta_\lambda(m) = a$ and hence $m = rp_\lambda$ for some $r \in \mathbb{Z}$. Thus, we find that S_λ is the disjoint union of $S_{\vartheta_\lambda(0),\lambda}$ and $\{rp_\lambda \colon r \in \mathbb{Z}\}$. In particular, we see that no element of $S_{\vartheta_\lambda(0),\lambda}$ is an integral multiple of p_λ. As $\xi^m_a = \sum_{\lambda \in L} \xi^m_{a,\lambda}$, we deduce from Lemma 9.3.6(iii) that

$$m \in S_\lambda \text{ if and only if } \xi^m_a \ge \xi^0_a \text{ for } a = \vartheta_\lambda(0). \tag{9.16}$$

This implies that $z^\lambda_a = \bigwedge_{m \in S_\lambda} \xi^m_a = \xi^0_a$ for $a = \vartheta_\lambda(0)$. Let us denote the restriction of f to the lower semi-lattice V generated by $\{\xi^m \colon 0 \le m < p\}$ by g. Then g is a lower semi-lattice homomorphism and therefore

$$g^{p_\lambda}(z^\lambda) = \bigwedge_{m \in S_\lambda} g^{p_\lambda}(\xi^m) = \bigwedge_{m \in S_\lambda} \xi^{m+p_\lambda} = \bigwedge_{k \in S_\lambda + p_\lambda} \xi^k = \bigwedge_{k \in S_\lambda} \xi^k = z^\lambda.$$

Thus z^λ is a periodic point of g whose period q divides p_λ. Now suppose that $q \neq p_\lambda$. Then, for $r \in \mathbb{Z}$,

$$z^\lambda = g^{rq}(z^\lambda) = \bigwedge_{m\in S_\lambda} g^{rq}(\xi^m) = \bigwedge_{m\in S_\lambda} \xi^{m+rq} = \bigwedge_{k\in S_\lambda+rq} \xi^k.$$

This implies that $\xi^k_a \geq z^\lambda_a = \xi^0_a$ for $a = \vartheta_\lambda(0)$ and $k \in S_\lambda + rq$. Thus $S_\lambda + rq \subseteq S_\lambda$ for all $r \in \mathbb{Z}$ by (9.16), and hence $S_\lambda = S_\lambda + rq$ for all $r \in \mathbb{Z}$. From this we deduce that $q = 0 + q \in S_\lambda$ and, since q divides p_λ and $q \neq p_\lambda$, we find that $q \in S_{\vartheta_\lambda(0),\lambda}$. From Lemma 9.3.10 we get that $q + q_\lambda \in S_{\vartheta_\lambda(0),\lambda} \subseteq S_\lambda$ and $q_\lambda = (q + q_\lambda) - q \in S_\lambda - q = S_\lambda$. But $q_\lambda \notin \{rp_\lambda : r \in \mathbb{Z}\}$, as there exists a prime power $m_\lambda^{\alpha_\lambda}$ that is a divisor of p_λ but not of q_λ. Thus, $q_\lambda \in S_{\vartheta_\lambda(0),\lambda}$ and hence $0 = q_\lambda - q_\lambda \in S_{\vartheta_\lambda(0),\lambda}$ by Lemma 9.3.10, which is a contradiction. This shows that $q = p_\lambda$.

Next we prove assertion (i). We first show that if $\zeta \in V$ and $\zeta < z^\lambda$, then $\zeta_a \leq z^\lambda_a - 1/2$ for $a = \vartheta_\lambda(0)$. There exists $J \subseteq \{0, 1, \ldots, p-1\}$ such that $\zeta = \bigwedge_{j\in J} \xi^j$. If $J \subseteq S_\lambda$, then $\zeta \geq z^\lambda$. But $\zeta < z^\lambda$, so there exists $j \in J$ such that $j \notin S_\lambda$. For this j we must have that $\xi^j_{a,\lambda} = 0$, and hence it follows from Lemma 9.3.6 that

$$\zeta_a \leq \xi^j_a \leq \xi^0_a - 1/2 = z^\lambda_a - 1/2,$$

since $a = \vartheta_\lambda(0)$.

To see that z^λ is irreducible in V, it suffices to construct $w \in V$ such that $\zeta \leq w \wedge z^\lambda < z^\lambda$ for all $\zeta \in V$ with $\zeta < z^\lambda$. Put $w = \xi^{q_\lambda} = g^{q_\lambda}(\xi^0)$, where q_λ is as in Lemma 9.3.10. Note that

$$w_a = \xi^{q_\lambda}_a = \sum_{\lambda'\in L} \xi^{q_\lambda}_{a,\lambda'}.$$

It follows from Claim 1 in the proof of Proposition 9.3.7 that $\xi^{q_\lambda}_{a,\lambda'} = \xi^0_{a,\lambda'}$ for $\lambda' \neq \lambda$, as $p_{\lambda'}$ is a divisor of q_λ whenever $\lambda' \neq \lambda$. Hence

$$w_a = \xi^{q_\lambda}_{a,\lambda} + \sum_{\lambda'\in L:\, \lambda'\neq\lambda} \xi^0_{a,\lambda'}. \tag{9.17}$$

Lemma 9.3.10 implies that $\xi^{q_\lambda}_{a,\lambda} = 1$ if and only if $\xi^0_{a,\lambda} = 1$. If $a = \vartheta_\lambda(0)$, then $\xi^0_{a,\lambda} = 1/2$ and hence $\xi^{q_\lambda}_{a,\lambda} \neq 1$. Since q_λ is not an integral multiple of p_λ, $\xi^{q_\lambda}_{a,\lambda} = 0$ and therefore

$$w_a = \left(\sum_{\lambda'\in L} \xi^0_{a,\lambda'}\right) - 1/2 = \xi^0_a - 1/2 \quad \text{for } a = \vartheta_\lambda(0)$$

by (9.17). If $a = \vartheta_\lambda(q_\lambda)$, then by the same argument $\xi^0_a < w_a$. If $\vartheta_\lambda(0) \neq a$ and $\vartheta_\lambda(q_\lambda) \neq a$, then neither $\xi^0_{a,\lambda}$ nor $\xi^{q_\lambda}_{a,\lambda}$ can be equal to $1/2$. Thus in that case

$\xi^0_{a,\lambda}, \xi^{q_\lambda}_{a,\lambda} \in \{0, 1\}$. As $\xi^{q_\lambda}_{a,\lambda} = 1$ if and only if $\xi^0_{a,\lambda} = 1$, we find that $\xi^0_{a,\lambda} = \xi^{q_\lambda}_{a,\lambda}$ in that case. From (9.17) it follows that $w_a = \xi^0_a$ if $\vartheta_\lambda(0) \neq a \neq \vartheta_\lambda(q_\lambda)$. From these inequalities we deduce that $(w \wedge z^\lambda)_a = z^\lambda_a$ for all $a \neq \vartheta_\lambda(0)$, as $z^\lambda \leq \xi^0$, and $(w \wedge z^\lambda)_a = w_a = z^\lambda_a - 1/2$ for $a = \vartheta_\lambda(0)$. We already saw that if $\zeta \in V$ and $\zeta < z^\lambda$, then $\zeta_a \leq z^\lambda_a - 1/2$ for $a = \vartheta_\lambda(0)$. Thus, $\zeta \leq w \wedge z^\lambda < z^\lambda$ for all $\zeta < z^\lambda$ in V, and hence z^λ is irreducible.

To prove assertion (iii) we note that $a = \vartheta_\lambda(0)$ is the only coordinate for which $(w \wedge z^\lambda)_a < z^\lambda_a$, and hence $I_V(z^\lambda) = \{\vartheta_\lambda(0)\}$. As z^λ is irreducible, we know that $g^k(z^\lambda)$ is also irreducible by Lemma 9.2.3(ii). Moreover, one can show, as before, that if $\zeta \in V$ and $\zeta < g^k(z^\lambda)$, then $\zeta \leq g^k(w) \wedge g^k(z^\lambda) < g^k(z^\lambda)$ and $a = \vartheta_\lambda(k)$ is the only coordinate such that $(g^k(w) \wedge g^k(z^\lambda))_a < g^k(z^\lambda)_a$. Thus, $I_V(g^k(z^\lambda)) = \{\vartheta_\lambda(k)\}$ for all $k \in \mathbb{Z}$.

It is convenient to prove assertion (vi) before (iv) and (v). To do so it is sufficient to show that for each $1 \leq a \leq n$ there exists $\lambda \in L$ such that $\xi^m_a \geq \xi^0_a$ for all $m \in S_\lambda$, since $z^\lambda \leq \xi^0$. If $\xi^0_a - \lfloor \xi^0_a \rfloor > 0$, then there exists $\lambda \in L$ with $\vartheta_\lambda(0) = a$. In that case it follows from (9.16) that $\xi^m_a \geq \xi^0_a$ for all $m \in S_\lambda$. Now assume that $\xi^0_a - \lfloor \xi^0_a \rfloor = 0$. If $\xi^0_a = 0$, there is nothing to prove, since $0 \leq z^\lambda \leq \xi^0$ for all $\lambda \in L$. On the other hand, if $\xi^0_a > 0$, then there exists $\lambda' \in L$ such that $\xi^0_{a,\lambda'} = 1$ and $\xi^0_{a,\lambda} = 0$ for all $\lambda > \lambda'$ in L. Remark that

$$\xi^0_a = \sum_{\lambda \in L} \xi^0_{a,\lambda} = \sum_{\lambda \leq \lambda'} \xi^0_{a,\lambda}$$

and $\xi^m_a = \sum_{\lambda \in L} \xi^m_{a,\lambda}$. It follows from Lemma 9.3.6(iii) that $\xi^m_a \geq \xi^0_a$ if $\xi^m_{a,\lambda'} = 1$. Thus, we need to show that there exists $\lambda \in L$ such that for all $m \in S_\lambda$, $\xi^m_{a,\lambda'} = 1$, as $z^\lambda = \bigwedge_{m \in S_\lambda} \xi^m$.

Since $\xi^0_{a,\lambda'} = 1$, we know there exist distinct $\lambda_1 < \ldots < \lambda_{r+1}$ in L, with $\lambda_1 = \lambda'$, and $\delta \in \mathbb{Z}$ such that $a = \vartheta_{\lambda'}(-\delta)$,

$$\vartheta_{\lambda_i}(s_i) = \vartheta_{\lambda_{i+1}}(t_i) \quad \text{for } 1 \leq i \leq r,$$

and $\sum_{i=1}^r (t_i - s_i) \equiv \delta \bmod \rho$, where $\rho = \gcd(\{p_{\lambda_i} : 1 \leq i \leq r+1\})$. Put $\lambda = \lambda_{r+1}$ and let $m \in S_\lambda$. Then $\xi^m_{b,\lambda} \geq \xi^0_{b,\lambda}$, where $b = \vartheta_\lambda(0)$. As $\xi^0_{b,\lambda} = \xi^0_{\vartheta_\lambda(0),\lambda} = 1/2$, either $\xi^m_{b,\lambda} = 1/2$ or $\xi^m_{b,\lambda} = 1$. In the first case we know that there exists $k \in \mathbb{Z}$ such that $m = kp_\lambda$. Put $s^*_i = s_i$ for $1 \leq i \leq r$, $t^*_i = t_i$ for $1 \leq i \leq r-1$, and $t^*_r = t_r + kp_\lambda$. Then $a = \vartheta_{\lambda'}(m - \delta^*)$, where $\delta^* = m + \delta$,

$$\vartheta_{\lambda_i}(s^*_i) = \vartheta_{\lambda_{i+1}}(t^*_i) \quad \text{for } 1 \leq i \leq r,$$

and $\sum_{i=1}^r (t^*_i - s^*_i) \equiv \delta^* \bmod \rho$. Therefore $\xi^m_{a,\lambda'} = 1$. On the other hand, if $\xi^m_{b,\lambda} = 1$, then there exist $\lambda_{r+1} < \ldots < \lambda_{r+s+1}$ in L and $\mu \in \mathbb{Z}$ such that $\vartheta_{\lambda_{r+1}}(m - \mu) = b$,

$$\vartheta_{\lambda_{r+i}}(s_{r+i}) = \vartheta_{\lambda_{r+i+1}}(t_{r+i}) \quad \text{for } 1 \le i \le s,$$

and $\sum_{i=1}^{s}(t_{r+i} - s_{r+i}) \equiv \mu \bmod \rho'$, where $\rho' = \gcd(\{p_{\lambda_{r+i}} : 1 \le i \le s+1\})$. As $\vartheta_{\lambda_{r+1}}(0) = b$ we may assume that $\mu = m$. Now we note that

$$\sum_{i=1}^{r+s}(t_i - s_i) \equiv \delta + \mu \bmod \rho^*, \quad \text{where } \rho^* = \gcd(\rho, \rho'),$$

and $\vartheta_{\lambda'}(m - (\delta + \mu)) = \vartheta_{\lambda'}(-\delta) = a$. This implies that $\xi^m_{a,\lambda'} = 1$. Therefore we conclude that $\xi^m_{a,\lambda'} = 1$ for all $m \in S_\lambda$, which shows that

$$\xi^0 = \bigvee_{\lambda \in L} z^\lambda. \tag{9.18}$$

It is now easy to prove that z^λ and $z^{\lambda'}$ are not comparable for all $\lambda \neq \lambda'$ in L. Indeed, it follows from (9.18) that if L' is a proper subset of L such that $\xi^0 = \bigvee_{\lambda \in L'} z^\lambda$, then

$$g^q(\xi^0) = \bigvee_{\lambda \in L'} g^q(z^\lambda) = \bigvee_{\lambda \in L'} z^\lambda = \xi^0$$

for $q = \operatorname{lcm}(\{p_\lambda : \lambda \in L'\})$. But, as ϑ is a minimal admissible array, we get that $q < p$, which contradicts the fact that ξ^0 has period p under g. Thus no two elements z^λ are comparable.

It remains to show that (v) holds. If $R(\vartheta_\lambda) \cap R(\vartheta_{\lambda'}) \neq \emptyset$ for $\lambda < \lambda'$, then there exist $s, t \in \mathbb{Z}$ such that $\vartheta_\lambda(s) = \vartheta_{\lambda'}(t)$. We claim that $S_{\lambda'} + (t - s) \subseteq S_{\vartheta_\lambda(0),\lambda}$. Indeed, if $\xi^m_{\vartheta_{\lambda'}(0),\lambda'} > 0$, then $\xi^{m+t}_{\vartheta_\lambda(s),\lambda'} = \vartheta^{m+t}_{\vartheta_{\lambda'}(t),\lambda'} > 0$ by Claim 1 in the proof of Proposition 9.3.7. As $\lambda < \lambda'$, this implies that $\xi^{m+t}_{\vartheta_\lambda(s),\lambda} = 1$ by Lemma 9.3.6(iii). Therefore $\xi^{m+t-s}_{\vartheta_\lambda(0),\lambda} = 1$, which shows that $m + (t - s) \in S_{\vartheta_\lambda(0),\lambda}$.

From the inclusion we deduce that

$$g^{(t-s)}(z^{\lambda'}) = \bigwedge_{m \in S_{\lambda'}+(t-s)} \xi^m \ge \bigwedge_{m \in S_{\vartheta_\lambda(0),\lambda}} \xi^m.$$

By definition $\xi^m_{\vartheta_\lambda(0),\lambda} = 1$ for all $m \in S_{\vartheta_\lambda(0),\lambda}$, so that $\xi^m_{\vartheta_\lambda(0),\lambda} > \xi^0_{\vartheta_\lambda(0),\lambda} = 1/2$ for all $m \in S_{\vartheta_\lambda(0),\lambda}$. This implies that

$$\bigwedge_{m \in S_{\vartheta_\lambda(0),\lambda}} \xi^m > \bigwedge_{m \in S_\lambda} \xi^m = z^\lambda$$

and hence $g^{(t-s)}(z^{\lambda'}) > z^\lambda$. As $h_V(g^m(y)) = h_V(y)$ for all $y \in V$ and $m \in \mathbb{Z}$, we conclude that $h_V(z^{\lambda'}) > h_V(z^\lambda)$, which completes the proof. □

From Theorem 9.3.11 it follows that if ϑ is a minimal admissible array on n symbols, then the admissible array induced by the periodic point ξ^0 of the sand-shift map $f_\vartheta\colon \mathbb{R}^n_+ \to \mathbb{R}^n_+$ defined in Definition 9.3.5 coincides with ϑ. Thus, every minimal admissible array on n symbols is the array of a periodic point of a map in $\mathcal{F}_2(n)$.

9.4 Computing periods of admissible arrays

For fixed $n \in \mathbb{N}$ the set of possible periods of admissible arrays on n symbols, denoted $Q(n)$, can be computed in finite time. To see this we first remark that

$$Q(n) = \{p \in \mathbb{N}\colon p \text{ is the period of a minimal array on } n \text{ letters}\}.$$

As the period of each ϑ_λ in an admissible array ϑ is at most n, there are only finitely many candidate sets $S \subseteq \{1, \ldots, n\}$ for the periods of the individual maps ϑ_λ to consider. Moreover, given a set $S \subseteq \{1, \ldots, n\}$ there are finitely many ways to order the elements of S, which corresponds to selecting the totally ordered set $(L, <)$, and there are only finitely many choices for each map ϑ_λ, as they are periodic. Thus, for a given n, we only need to decide for finitely many arrays ϑ if they are admissible or not. To verify if an array is admissible one needs to check the two conditions from Definition 9.3.1. The first condition is automatically fulfilled by choosing the maps ϑ_λ appropriately. To check the second condition we only need to consider $s_i, t_i \in \mathbb{Z}$ with $0 \leq s_i, t_i \leq \mathrm{lcm}(\{p_\lambda\colon \lambda \in L\})$ for all $1 \leq i \leq r$. From these observations it follows that the set $Q(n)$ can in theory be computed in finite time. However, the computation of $Q(n)$ becomes rapidly harder when n gets large. A detailed analysis of the set $Q(n)$ for $1 \leq n \leq 50$ has been given in [171]. In this section we discuss some basic properties of $Q(n)$ that allow us to quickly compute $Q(n)$ for relatively small n.

To simplify matters it is useful to first identify a large set $P(n)$ such that $P(n) \subseteq Q(n)$ for all $n \in \mathbb{N}$.

Definition 9.4.1 For each $n \in \mathbb{N}$ the set $P(n)$ is defined inductively by $P(1) = \{1\}$, and for $n > 1$, $p \in P(n)$ if either

(A) $p = \mathrm{lcm}(p_1, p_2)$, where $p_1 \in P(n_1)$, $p_2 \in P(n_2)$ and $n = n_1 + n_2$, or
(B) $p = r\mathrm{lcm}(p_1, p_2, \ldots, p_r)$, where $r > 1$ and there exist $m \geq 1$ such that $p_i \in P(m)$ for $1 \leq i \leq r$ and $n = mr$.

It is easy to see that $P(n)$ contains the set

$$L(n) = \{\mathrm{lcm}(p_1, \ldots, p_m)\colon p_1 + \cdots + p_m \leq n \text{ and } p_1, \ldots, p_m \in \mathbb{N}\}, \quad (9.19)$$

which is the set of orders of permutations on n letters. In general $P(n)$ is much larger than $L(n)$, e.g., $12 \in P(6) \setminus L(6)$. It can be shown that $P(n) \subseteq Q(n)$ for all $n \in \mathbb{N}$. Instead of proving this inclusion directly, we show that $P(n) \subseteq P_1(n)$ for all $n \in \mathbb{N}$, which is a slightly stronger statement; see Theorem 9.3.3.

Theorem 9.4.2 *For each $n \in \mathbb{N}$, $P(n) \subseteq P_1(n)$.*

Proof We argue by induction on n. Clearly $P(1) \subseteq P_1(n)$. Now suppose that $P(m) \subseteq P_1(m)$ for all $m < n$. We need to show that $P(n) \subseteq P_1(n)$. So let $p \in P(n)$. By definition there are two cases, (A) and (B). If we are in case (A), then $p = \operatorname{lcm}(p_1, p_2)$, where $p_1 \in P(n_1)$, $p_2 \in P(n_2)$, and $n_1 + n_2 = n$. By induction we get that $p_1 \in P_1(n_1)$ and $p_2 \in P_1(n_2)$, so that there exist $f \in \mathcal{F}_1(n_1)$ and $g \in \mathcal{F}_1(n_2)$, which have periodic points $\xi \in \mathbb{R}^{n_1}_+$ and $\eta \in \mathbb{R}^{n_2}_+$ with periods p_1 and p_2, respectively. Define $h\colon \mathbb{R}^n_+ \to \mathbb{R}^n_+$ by $h(x, y) = (f(x), g(y))$ for all $(x, y) \in \mathbb{R}^{n_1}_+ \times \mathbb{R}^{n_2}_+$. It is easy to verify that $h \in \mathcal{F}_1(n)$ and that (ξ, η) is a periodic point of h with period $\operatorname{lcm}(p_1, p_2) = p$. Thus $p \in P_1(n)$ in case (A).

In case (B) the induction hypothesis gives $p_i \in P_1(m)$ for $1 \leq i \leq r$. By Theorem 9.2.7 we know that there exists $f \in \mathcal{F}_1(m)$ such that f has periodic points $\xi^i \in \mathbb{R}^m_+$ with period p_i for $1 \leq i \leq r$ and $\mathcal{O}(\xi^i) \cap \mathcal{O}(\xi^j) = \emptyset$ for all $i \neq j$. Define $h\colon \mathbb{R}^n_+ \to \mathbb{R}^n_+$ by

$$h(x^1, x^2, \ldots, x^r) = (f(x^r), x^1, x^2, \ldots, x^{r-1}),$$

where $x^i \in \mathbb{R}^m_+$ for $1 \leq i \leq r$. Clearly $h \in \mathcal{F}_1(n)$. Let $\xi = (\xi^1, \ldots, \xi^r) \in \mathbb{R}^n_+$. As $\mathcal{O}(\xi^i) \cap \mathcal{O}(\xi^j) = \emptyset$ for all $i \neq j$, $h^q(\xi) = \xi$ only if r divides q. Moreover $h^{kr}(x) = (f^k(x^1), f^k(x^2), \ldots, f^k(x^r))$ for each $k \in \mathbb{N}$. Therefore, $h^{kr}(\xi) = \xi$ if and only if p_i divides k for all $1 \leq i \leq r$. The smallest such k is equal to $\operatorname{lcm}(p_1, p_2, \ldots, p_r)$ and hence ξ has period $r\operatorname{lcm}(p_1, p_2, \ldots, p_r)$ under h. This shows that $p \in P_1(n)$. □

The set $P(n)$ is fairly easy to compute, certainly in comparison to $Q(n)$. Using the inclusion $P(n) \subseteq Q(n)$ it is straightforward to compute $Q(n)$ for small n. For example, $Q(1) = \{1\}$ and $Q(2) = \{1, 2\}$. We know that $Q(3) \subseteq \{1, 2, 3, 6\}$, but $6 \notin Q(3)$. Indeed, if $6 \in Q(3)$, there would exist an admissible array $\vartheta = (\vartheta_\lambda\colon \mathbb{Z} \to \{1, 2, 3\} \mid \lambda = 1, 2)$ with period 6 such that ϑ_1 has period 2 and ϑ_2 has period 3. This implies that $R(\vartheta_1) \cap R(\vartheta_2)$ is non-empty and hence there exist $s, t \in \mathbb{Z}$ such that $\vartheta_1(s) = \vartheta_2(t)$. But this contradicts the fact that $s - t \not\equiv 0 \bmod 1$ and therefore $6 \notin Q(3)$. The reasoning above exemplifies the following basic property of admissible arrays.

Lemma 9.4.3 *If $\vartheta = (\vartheta_\lambda \colon \mathbb{Z} \to \Sigma \mid \lambda \in L)$ is an admissible array on n symbols and there exists $L' \subseteq L$ such that* $\gcd(p_\lambda, p_{\lambda'}) = 1$ *for all $\lambda \neq \lambda'$ in L', then $\sum_{\lambda \in L'} p_\lambda \leq n$.*

Proof If $\sum_{\lambda \in L'} p_\lambda > n$, then there exists $\lambda \neq \lambda'$ in L' such that $R(\vartheta_\lambda) \cap R(\vartheta_{\lambda'}) \neq \emptyset$. This implies that $\vartheta_\lambda(s) = \vartheta_{\lambda'}(t)$ for some $s, t \in \mathbb{Z}$, which contradicts the fact that $s - t \not\equiv 0 \bmod 1$. □

As a consequence of Lemma 9.4.3, we deduce that $6, 12 \notin Q(4)$ and hence $Q(4) = P(4) = \{1, 2, 3, 4\}$. To compute $Q(5)$ quickly it is convenient to use another basic property of $Q(n)$.

Lemma 9.4.4 *If $p \in Q(n)$ and q divides p, then $q \in Q(n)$.*

Proof Clearly it is true that if $p \in P_2(n)$ and q divides p, then $q \in P_2(n)$. As $P_2(n) = Q(n)$ by Theorem 9.3.8, the set $Q(n)$ also has this property. □

Note that the property in Lemma 9.4.4 is not immediately clear from the definition of the admissible arrays. To compute $Q(5)$ we first remark that $Q(5) \subseteq \{1, 2, 3, 4, 5, 6, 10, 12, 15, 20, 30, 60\}$. But 10, 12, and 15 are not in $Q(5)$ by Lemma 9.4.3, so that $Q(5) = P(5) = \{1, 2, 3, 4, 5, 6\}$ by Lemma 9.4.4. Similarly we know that $Q(6) \subseteq \{1, 2, 3, 4, 5, 6, 10, 12, 15, 20, 30, 60\}$ and that 10 and 15 are not in $Q(6)$ by Lemma 9.4.3. As $\{1, 2, 3, 4, 5, 6, 12\} = P(6) \subseteq Q(6)$, it follows from Lemma 9.4.4 that $Q(6) = \{1, 2, 3, 4, 5, 6, 12\}$. To construct an admissible array on 6 symbols with period 12 we cannot use periods 3 and 4, but instead we have to use periods 6 and 4. An example of an admissible array on 6 with period 12 is given by

...	1	2	3	4	5	6	1	2	3	4	5	6	...
...	2	3	4	5	2	3	4	5	2	3	4	5	...

Using the same type of arguments, the set $Q(n)$ can be computed for $1 \leq n \leq 10$. See Table 9.1 below. It turns out that $P(n) = Q(n)$ for all $1 \leq n \leq 10$. In fact, it has been shown in [171], with the aid of a computer, that $P(n) = Q(n)$ for $1 \leq n \leq 50$. From this equality one might conjecture that $Q(n)$ is equal to $P(n)$ for all n, but it has been shown in [171, theorem 7.10 and corollary 7.11] that if p_1 and p_2 are twin primes, so $p_1 = p_2 + 2$, with $p_2 \geq 11$ and $p_2 \neq 41$, then $q = 2^3 7^2 p_1 p_2 \in Q(56 + 2p_2)$, but $q \notin P(56 + 2p_2)$. In particular, 56 056 is in $Q(78)$, but not in $P(78)$. In connection with Problem 9.3.9 we remark that $P(n) \subseteq P_1(n) \subseteq Q(n)$ by Theorems 9.3.3 and 9.4.2, so that $P_1(n) = Q(n)$ for $1 \leq n \leq 50$. Moreover, it is known that $56\,056 \in P_1(78)$ and therefore $P(n) \neq P_1(n)$ in general.

Table 9.1 *The elements of $Q(n)$ for $1 \leq n \leq 10$.*

n	Elements of $Q(n)$
1	1
2	1, 2
3	1, 2, 3,
4	1, 2, 3, 4
5	1, 2, 3, 4, 5, 6
6	1, 2, 3, 4, 5, 6, 12
7	1, 2, 3, 4, 5, 6, 7, 10, 12
8	1, 2, 3, 4, 5, 6, 7, 8, 10, 12, 15, 24
9	1, 2, 3, 4, 5, 6, 7, 8, 9, 10, 12, 14, 15, 18, 20, 24
10	1, 2, 3, 4, 5, 6, 7, 8, 9, 10, 12, 14, 15, 18, 20, 21, 24, 30, 40, 60

In view of the computational difficulties in determining $Q(n)$, the following question is interesting.

Problem 9.4.5 *Does there exist a different characterization of $Q(n)$ that allows it to be computed more easily?*

The computation of $Q(n)$ for $1 \leq n \leq 50$ requires more sophisticated properties of admissible arrays. For the interested reader we discuss some of these properties at the end of this section, but first we give an asymptotic estimate for the largest element of $Q(n)$.

Let $g(n) = \max\{p \colon p \in L(n)\}$, so $g(n)$ is the maximum order of a permutation on n letters. Landau [118] has proved the following classical result concerning the asymptotics of $g(n)$:

$$\lim_{n\to\infty} \frac{\log g(n)}{\sqrt{n \log n}} = 1. \tag{9.20}$$

A proof of this result can be obtained using the prime number theorem; see Miller [147]. More precise estimates for $g(n)$ have been obtained by Massias [141]. We note that as $L(n) \subseteq P(n) \subseteq P_1(n) \subseteq Q(n)$, $g(n)$ is a lower bound for the largest element of $Q(n)$. It turns out that on a log scale they have the same asymptotics.

Theorem 9.4.6 *If $\gamma(n) = \max\{p \colon p \in Q(n)\}$, then*

$$\lim_{n\to\infty} \frac{\log \gamma(n)}{\sqrt{n \log n}} = 1.$$

Proof Let $\gamma(n) = \prod_{i=1}^{s} p_i^{\alpha_i}$, with $p_1 < p_2 < \ldots < p_s$, be the prime factorization of $\gamma(n)$. Since $\gamma(n) \in Q(n)$, there exists an admissible array $\vartheta = (\vartheta_\lambda \colon \mathbb{Z} \to \Sigma \mid \lambda \in L)$ on n symbols such that $\gamma(n) = \operatorname{lcm}(\{q_\lambda \colon \lambda \in L\})$, where q_λ is the period ϑ_λ. This implies that for each $1 \leq i \leq s$ there exists $\lambda \in L$ such that $p_i^{\alpha_i}$ is a divisor of q_λ. As $q_\lambda \leq n$, we deduce that

$$p_i^{\alpha_i} \leq n \quad \text{for } 1 \leq i \leq s. \tag{9.21}$$

Let $0 \leq r \leq s$ be the integer such that $p_r \leq \sqrt{n}$ and $p_{r+1} > \sqrt{n}$, where we define $r = 0$ if $p_1 > \sqrt{n}$, and $r = s$ if $p_s \leq \sqrt{n}$. Put

$$\gamma_1(n) = \prod_{i=1}^{r} p_i^{\alpha_i} \quad \text{and} \quad \gamma_2(n) = \prod_{i=r+1}^{s} p_i^{\alpha_i},$$

where $\gamma_1(n) = 1$ if $r = 0$, and $\gamma_2(n) = 1$ if $r = s$. From Lemma 9.4.4 it follows that $\gamma_1(n) \in Q(n)$ and $\gamma_2(n) \in Q(n)$. Let us first analyze $\gamma_2(n)$. Remark that $\gamma_2(n) = \prod_{i=r+1}^{s} p_i^{\alpha_i}$ and $p_i > \sqrt{n}$ for $i \geq r+1$. Therefore $\alpha_i = 1$ for all $i \geq r+1$ by (9.21). Since $\gamma_2(n) \in Q(n)$, there exists an admissible array $\psi = (\psi_\lambda : \mathbb{Z} \to \Sigma \mid \lambda \in L)$ on n symbols such that $\gamma_2(n) = \operatorname{lcm}(\{q'_\lambda : \lambda \in L'\})$, where q'_λ is the period of ψ_λ for $\lambda \in L'$. Moreover for each $r+1 \leq i \leq s$ there exists $\lambda_i \in L'$ such that p_i divides q'_{λ_i}. As $q'_{\lambda_i} \leq n$, p_i is the only prime among $p_{r+1}, \ldots, p_s$ that divides q'_{λ_i}. From this it follows that $p_i = q'_{\lambda_i}$, since $\gamma_2(n) = \operatorname{lcm}(\{q'_\lambda : \lambda \in L'\})$. Thus, $\gcd(q'_\lambda, q'_{\lambda'}) = 1$ for all $\lambda \neq \lambda'$ in L' and hence $\sum_{\lambda \in L'} q'_\lambda \leq n$ by Lemma 9.4.3. Therefore

$$\gamma_2(n) = \operatorname{lcm}(\{q'_\lambda : \lambda \in L'\}) \leq g(n). \tag{9.22}$$

To estimate $\gamma_1(n)$ we let $\pi(\sqrt{n})$ denote the number of primes p such that $p \leq \sqrt{n}$. As $p_i^{\alpha_i} \leq n$, we get that

$$\gamma_1(n) = \prod_{i=1}^{r} p_i^{\alpha_i} \leq n^{\pi(\sqrt{n})}. \tag{9.23}$$

Combining equations (9.22) and (9.23) gives

$$\log g(n) \leq \log \gamma(n) = \log \gamma_1(n) + \log \gamma_2(n) \leq \pi(\sqrt{n}) \log n + \log g(n). \tag{9.24}$$

Now using the prime number theorem, which says that

$$\lim_{n\to\infty} \frac{\pi(n) \log n}{n} = 1,$$

we see that

$$\lim_{n\to\infty} \frac{\pi(\sqrt{n}) \log n}{\sqrt{n \log n}} = \lim_{n\to\infty} \left(\frac{\pi(\sqrt{n}) \log \sqrt{n}}{\sqrt{n}}\right)\left(\frac{2}{\sqrt{\log n}}\right) = 0,$$

so that (9.20) and (9.24) yield

$$\lim_{n\to\infty} \frac{\log \gamma(n)}{\sqrt{n \log n}} = 1.$$

□

Although on a log scale there is no difference between $g(n)$ and $\gamma(n)$, it can be shown that $L(n)$ is strictly contained in $Q(n)$ for all $n \geq 6$. A proof of this result and many other properties of $\gamma(n)$ can be found in [172].

We conclude this section with some more sophisticated properties of admissible arrays that are useful in the computation of $Q(n)$. This material is not essential for the remainder of the book, but is included for the curious reader.

Theorem 9.4.7 *Let $L = \{1, \ldots, m+1\}$ be equipped with the usual ordering and $m \geq 1$. Suppose that $\vartheta = (\vartheta_i : \mathbb{Z} \to \Sigma \mid i \in L)$ is an admissible array on n symbols. Let R_i denote the range of ϑ_i and let p_i be the period of ϑ_i for $i \in L$. If the following conditions hold:*

(i) $R_i \cap R_{i+1} \neq \emptyset$ for all $1 \leq i \leq m$,
(ii) there exists an integer $r > 1$ such that $\gcd(p_i, p_{i+1})$ divides r for all $1 \leq i \leq m$, and
(iii) there exist $1 \leq k \leq m+1$ and integers $r_1 > 1$ and $r_2 \geq 1$ such that $r = r_1 r_2$, and $\gcd(p_{k-1}, p_k)$ and $\gcd(p_k, p_{k+1})$ are divisors of r_1, where we let $p_0 = p_{m+2} = 1$,

then

$$m + 1 = |L| \leq r_1 r_2 - r_2 + 1.$$

Proof We first note that if k is as in the statement of the theorem, then we may assume that $k > 1$. Indeed, if $k = 1$, then we can consider the reversed array $\vartheta' = (\vartheta_i' : \mathbb{Z} \to \Sigma \mid i \in L')$, where $L' = L$ and $<'$ is the reverse ordering on $L' = L$, so $a <' b$ if and only if $b < a$. It is clear that ϑ' is also an admissible array on n symbols.

As $R_i \cap R_{i+1} \neq \emptyset$ for all $1 \leq i \leq m$, there exist $s_i, t_i \in \mathbb{Z}$ such that

$$\vartheta_i(s_i) = \vartheta_{i+1}(t_i) \quad \text{for all } 1 \leq i \leq m.$$

For $1 \leq j \leq m$ we let

$$\eta_j = \sum_{i=1}^{j} (s_i - t_i).$$

Now for $1 \leq j \leq m$ and $j \neq k-1$ we claim that

$$\eta_{k-1} \not\equiv \eta_j \bmod r_1. \tag{9.25}$$

To prove this, note that for $1 \leq j < k-1$ we have that

$$\eta_{k-1} - \eta_j = \sum_{i=j+1}^{k-1} (s_i - t_i).$$

It follows from the assumptions that $\gcd(\{p_i : j+1 \le i \le k\})$ divides r_1, so that $\eta_{k-1} - \eta_j \not\equiv 0 \bmod r_1$, as ϑ is an admissible array. On the other hand, if $k-1 < j \le m$, then

$$\eta_j - \eta_{k-1} = \sum_{i=k}^{j} (s_i - t_i).$$

Again the assumptions imply that $\gcd(\{p_i : k \le i \le j+1\})$ divides r_1, so that $\eta_j - \eta_{k-1} \not\equiv 0 \bmod r_1$. This proves the claim.

Let $\pi : \mathbb{Z} \to \mathbb{Z}/r\mathbb{Z}$ be the quotient map that maps $a \in \mathbb{Z}$ to its equivalence class modulo r. Define

$$T = \{\eta_j : 1 \le j \le m \text{ and } j \ne k-1\} \cup \{\eta_{k-1} + sr_1 : 0 \le s < r_2\}.$$

As $\eta_l - \eta_j = \sum_{i=j+1}^{l} s_i - t_i \not\equiv 0 \bmod r$ for $1 \le j < l \le m$, it follows from (9.25) that $|T| = m - 1 + r_2$ and π restricted to T is one-to-one. As a final step we show that $\pi(\eta) \ne 0$ for all $\eta \in T$. Clearly $\pi(\eta_j) \ne 0$ for all $1 \le j \le m$ and $j \ne k-1$; otherwise $\eta_j = \sum_{i=1}^{j} s_i - t_i \equiv 0 \bmod r$, which contradicts the fact that ϑ is an admissible array. But also if $\pi(\eta_{k-1} + sr_1) \equiv 0 \bmod r$ for some $0 \le s < r_2$, then $\eta_{k-1} \equiv 0 \bmod r_1$. This implies that

$$\eta_{k-1} = \sum_{i=1}^{k-1} (s_i - t_i) \equiv 0 \bmod r_1,$$

which contradicts the fact that ϑ is an admissible array, as $\gcd(\{p_i : 1 \le i \le k\})$ divides r_1. Thus, π is a one-to-one map of T into $(\mathbb{Z}/r\mathbb{Z}) \setminus \{0\}$, and hence

$$|T| = m - 1 + r_2 \le r - 1,$$

which completes the proof. □

The following simple consequence of Theorem 9.4.7 is often easier to apply.

Corollary 9.4.8 *If $\vartheta = (\vartheta_i : \mathbb{Z} \to \Sigma \mid i \in L)$ is an admissible array on n symbols, where ϑ_i has period p_i and range R_i for $i \in L$, then the following conditions hold for $S = \{p_i : i \in L\}$:*

A. There does not exist $R \subseteq S$ such that

(i) $|R| = r + 1$, where $r > 1$ and $\gcd(p_i, p_j)$ divides r for all $p_i, p_j \in R$ with $p_i \ne p_j$, and

(ii) $R_i \cap R_j \ne \emptyset$ for all $p_i, p_j \in R$.

B. There does not exist $R \subseteq S$ such that

(i) $|R| = r_1 r_2 - r_2 + 2$, where $r_1 > 1$ and $r_2 \ge 1$ are integers,

(ii) $\gcd(p_i, p_j)$ divides $r = r_1 r_2$ for all $p_i \ne p_j$ in R,

(*iii*) *there exists* $p_k \in R$ *such that* $\gcd(p_k, p_i)$ *divides* r_1 *for all* $p_i \neq p_k$ *in* R, *and*

(*iv*) $R_i \cap R_j \neq \emptyset$ *for all* $p_i, p_j \in R$.

Proof Obviously assertion A is a special case of assertion B, where $r_2 = 1$. To prove assertion B, we simply consider the sub-array $\vartheta' = (\vartheta_i : \mathbb{Z} \to \Sigma \mid i \in L')$ of ϑ, where L' consists of those $i \in L$ such that $p_i \in R$, and apply Theorem 9.4.7. □

9.5 Maps on the whole space

Instead of looking at maps on the positive cone we can consider maps $f : \mathbb{R}^n \to \mathbb{R}^n$ that are non-expansive under the ℓ_1-norm. We know from Corollary 4.2.5 and Lemma 4.2.6 that the set $R(n)$, consisting of those $p \in \mathbb{N}$ for which there exists an ℓ_1-norm non-expansive map $f : \mathbb{R}^n \to \mathbb{R}^n$ that has a periodic point with period p, is finite. In view of Theorem 9.3.8 we may wonder if there exists a characterization of $R(n)$ in terms of arithmetical and combinatorial constraints. Surprisingly, such a characterization exists and is closely related to the set $Q(n)$. A key result in our analysis is the following lemma, where $\mathbb{E}^{2n} = \{(x, y) \in \mathbb{R}^n_+ \times \mathbb{R}^n_+ : x \wedge y = 0\}$.

Lemma 9.5.1 *An integer* p *is an element of* $R(n)$ *if and only if there exists a map* $f \in \mathcal{F}_3(2n)$ *such that* f *has a periodic point* $\xi \in \mathbb{E}^{2n}$ *with period* p *and* $\mathcal{O}(\xi) \subseteq \mathbb{E}^{2n}$.

Proof If $p \in R(n)$, there exist an ℓ_1-norm non-expansive map $f : \mathbb{R}^n \to \mathbb{R}^n$ and a periodic point $\xi \in \mathbb{R}^n$ of f with period p. As $\Omega_f \neq \emptyset$, we know that f has a fixed point, $\eta \in \mathbb{R}^n$, by Proposition 3.2.4. Define $h : \mathbb{R}^n \to \mathbb{R}^n$ by $h(x) = f(x + \eta) - \eta$ for all $x \in \mathbb{R}^n$. Clearly h is also non-expansive under the ℓ_1-norm and $h(0) = 0$.

Let $J : \mathbb{R}^n \to \mathbb{E}^{2n}$ be defined by $J(x) = (x \vee 0, (-x) \vee 0)$ for all $x \in \mathbb{R}^n$. It is easy to verify that J is an ℓ_1-norm isometry that maps $\mathbb{R}^n$ onto $\mathbb{E}^{2n}$. Therefore the inverse, J^{-1}, is also an ℓ_1-norm isometry on $\mathbb{E}^{2n}$. Let $R : \mathbb{R}^{2n}_+ \to \mathbb{E}^{2n}$ be the retraction onto $\mathbb{E}^{2n}$ given by

$$R(x, y) = (x - (x \wedge y), y - (x \wedge y)) \quad \text{for all } (x, y) \in \mathbb{R}^n_+ \times \mathbb{R}^n_+.$$

It is not hard to show that R is ℓ_1-norm non-expansive.

Finally let $g : \mathbb{R}^{2n}_+ \to \mathbb{R}^{2n}_+$ be defined by

$$g(z) = (J \circ h \circ J^{-1} \circ R)(z) \quad \text{for all } z \in \mathbb{R}^{2n}_+.$$

By construction, $g(0) = 0$ and g is non-expansive under the ℓ_1-norm, so that $g \in \mathcal{F}_3(2n)$. Moreover, $\zeta = J(\xi - \eta)$ is a periodic point of g with period p and $g^k(\zeta) \in \mathbb{E}^{2n}$ for all $k \in \mathbb{N}$.

On the other hand, if $g \in \mathcal{F}_3(2n)$ and $\xi \in \mathbb{E}^{2n}$ is a periodic point of g with period p and $g^k(\xi) \in \mathbb{E}^{2n}$ for all $k \in \mathbb{N}$, then $f\colon \mathbb{R}^n \to \mathbb{R}^n$, given by $f(x) = (J^{-1} \circ R \circ g \circ J)(x)$ for all $x \in \mathbb{R}^n$, is ℓ_1-norm non-expansive and has $J^{-1}(\xi)$ as a periodic point with period p. Therefore $p \in R(n)$ in that case. □

Note that Lemma 9.5.1 implies $R(n) \subseteq P_3(2n)$, which gives the following inclusion by Theorem 9.3.8.

Corollary 9.5.2 *For each $n \in \mathbb{N}$, $R(n) \subseteq Q(2n)$.*

In general $R(n) \neq Q(2n)$, but, as we shall see, there exists an additional constraint on the admissible arrays on $2n$ symbols such that their periods characterize the set $R(n)$. To formulate this additional constraint it is convenient to introduce the following notion.

Definition 9.5.3 Given an admissible array $\vartheta = (\vartheta_\lambda\colon \mathbb{Z} \to \Sigma \mid \lambda \in L)$, we say that a symbol $a \in \Sigma$ is *reachable* from $q \in \mathbb{Z}$ if either

(i) $a = \vartheta_\lambda(q)$ for some $\lambda \in L$, or
(ii) there exist distinct $\lambda_1 < \lambda_2 < \ldots < \lambda_{r+1}$ in L and $\delta \in \mathbb{Z}$ such that $a = \vartheta_{\lambda_1}(q - \delta)$,

$$\vartheta_{\lambda_i}(s_i) = \vartheta_{\lambda_{i+1}}(t_i) \quad \text{for } 1 \leq i \leq r,$$

and

$$\sum_{i=1}^{r}(t_i - s_i) \equiv \delta \bmod \rho,$$

where $\rho = \gcd(\{p_{\lambda_i} : 1 \leq i \leq r + 1\})$.

We denote the set of all symbols that are reachable from q by $\mathcal{R}(q, \vartheta)$; so,

$$\mathcal{R}(q, \vartheta) = \{a \in \Sigma\colon\ a \text{ is reachable from } q\}.$$

If $\Sigma = \{1, 2, \ldots, 2n\}$ and $a \in \Sigma$, we write $a^+ = a + n$ if $1 \leq a \leq n$ and $a^+ = a - n$ otherwise. Further, for $S \subseteq \Sigma$, we let $S^\bullet = \{a \in \Sigma\colon a \in S \text{ or } a^+ \in S\}$. Using these notions we can now formulate the additional constraint on the admissible arrays on $2n$ symbols.

Definition 9.5.4 An admissible array ϑ on $2n$ symbols is called a *restricted admissible array on* $2n$ *symbols* if

$$\{a, a^+\} \not\subseteq \mathcal{R}(q, \vartheta) \quad \text{for all } a \in \Sigma \text{ and } q \in \mathbb{Z}.$$

Note that if p is the period of ϑ, then $\mathcal{R}(q,\vartheta) = \mathcal{R}(q+p,\vartheta)$ for all $q \in \mathbb{Z}$, and hence we can decide in finite time whether an admissible array on $2n$ symbols is restricted or not. Define, for each $n \in \mathbb{N}$,

$$Q'(2n) = \{p \in \mathbb{N}\colon\ p \text{ is the period of a restricted admissible array on } 2n \text{ symbols}\}.$$

The characterization of $R(n)$ can now be stated as follows.

Theorem 9.5.5 *For each $n \in \mathbb{N}$, $R(n) = Q'(2n)$.*

We first prove that $Q'(2n) \subseteq R(n)$ for each $n \in \mathbb{N}$. As we shall see, the additional constraint on the admissible ϑ ensures that the orbit of the periodic point $\xi^0 \in \mathbb{R}^{2n}_+$ defined in Equation (9.15) is contained in $\mathbb{E}^{2n}$ under the sand-shift map $f_\vartheta\colon \mathbb{R}^{2n}_+ \to \mathbb{R}^{2n}_+$. So we can apply Lemma 9.5.1 to derive the following inclusion.

Proposition 9.5.6 *For each $n \in \mathbb{N}$, $Q'(2n) \subseteq R(n)$.*

Proof If $p \in Q'(2n)$, then there exists a restricted admissible array $\vartheta = (\vartheta_\lambda\colon \mathbb{Z} \to \Sigma \mid \lambda \in L)$ on $2n$ symbols with period p. Let $f_\vartheta\colon \mathbb{R}^{2n}_+ \to \mathbb{R}^{2n}_+$ be the sand-shift map with rule γ_ϑ defined in Definition 9.3.4 and let $\xi^q \in \mathbb{R}^{2n}_+$ for $q \in \mathbb{Z}$ be given by (9.15). It follows from Proposition 9.3.7 that ξ^0 is a periodic point of f_ϑ with period p and $f^q_\vartheta(\xi^0) = \xi^q$ for all $q \geq 0$. Moreover, we get from Definition 9.3.5 that $\xi^q_a > 0$ if and only if $a \in \mathcal{R}(q,\vartheta)$. As ϑ is a restricted admissible array, we deduce that $\xi^q_a \xi^q_{a^+} = 0$ for all $a \in \{1,2,\ldots,2n\} = \Sigma$ and $q \geq 0$. Thus, Lemma 9.5.1 implies that $p \in R(n)$. □

The proof of the opposite inclusion, $R(n) \subseteq Q'(2n)$, relies on the following proposition.

Proposition 9.5.7 *Let $W \subseteq \mathbb{E}^{2n}$ be a lower semi-lattice and let $g\colon W \to W$ be a lower semi-lattice homomorphism that has a periodic point $\xi \in W$ with period p. Let (a_{ij}), where $1 \leq i \leq m$ and $j \in \mathbb{Z}$, be an array of ξ. Put $\Sigma = \{1,2,\ldots,2n\}$ and let $L = \{1,\ldots,m\}$ be equipped with the usual ordering. If $\vartheta = (\vartheta_\lambda\colon \mathbb{Z} \to \Sigma \mid \lambda \in L)$ is defined by*

$$\vartheta_\lambda(j) = a_{\lambda,j} \quad \textit{for all } \lambda \in L \textit{ and } j \in \mathbb{Z}, \tag{9.26}$$

then ϑ is a restricted admissible array on $2n$ symbols with period p.

Proof Let $V \subseteq W$ be the lower semi-lattice by $\mathcal{O}(\xi)$ and write f to denote the restriction of g to V. Assume that (a_{ij}), where $1 \leq i \leq m$ and $j \in \mathbb{Z}$, is an array of ξ and suppose that $(y^i)_{i=1}^m$ is a complete sequence for ξ that induces this array. Further let p_i denote the period y^i under f for $1 \leq i \leq m$. Put $\Sigma = \{1, 2 \ldots, 2n\}$ and let $L = \{1, \ldots, m\}$ be equipped with the usual ordering. Now define $\vartheta = (\vartheta_\lambda : \mathbb{Z} \to \Sigma \mid \lambda \in L)$ by (9.26).

It follows from Proposition 9.3.2 that ϑ is an admissible array on $2n$ symbols with period p. Therefore it remains to show that

$$\{a, a^+\} \not\subseteq \mathcal{R}(q, \vartheta) \quad \text{for all } a \in \Sigma \text{ and } q \in \mathbb{Z}. \tag{9.27}$$

If $|L| = 1$, then it follows from Definition 9.5.3 that $|\mathcal{R}(q, \vartheta)| = 1$ for all $a \in \Sigma$ and $q \in \mathbb{Z}$, and hence (9.27) holds. To prove (9.27) if $|L| > 1$, we use the following claim: if $|L| > 1$ and $a \in \mathcal{R}(q, \vartheta)$, then there exists $y \in V$ such that $y_a > 0$ and $y \leq f^q(\xi)$.

It suffices to prove the claim, as $\{a, a^+\} \subseteq \mathcal{R}(q, \vartheta)$ implies that there exist $y, y' \in V$ with $y_a > 0$ and $y'_{a^+} > 0$, such that $y \leq f^q(\xi)$ and $y' \leq f^q(\xi)$, which contradicts the fact that $f^q(\xi) \in \mathbb{E}^{2n}$.

So, let $|L| > 1$ and $a \in \mathcal{R}(q, \vartheta)$. As $|L| > 1$ it follows from Definition 9.2.8 that $y^\lambda > \inf_V(V)$ for all $\lambda \in L$. This implies that $S_{f^k(y^\lambda)}$ is non-empty for all $\lambda \in L$ and $k \in \mathbb{Z}$, because f is order-preserving and y^λ is a periodic point of f. It follows from Lemma 9.2.3 that for each $\lambda \in L$ and $k \in \mathbb{Z}$ we have that

$$f^k(y^\lambda)_a > 0 \quad \text{for all } a \in I_V(f^k(y^\lambda)), \tag{9.28}$$

as $f^k(y^\lambda)_a > \sup_V(S_{f^k(y^\lambda)}) \geq 0$. Since $a \in \mathcal{R}(q, \vartheta)$, there are two cases to consider. We begin with the first one: $a = \vartheta_\lambda(q)$ for some $\lambda \in L$. By construction $a = \vartheta_\lambda(q) = a_{\lambda,q} \in I_V(f^q(y^\lambda))$. Hence $f^q(y^\lambda)_a > 0$ by (9.28). As $y^\lambda \leq \xi$ by Definition 9.2.8, we also get that $f^q(y^\lambda) \leq f^q(\xi)$ and hence the proof of the claim is complete in this case.

In the other case there exist distinct $\lambda_1 < \lambda_2 < \ldots < \lambda_{r+1}$ in L and $\delta \in \mathbb{Z}$ such that $a = \vartheta_{\lambda_1}(q - \delta)$,

$$\vartheta_{\lambda_i}(s_i) = \vartheta_{\lambda_{i+1}}(t_i) \quad \text{for } 1 \leq i \leq r,$$

and

$$\sum_{i=1}^{r} (t_i - s_i) \equiv \delta \bmod \rho,$$

where $\rho = \gcd(\{p_{\lambda_i} : 1 \leq i \leq r + 1\})$. The remainder of the argument is similar to the one used in the proof of Proposition 9.3.2. First remark that there exist integers $A_1, A_2, \ldots, A_{r+1}$ such that $\rho = \sum_{i=1}^{r+1} A_i p_{\lambda_i}$. As $\sum_{i=1}^{r} t_i - s_i \equiv \delta \bmod \rho$, there exist integers $B_1, B_2, \ldots, B_{r+1}$ such that

$$\sum_{i=1}^{r} (t_i - s_i) - \sum_{i=1}^{r+1} B_i p_{\lambda_i} = \delta. \tag{9.29}$$

Because $\vartheta_{\lambda_i}(s_i) = \vartheta_{\lambda_{i+1}}(t_i)$, $f^{s_i}(y^{\lambda_i})$ and $f^{t_i}(y^{\lambda_{i+1}})$ are comparable by Lemma 9.2.3(iii). From this we deduce that $f^{s_i}(y^{\lambda_i}) \leq f^{t_i}(y^{\lambda_{i+1}})$, since $\lambda_i < \lambda_{i+1}$. This implies that

$$y^{\lambda_i} \leq f^{t_i - s_i - B_i p_{\lambda_i}}(y^{\lambda_{i+1}}) \quad \text{for } 1 \leq i \leq r, \tag{9.30}$$

because y^{λ_i} has period p_{λ_i} under f. Applying (9.30) iteratively gives

$$y^{\lambda_1} \leq f^{\nu}(y^{\lambda_{r+1}}), \quad \text{where } \nu = \sum_{i=1}^{r}(t_i - s_i) - \sum_{i=1}^{r} B_i p_{\lambda_i}. \tag{9.31}$$

Put $\mu = B_{r+1} p_{\lambda_{r+1}}$ and remark that $f^{\mu}(y^{\lambda_{r+1}}) = y^{\lambda_{r+1}}$. Therefore

$$y^{\lambda_1} \leq f^{\nu+\mu}(y^{\lambda_{r+1}}) = f^{\delta}(y^{\lambda_{r+1}}) \tag{9.32}$$

by (9.29). Now we can use Definition 9.2.8(i) and the fact that f is order-preserving to deduce that

$$f^{q-\delta}(y^{\lambda_1}) \leq f^{q}(y^{\lambda_{r+1}}) \leq f^{q}(\xi).$$

Furthermore, it follows from (9.28) that $f^{q-\delta}(y^{\lambda_1})_a > 0$, as $a = \vartheta_{\lambda_1}(q-\delta) = a_{\lambda_1, q-\delta} \in I_V(f^{q-\delta}(y^{\lambda_1}))$. This proves the claim in the second case. □

Using this proposition it is now easy to show the other inclusion.

Proposition 9.5.8 *For each $n \in \mathbb{N}$, $R(n) \subseteq Q'(2n)$.*

Proof Let $p \in R(n)$. By Lemma 9.5.1 there exists a map $f \in \mathcal{F}_3(2n)$ which has a periodic point ξ with period p and $\mathcal{O}(\xi) \subseteq \mathbb{E}^{2n}$. Let V be the lower semi-lattice generated by the orbit of ξ and let g be the restriction of f to V. It follows from Proposition 9.1.3 that g is a lower semi-lattice homomorphism that maps V onto itself. Moreover V is contained in $\mathbb{E}^{2n}$, as $\mathbb{E}^{2n}$ is a lower semi-lattice. Thus it follows from Proposition 9.5.7 that $p \in Q'(2n)$. □

A combination of Propositions 9.5.6 and 9.5.8 yields $R(n) = Q'(2n)$ for all $n \in \mathbb{N}$, which proves Theorem 9.5.5. The additional constraint on the admissible arrays complicates the computation of $R(n)$. For small n, however, it is still feasible to determine $R(n)$. For $1 \leq n \leq 10$ a list of the elements of $R(n)$ is given in Table 9.2. The interested reader is referred to [122] for details. We conclude this section with an analysis of the asymptotics of the largest element of $R(n)$.

Theorem 9.5.9 *If $\psi(n) = \max\{p \colon p \in R(n)\}$ for $n \in \mathbb{N}$, then*

$$\lim_{n\to\infty} \frac{\log \psi(n)}{\sqrt{2n \log n}} = 1.$$

Table 9.2 *The elements of $R(n)$ for $1 \leq n \leq 10$.*

n	Elements of $R(n)$
1	1, 2
2	1, 2, 3, 4
3	1, 2, 3, 4, 5, 6
4	1, 2, 3, 4, 5, 6, 7, 8, 10, 12
5	1, 2, 3, 4, 5, 6, 7, 8, 9, 10, 12, 14, 15, 20
6	1, 2, 3, 4, 5, 6, 7, 8, 9, 10, 11, 12, 14, 15, 18, 20, 21, 24, 28, 30
7	1, 2, 3, 4, 5, 6, 7, 8, 9, 10, 11, 12, 13, 14, 15, 18, 20, 21, 22, 24, 28, 30, 35, 36, 40, 42, 60
8	1, 2, 3, 4, 5, 6, 7, 8, 9, 10, 11, 12, 13, 14, 15, 16, 18, 20, 21, 22, 24, 26, 28, 30, 33, 35, 36, 40, 42, 44, 45, 48, 56, 60, 70, 84
9	1, 2, 3, 4, 5, 6, 7, 8, 9, 10, 11, 12, 13, 14, 15, 16, 17, 18, 20, 21, 22, 24, 26, 28, 30, 33, 35, 36, 39, 40, 42, 44, 45, 48, 52, 55, 56, 60, 63, 66, 70, 72, 84, 90, 105, 120, 140
10	1, 2, 3, 4, 5, 6, 7, 8, 9, 10, 11, 12, 13, 14, 15, 16, 17, 18, 19, 20, 21, 22, 24, 26, 28, 30, 33, 34, 35, 36, 39, 40, 42, 44, 45, 48, 52, 55, 56, 60, 63, 65, 66, 70, 72, 77, 78, 80, 84, 88, 90, 105, 110, 120, 126, 132, 140, 168, 180, 210

Proof Let $g(n) = \max\{p \colon p \in L(n)\}$ and let $\pi(n)$ denote the number of primes not exceeding n. We first prove the following inequality:

$$g(2n - \pi(2n)) \leq \psi(n) \quad \text{for all } n \in \mathbb{N}. \tag{9.33}$$

To derive this inequality it suffices to find for each $q \in L(2n - \pi(2n))$ a restricted admissible array on $2n$ symbols with period q. Let $\prod_{i=1}^{k} q_i^{\alpha_i}$ be the prime factorization of q. As $q \in L(2n-\pi(2n))$, there exists a permutation μ on $2n - \pi(2n)$ letters which has a disjoint cycle representation $\mu = \mu_1\mu_2 \ldots \mu_k$, where the order of μ_i equals $q_i^{\alpha_i}$ for $1 \leq i \leq k$. Moreover, $\sum_{i=1}^{k} q_i^{\alpha_i} \leq 2n - \pi(2n)$. Let D_i denote the domain of μ_i for $1 \leq i \leq k$; so, $|D_i| = q_i^{\alpha_i}$ and put $D = \cup_i D_i$. Since $k \leq \pi(2n)$, we know that $\sum_{i=1}^{k}(q_i^{\alpha_i} + 1) \leq 2n$. So, we can rename the elements of D such that $D \subseteq \{1, 2, \ldots, 2n\}$ and the sets D_i satisfy $|D_i^\bullet| \leq |D_i| + 1$ and $D_i^\bullet \cap D_j^\bullet = \emptyset$ for all $i \neq j$. Now let a_i be the smallest element of D_i and let $\mu_i^j(a_i)$ denote the j-th iterate of a_i under μ_i. We define an array $\vartheta = (\vartheta_i \colon \mathbb{Z} \to \{1, 2, \ldots, 2n\} \mid 1 \leq i \leq k)$ by $\vartheta_i(j) = \mu_i^j(a_i)$ for all $1 \leq i \leq k$ and $j \in \mathbb{Z}$. As $D_i^\bullet \cap D_j^\bullet = \emptyset$ for all $i \neq j$, ϑ is a restricted admissible array on $2n$ synbols with period $q = \prod_{i=1}^{k} q_i^{\alpha_i}$, and hence inequality (9.33) holds.

If we write $\gamma(n) = \max\{p\colon p \in Q(n)\}$, then we know that $\psi(n) \leq \gamma(2n)$ for all $n \in \mathbb{N}$, as $R(n) \subseteq Q(2n)$. Thus,

$$\frac{\log g(2n - \pi(2n))}{\sqrt{2n \log n}} \leq \frac{\log \psi(n)}{\sqrt{2n \log n}} \leq \frac{\log \gamma(2n)}{\sqrt{2n \log n}} \tag{9.34}$$

for all $n \in \mathbb{N}$ by (9.33). It follows from Theorem 9.4.6 that

$$\lim_{n\to\infty} \frac{\log \gamma(n)}{\sqrt{n \log n}} = 1$$

and hence

$$\lim_{n\to\infty} \frac{\log \gamma(2n)}{\sqrt{2n \log n}} = 1. \tag{9.35}$$

On the other hand, Landau [118] proved that

$$\lim_{n\to\infty} \frac{\log g(n)}{\sqrt{n \log n}} = 1,$$

so that we can use the prime number theorem, $\lim_{n\to\infty}(\pi(n)\log n)/n = 1$, to deduce that

$$\lim_{n\to\infty} \frac{\log g(2n - \pi(2n))}{\sqrt{2n \log n}} = 1. \tag{9.36}$$

Combining equations (9.34), (9.35), and (9.36) yields

$$\lim_{n\to\infty} \frac{\log \psi(n)}{\sqrt{2n \log n}} = 1.$$

□

Appendix A

The Birkhoff–Hopf theorem

In Section 2.1 we briefly discussed a classical result by Birkhoff [25, 26] (see page 31) concerning the contraction ratio of linear maps that leave a closed cone invariant. A closely related theorem was proved by Hopf [90, 91], who was apparently unaware of Birkhoff's work. The Birkhoff–Hopf results have stimulated various authors to generalize and sharpen the original theorems and to elucidate their connections. A partial list of contributors includes Bauer [17], Ostrowski [175, 176], and Bushell [43, 44, 46]. Related theorems have been obtained by Krasnosel'skii, Pokornyi, Sobolev, and Zabreiko [109, 111, 230] [110], who also seem to have been unaware of the work by Birkhoff and Hopf.

The purpose of this appendix is to discuss a generalization of the theorems of Birkhoff, Hopf, Bauer, Ostrowski, Bushell, and others. We shall call the cumulative result the Birkhoff–Hopf theorem. Our presentation roughly follows [61]. Although the focus of this book is on finite-dimensional cones, we will work in infinite dimensions here, as there are no additional complications. In fact, our approach will be to show that the general infinite-dimensional version of the Birkhoff–Hopf theorem can be deduced from a special case where the linear map leaves $\mathbb{R}^2_+$ invariant.

A.1 Preliminaries

Let us start by introducing some new definitions and generalizing several earlier concepts to the context of infinite-dimensional spaces. All of this will be rather straightforward. A subset K of a real vector space V is called a *wedge* if K is convex and $\lambda K \subseteq K$ for all $\lambda \geq 0$. A wedge is called a *cone* if, in addition, $K \cap (-K) = \{0\}$. As before a cone induces a partial ordering $\leq_K$ on V by $x \leq_K y$ if $y - x \in K$.

Recall that a vector space V is called a *topological vector space* if V is endowed with a topology in which every point is a closed set and the multiplication and addition operations are continuous. Such a topology is always Hausdorff; see [190, chapter 1]. In the version of the Birkhoff–Hopf theorem we shall present here, however, topology will play a limited role, and we shall not need to assume that V is a topological vector space.

Let K be a cone in a real vector space V. For $y \in K$ and $x \in V$ we say that y *dominates* x if there exist $\alpha, \beta \in \mathbb{R}$ such that $\alpha y \leq x \leq \beta y$. In that case we define

$$M(x/y; K) = \inf\{\beta \in \mathbb{R} \colon x \leq \beta y\}$$

and

$$m(x/y; K) = \sup\{\alpha \in \mathbb{R} \colon \alpha y \leq x\}.$$

Moreover, we let

$$\omega(x/y; K) = M(x/y; K) - m(x/y; K). \tag{A.1}$$

As usual we shall drop K in the notation above when there is no danger of confusion. Obviously $M(x/y) \geq m(x/y)$, so that $\omega(x/y)$ is nonnegative and finite if y dominates x. If $y = 0$ and y dominates x, then $x = 0$ and we put $\omega(0/0; K) = 0$. If $y \in K \setminus \{0\}$ and $x \in V$ is dominated by y, then it is easy to verify that ty also dominates $\lambda x + \mu y$ for all $\lambda, \mu \in \mathbb{R}$ and $t > 0$, and

$$\omega((\lambda x + \mu y)/ty) = \frac{|\lambda|}{t}\omega(x/y). \tag{A.2}$$

Moreover, if $x_1, x_2 \in V$ are dominated by $y \in K \setminus \{0\}$, then y also dominates $x_1 + x_2$ and

$$\omega((x_1 + x_2)/y) \leq \omega(x_1/y) + \omega(x_2/y). \tag{A.3}$$

These observations yield the following lemma.

Lemma A.1.1 *Let K be a cone in a real vector space V. If $y \in K \setminus \{0\}$ and $V_y = \{x \in V \colon y \text{ dominates } x\}$, then V_y is a linear subspace of V. Moreover, if we let $p_y(x) = \omega(x/y; K)$ for $x \in V_y$, then $p_y(\cdot)$ is a semi-norm on V_y, i.e., $p_y(x + z) \leq p_y(x) + p_y(z)$ and $p_y(\lambda x) = |\lambda| p_y(x)$ for all $x, z \in V_y$ and $\lambda \in \mathbb{R}$.*

If S is a compact Hausdorff space and V is the space of continuous real-valued functions on S, then V contains a natural cone K consisting of nonnegative functions $x: S \to \mathbb{R}_{\geq 0}$. If y is a strictly positive, continuous function, so $y(s) > 0$ for all $s \in S$, then y dominates each x in V. In that case

$$\omega(x/y) = \max_{s \in S} \frac{x(s)}{y(s)} - \min_{s \in S} \frac{x(t)}{y(t)}.$$

Because of this example, $\omega(x/y)$ is sometimes called the *oscillation* of x/y. In particular, we see that if $S = \{1, \ldots, n\}$, $V = \mathbb{R}^n$, and $K = \mathbb{R}^n_+$, then

$$\omega(x/y) = \mathsf{t}(x/y) - \mathsf{b}(x/y),$$

for all $x \in \mathbb{R}^n$ and y positive. (Here x/y denotes the vector with coordinates x_i/y_i.)

As in finite-dimensional spaces we have an equivalence relation $\sim_K$ on K by $x \sim_K y$ if x dominates y and vice versa. The equivalence classes are called *parts* of K. Clearly $\{0\}$ is a part of K, and if $x \sim_K y$ with $y \neq 0$, then there exist $\alpha, \beta > 0$ such that $\alpha y \leq x \leq \beta y$. For $x, y \in K$, with $x \sim_K y$ and $y \neq 0$, we let *Hilbert's metric* be given by

$$d_H(x, y) = \log \frac{M(x, y)}{m(x/y)}.$$

For consistency we put $d_H(0, 0) = 0$. It is straightforward to verify that the assertions in Proposition 2.1.1 also hold when K is a cone in a real vector space V.

Although we do not need to assume V to be a Hausdorff topological vector space in the Birkhoff–Hopf theorem, point set topology will play a small part in the proof. In particular, it is useful to recall the following basic fact; see [190, chapter 1].

Lemma A.1.2 *If E is a finite-dimensional real vector space, then there exists a unique topology τ that turns E into a Hausdorff topological vector space. In particular, all norms on E yield the same topology, and if* $\dim(E) = n$*, then each linear isomorphism of $\mathbb{R}^n$ into E is a homeomorphism.*

If V is a real vector space and $E \subseteq V$ is a finite-dimensional linear subspace, we shall always assume that E is endowed with the natural Hausdorff topology τ of Lemma A.1.2. Moreover, if K is a cone in V, then $\overline{K \cap E}$ will denote the closure of $K \cap E$ in E under the Hausdorff topology τ.

The following basic fact concerning convex sets will also be useful in our subsequent work; see [58, p. 413].

Lemma A.1.3 *Let C be a convex set in a topological Hausdorff vector space V. If $x \in$* $\operatorname{int}(C)$ *and $y \in C$, then $\lambda x + (1 - \lambda)y \in$* $\operatorname{int}(C)$ *for all* $0 < \lambda \leq 1$.

A.2 Almost Archimedean cones

To clarify some pathological cases in the proof of the Birkhoff–Hopf theorem, it is useful to introduce the notion of an almost Archimedean cone, due to Bonsall [29].

Definition A.2.1 A cone K in a real vector space V is called *almost Archimedean* if for each $z \in V$ we have that $z = 0$ whenever there exists $y \in V$ such that $-\varepsilon y \leq_K z \leq_K \varepsilon y$ for all $\varepsilon > 0$.

Not all cones are almost Archimedean. For example, the cone $K_{\text{lex}} = \{x \in \mathbb{R}^2 : x_1 > 0, \text{ or } x_1 = 0 \text{ and } x_2 \geq 0\}$. The partial ordering $\leq$ corresponding to K_{lex} corresponds to the lexicographic ordering on $\mathbb{R}^2$. Indeed, $x \leq y$ if and only if $x_1 < y_1$, or $x_1 = y_1$ and $x_2 \leq y_2$. Thus, $z = (0, 1)$ and $y = (1, 0)$ satisfy $-\varepsilon y \leq z \leq \varepsilon y$ for all $\varepsilon > 0$. The reason that K_{lex} is not almost Archimedean is that the closure of K_{lex} is not a cone, but a wedge; see Figure A.1.

In fact, the following proposition holds; cf. [93].

Proposition A.2.2 *If K is a cone in a real vector space V, then K is almost Archimedean if and only if for each linear subspace $E \subseteq V$, with* $\dim(E) \leq 2$, *we have that $\overline{E \cap K}$ is a cone.*

Proof Let K be an almost Archimedean cone in V and let $E \subseteq V$ be a linear subspace with $\dim(E) \leq 2$. It is clear that $\overline{E \cap K}$ is a wedge. To prove that $\overline{E \cap K}$ is a cone it remains to show that if $z \in \overline{E \cap K}$ and $-z \in \overline{E \cap K}$, then $z = 0$. If there exist no $u, v \in E \cap K$ such that u and v are linearly independent, then $E \cap K = \{\lambda w : \lambda \geq 0\}$ for some $w \in E \cap K$. In that case $\overline{E \cap K}$ is a cone, so $z = 0$. On the other hand, if there exist $u, v \in E \cap K$ such that u and v are linearly independent, then the linear isomorphism $(a, b) \mapsto au + bv$ from $\mathbb{R}^2$ to E is a homeomorphism by Lemma A.1.2. As $au + bv \in E \cap K$ for all $a, b \geq 0$, we deduce that $w = a_0 u + b_0 v \in \text{int}(E \cap K)$ for $a_0, b_0 > 0$. (Here $\text{int}(E \cap K)$ is relative to E.) Lemma A.1.3 implies that $\pm z + \varepsilon w \in \text{int}(E \cap K)$ for all $\varepsilon > 0$. Thus, $-\varepsilon w \leq z \leq \varepsilon w$ for all $\varepsilon > 0$, which implies that $z = 0$.

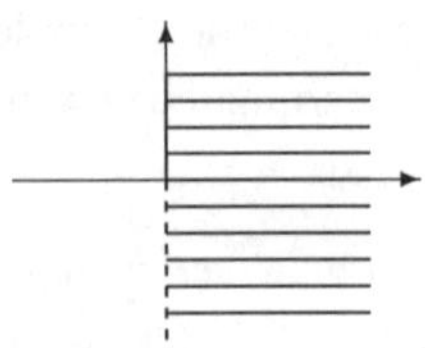

Figure A.1 Lexicographic cone.

To prove the opposite implication let $y, z \in V$ and suppose that $-\varepsilon y \leq z \leq \varepsilon y$ for all $\varepsilon > 0$. Let E denote the linear subspace spanned by y and z. Note that $z + \varepsilon y \in E \cap K$ and $-z + \varepsilon y \in E \cap K$ for all $\varepsilon > 0$. Clearly $z + \varepsilon y \to z$ and $-z + \varepsilon y \to -z$ as $\varepsilon \to 0^+$, in the Hausdorff vector space topology on E. It follows that z and $-z$ are in $\overline{E \cap K}$, which implies that $z = 0$. □

For $x, y \in V$, we write $V(x, y)$ to denote the linear span of x and y in V, and let $K(x, y) = K \cap V(x, y)$, where K is a cone in V. A key property of almost Archimedean cones is the following.

Lemma A.2.3 *If K is an almost Archimedean cone in a real vector space V and $x, y \in K$ are such that $x \sim_K y$ and $d_H(x, y) = 0$, then $x = \beta y$ for some $\beta \geq 0$.*

Proof Let $E = V(x, y)$ and $\beta = M(x/y)$. Note that $\beta \geq 0$, as $x, y \in K$. If $d_H(x, y) = 0$, then either $x = y = 0$ or $\beta = m(x/y)$. In the first case we are done. The second case implies that $\beta y - x \in \overline{E \cap K}$ and $x - \beta y \in \overline{E \cap K}$. As K is almost Archimedean, $\overline{E \cap K}$ is a cone by Proposition A.2.2, and hence $\beta y - x = x - \beta y = 0$, which shows that $x = \beta y$. □

A.3 Projective diameter

To formulate the Birkhoff–Hopf theorem we need to introduce the projective diameter of a linear map $L: V \to W$ that maps a cone K into a cone K'. The basic idea is to measure the diameter of its range $L(K) \subseteq K'$ under Hilbert's metric. There are several natural ways to do this. The first one is the following.

Definition A.3.1 Let K be a cone in a real vector space V and suppose that $A \subseteq K$. If $A \setminus \{0\}$ is contained in a part of K, we define

$$\mathrm{diam}_1(A; K) = \sup\{d_H(x, y): x, y \in A \setminus \{0\}\}.$$

Otherwise, we put $\mathrm{diam}_1(A; K) = \infty$.

The second definition involves the convex hull of a set A in V, which will be denoted by $\mathrm{co}(A)$.

Definition A.3.2 If K is a cone in a real vector space V and $A \subseteq K$, we define

$$\mathrm{diam}_2(A; K) = \sup\{d_H(x, y): x, y \in \mathrm{co}(A) \text{ and } x \sim_K y\}.$$

The two definitions are in general not equivalent. Indeed, if we consider K_{lex}, then $(0, 1)$ and $(1, 0)$ are in different parts of the cone K_{lex}, so that

$\mathrm{diam}_1(K_{\mathrm{lex}}; K_{\mathrm{lex}}) = \infty$. On the other hand, if $x, y \in K_{\mathrm{lex}}$ and $x \sim_{K_{\mathrm{lex}}} y$, then $d_H(x, y) = 0$, so that $\mathrm{diam}_2(K_{\mathrm{lex}}; K_{\mathrm{lex}}) = 0$.

Although Definition A.3.2 may seem more contrived, it appears to be the most suitable one for our purposes. We define the *projective diameter* of a linear map $L: V \to W$, where $L(K) \subseteq K'$, by

$$\Delta(L; K, K') = \mathrm{diam}_2(L(K); K'). \tag{A.4}$$

It follows that $\Delta(L; K, K') = \sup\{d_H(Lx, Ly) : x, y \in K \text{ with } Lx \sim_K Ly\}$. For the analysis it is convenient to establish the equivalence between the two definitions if K is almost Archimedean. To prove this equivalence we first consider a special case.

Lemma A.3.3 *If K is an almost Archimedean cone in a real vector space V and $A \subseteq K$ is convex, then* $\mathrm{diam}_1(A; K) = \mathrm{diam}_2(A; K)$. *If, in addition,* $\mathrm{diam}_2(A; K) < \infty$, *then $A \setminus \{0\}$ is contained in a part of K.*

Proof Let us suppose for the moment that the second assertion is true. If $\mathrm{diam}_1(A; K) < \infty$, then $u \sim_K v$ for all $u, v \in A \setminus \{0\}$, so that $\mathrm{diam}_1(A; K) = \mathrm{diam}_2(A; K)$. If $\mathrm{diam}_2(A; K) = \infty$ and $u \sim_K v$ for all $u, v \in A \setminus \{0\}$, then again $\mathrm{diam}_1(A; K) = \mathrm{diam}_2(A; K)$. If $\mathrm{diam}_2(A; K) = \infty$ and there exist $u, v \in A \setminus \{0\}$ with $u \not\sim_K v$, we get that $\mathrm{diam}_1(A; K) = \infty$ by definition, and hence $\mathrm{diam}_1(A; K) = \mathrm{diam}_2(A; K)$. Therefore it suffices to prove the second assertion.

For the sake of contradiction assume that $\mathrm{diam}_2(A; K) < \infty$ and there exist $u, v \in A \setminus \{0\}$ with $u \not\sim_K v$. This means that either there exists no $\alpha > 0$ such that $v \leq \alpha u$, or there exists no $\beta > 0$ such that $u \leq \beta v$. By symmetry we may assume that there exists no $\beta > 0$ such that

$$u \leq \beta v. \tag{A.5}$$

For $0 < \varepsilon < 1$, we let $u_\varepsilon = u + \varepsilon v$ and $v_\varepsilon = v + \varepsilon u$. We have that

$$v_\varepsilon = v + \varepsilon u \leq \varepsilon^{-1}(u + \varepsilon v) = \varepsilon^{-1} u_\varepsilon$$

and

$$u_\varepsilon = u + \varepsilon v \leq \varepsilon^{-1}(v + \varepsilon u) = \varepsilon^{-1} v_\varepsilon.$$

It follows that $M(v_\varepsilon/u_\varepsilon) \leq \varepsilon^{-1}$ and $M(u_\varepsilon/v_\varepsilon) \leq \varepsilon^{-1}$, so that $m(v_\varepsilon/u_\varepsilon) \geq \varepsilon$. If $M(u_\varepsilon/v_\varepsilon) < \gamma < \varepsilon^{-1}$, then we find that

$$(1 - \varepsilon\gamma)u \leq (\gamma - \varepsilon)v,$$

which contradicts (A.5). Thus, we may assume that

$$M(u_\varepsilon/v_\varepsilon) = m(v_\varepsilon/u_\varepsilon)^{-1} = \varepsilon^{-1}.$$

Let us now distinguish two cases: (1) there exists no $\alpha > 0$ such that $v \leq \alpha u$, and (2) there exists $\beta > 0$ such that $v \leq \beta u$. In the first case the same argument as above shows that $M(v_\varepsilon/u_\varepsilon) = \varepsilon^{-1}$. As A is convex, $(1+\varepsilon)^{-1}u_\varepsilon \in A$ and $(1+\varepsilon)^{-1}v_\varepsilon \in A$, so that

$$\mathrm{diam}_2(A; K) \geq d_H((1+\varepsilon)^{-1}u_\varepsilon, (1+\varepsilon)^{-1}v_\varepsilon) = d_H(u_\varepsilon, v_\varepsilon) = \log \varepsilon^{-2}.$$

This contradicts the assumptions that $\mathrm{diam}_2(A; K) < \infty$.

In the second case we know that $M(v/u) = \beta$ for some $0 \leq \beta < \infty$. Let $E = V(u, v)$ and note that $\beta v - u \in \overline{E \cap K}$. Since K is almost Archimedean, we get that $\beta > 0$. Note that $M(v_\varepsilon/u) = \beta + \varepsilon$, so that

$$d_H((1+\varepsilon)^{-1}v_\varepsilon, u) = d_H(v_\varepsilon, u) = \log \frac{\beta + \varepsilon}{\varepsilon}.$$

As $\beta > 0$, $\log((\beta+\varepsilon)/\varepsilon) \to \infty$ when $\varepsilon \to 0^+$, which violates $\mathrm{diam}_2(A; K) < \infty$. □

In the next lemma we prove a few more facts about $\mathrm{diam}_i(A; K)$ for $i = 1, 2$. In particular, we show that

$$\mathrm{diam}_1(\mathrm{co}(A); K) = \mathrm{diam}_1(A; K).$$

Combining this equation with Lemma A.3.3 shows that if K is an almost Archimedean cone, and $A \subseteq K$, then

$$\mathrm{diam}_1(A; K) = \mathrm{diam}_1(\mathrm{co}(A); K) = \mathrm{diam}_2(A; K), \tag{A.6}$$

and hence Definitions A.3.1 and A.3.2 are equivalent for almost Archimedean cones.

Lemma A.3.4 *Let K be a cone in a real vector space V and suppose that $A \subseteq K$.*

(i) If $\Gamma = \{tx : t \geq 0$ and $x \in A\}$, then

$$\mathrm{diam}_i(\Gamma; K) = \mathrm{diam}_i(A; K) \quad \textit{for } i = 1, 2.$$

(ii) $\mathrm{diam}_1(\mathrm{co}(A); K) = \mathrm{diam}_1(A; K)$.

(iii) If V is a Hausdorff topological vector space and K is closed, then

$$\mathrm{diam}_i(\mathrm{cl}(A); K) = \mathrm{diam}_i(A; K) \quad \textit{for } i = 1, 2.$$

Proof To prove (i) we first note that if $A \setminus \{0\}$ is not contained in a single part of K, then $\mathrm{diam}_1(A; K) = \mathrm{diam}_1(\Gamma; K) = \infty$. On the other hand, if A is contained in a part of K, then $u \sim_K v$ for all $u, v \in \Gamma \setminus \{0\}$. As $d_H(\alpha x, \beta y) = d_H(x, y)$ for all $\alpha, \beta > 0$ and $x, y \in A$, we get that

$\mathrm{diam}_1(A; K) = \mathrm{diam}_1(\Gamma; K)$. To show that $\mathrm{diam}_2(\Gamma; K) = \mathrm{diam}_2(A; K)$, it suffices to prove that

$$\mathrm{diam}_2(\mathrm{co}(\Gamma); K) \leq \mathrm{diam}_2(\mathrm{co}(A); K).$$

For $x \in \mathrm{co}(\Gamma) \setminus \{0\}$ we have that

$$x = \sum_{k=1}^{m} a_k x_k,$$

where $a_k > 0$ and $x_k \in A \setminus \{0\}$ for $1 \leq k \leq m < \infty$. Suppose that $y \in \mathrm{co}(\Gamma) \setminus \{0\}$ and $x \sim_K y$. So, $y = \sum_{k=1}^{n} b_k y_k$, where $b_k > 0$ and $y_k \in A \setminus \{0\}$ for $1 \leq k \leq n < \infty$. Put $a = \sum_k a_k$ and $b = \sum_k b_k$. Remark that if we let $\xi = x/a$ and $\eta = y/b$, then $\xi, \eta \in \mathrm{co}(A) \setminus \{0\}$ and $d_H(x, y) = d_H(\xi, \eta)$. Thus, $\mathrm{diam}_2(\mathrm{co}(\Gamma); K) \leq \mathrm{diam}_2(\mathrm{co}(A); K)$.

To prove the second assertion we first remark that the equality holds if $\mathrm{diam}_1(A; K) = \infty$. So, assume that $\mathrm{diam}_1(A; K) < \infty$. Then A is contained in a part of K. If $A = \{0\}$, we are done. Suppose that $A \neq \{0\}$. Clearly $\mathrm{diam}_1(\mathrm{co}(A); K) \leq \mathrm{diam}_1(\mathrm{co}(A \cup \{0\}); K)$. It therefore suffices to prove that

$$\mathrm{diam}_1(\mathrm{co}(A \cup \{0\}); K) \leq \mathrm{diam}_1(A; K).$$

Let $R > \mathrm{diam}_1(A; K)$. For $u \in \mathrm{co}(A \cup \{0\})$ and $u \neq 0$ we have that

$$u = s \sum_{k=1}^{m} \lambda_k u_k,$$

where $0 < s < 1$, $\lambda_k > 0$, and $u_k \in A$ for $1 \leq k \leq m$, and $\sum_k \lambda_k = 1$. If $v \in A$, then there exist $\alpha_k, \beta_k > 0$ such that

$$\alpha_k v < u_k < \beta_k v$$

and $\log(\beta_k/\alpha_k) < R$ for all k. These inequalities imply for $0 < s < 1$ that

$$s\Big(\sum_k \lambda_k \alpha_k\Big) v \leq u \leq s\Big(\sum_k \lambda_k \beta_k\Big) v,$$

which yields

$$d_H(u, v) \leq \log\left(\frac{\sum_k \lambda_k \beta_k}{\sum_k \lambda_k \alpha_k}\right) < \log\left(\frac{\sum_k \lambda_k \alpha_k e^R}{\sum_k \lambda_k \alpha_k}\right) = R.$$

Now suppose that $u, v \in \mathrm{co}(A \cup \{0\}) \setminus \{0\}$, so $v = t \sum_{k=1}^{n} \gamma_k v_k$, where $0 < t < 1$, $\gamma_k > 0$, and $v_k \in A$ for $1 \leq k \leq n$, and $\sum_k \gamma_k = 1$. We have already shown that $d_H(u, v_k) < R$, so there exist $\delta_k, \rho_k > 0$ such that

$$\delta_k u \leq v_k \leq \rho_k u$$

and $\log \delta_k/\rho_k < R$ for all k. The same argument shows that

$$t\Big(\sum_k \delta_k \gamma_k\Big)u \leq v \leq t\Big(\sum_k \rho_k \gamma_k\Big)u \leq te^R\Big(\sum_k \delta_k \gamma_k\Big)u.$$

This implies that $d_H(u, v) \leq R$. As R was arbitrary, we get that $d_H(u, v) \leq \operatorname{diam}_1(A; K)$, and hence

$$\operatorname{diam}_1(\operatorname{co}(A \cup \{0\}); K) \leq \operatorname{diam}_1(A; K).$$

Using Lemma A.3.3 and statement (ii) of this lemma, we have now proved that if K is almost Archimedean, Equation (A.6) is satisfied.

Now suppose that K is a closed cone in a Hausdorff topological vector space V. Clearly K is almost Archimedean; so, Equation (A.6) implies that

$$\operatorname{diam}_1(A; K) = \operatorname{diam}_2(A; K).$$

It is easy to show that $\operatorname{cl}(\operatorname{co}(A)) \supseteq \operatorname{co}(\operatorname{cl}(A))$ and hence it suffices to prove that

$$\operatorname{diam}_2(\operatorname{cl}(A); K) \leq \operatorname{diam}_2(A; K)$$

when $A \subseteq K$ is convex. The inequality is trivial if $\operatorname{diam}_2(A; K) = \infty$, so we may assume that $\operatorname{diam}_2(A; K) < \infty$. Lemma A.3.3 implies that A is contained in a part of K. If $A = \{0\}$, then the inequality is trivial. So suppose that $A \neq \{0\}$. Let $u, v \in \operatorname{cl}(A)$ be such that $u \sim_K v$. Write $b = M(v/u)$ and $a = m(v/u)$. Let $a' > a$ and $b' < b$. By definition we have that $b'u - v$ and $v - a'u$ are not in K. As K is closed, there exist neighborhoods $\mathcal{U}_1$ of u and $\mathcal{V}_1$ of v such that $0 \notin \mathcal{U}_1 \cup \mathcal{V}_1$, and $b'u_1 - v_1$ and $v_1 - a'u_1$ are not in K for all $u_1 \in \mathcal{U}_1$ and $v_1 \in \mathcal{V}_1$. As $u, v \in \operatorname{cl}(A)$, there exist $u^* \in \mathcal{U}_1 \cap A$ and $v^* \in \mathcal{V}_1 \cap A$. Remark that $u^* \sim_K v^*$, and that $b'u^* - v^*$ and $v^* - a'u^*$ are not in K. This implies that $M(v^*/u^*) \geq b'$ and $m(v^*/u^*) \leq a'$, so that $d_H(u^*, v^*) \geq \log(b'/a')$. If we let $b' \to b$ and $a' \to a$, we see that $\operatorname{diam}_2(A; K) \geq d_H(u, v)$ for $u \sim_K v$ arbitrary in $\operatorname{cl}(A)$, which completes the proof. □

A.4 The Birkhoff–Hopf theorem: reduction to two dimensions

Let V and W be real vector spaces and let $K \subseteq V$ and $K' \subseteq W$ be cones. Suppose that $L: V \to W$ is a linear map such that $L(K) \subseteq K'$. The *Hopf oscillation ratio* of L is defined by

$$N(L; K, K') = \inf\{\mu \geq 0\colon \omega(Lx/Ly) \leq \mu\omega(x/y) \text{ for all } x \sim_K y \text{ in } K\}. \tag{A.7}$$

Furthermore, we define the *Birkhoff contraction ratio* of L by

$$k(L; K, K') = \inf\{\lambda \geq 0\colon d_H(Lx, Ly) \leq \lambda d_H(x, y) \text{ for all } x \sim_K y \text{ in } K\}. \tag{A.8}$$

If K and K' are clear from the context we simply write $N(L)$ and $k(L)$. Of course, if $x, y \in K \setminus \{0\}$ and $x \sim_K y$, then as in Proposition 2.1.3 we have that

$$M(Lx/Ly) \leq M(x/y) \quad \text{and} \quad m(x/y) \leq m(Lx/Ly).$$

Thus, $\omega(Lx/Ly) \leq \omega(x/y)$ for all $x \sim_K y$, so that $N(L; K, K') \leq 1$. In the same way we find that $k(L; K, K') \leq 1$.

We note that if L is one-to-one and $L(K) = K$, then the same argument applies to L^{-1} and we get that

$$M(Lx/Ly) = M(x/y) \quad \text{and} \quad m(Lx/Ly) = m(x/y) \tag{A.9}$$

for all $x \sim_K y$ in K. In that case we also have that

$$\omega(Lx/Ly) = \omega(x/y) \tag{A.10}$$

and

$$d_H(Lx, Ly) = d_H(x, y) \tag{A.11}$$

for all $x \sim_K y$ in K. After these preliminary observations we now state the Birkhoff–Hopf theorem.

Theorem A.4.1 (Birkhoff–Hopf) *Let K be a cone in a real vector space V and let K' be a cone in real vector space W. If $L\colon V \to W$ is a linear map with $L(K) \subseteq K'$ and $\Delta(L)$ is defined by (A.4), then*

$$N(L) = k(L) = \tanh(\frac{1}{4}\Delta(L)). \tag{A.12}$$

Of course $\tanh(y) = (e^y - e^{-y})/(e^y + e^{-y})$, and we put $\tanh(\infty) = 1$, so that $\Delta(L) = \infty$ in Equation (A.12) gives $N(L) = k(L) = 1$.

Birkhoff [25] basically established

$$k(L) = \tanh(\frac{1}{4}\Delta(L))$$

in case $V = W$ is a Banach space, $K = K'$ is a closed cone, and $L\colon V \to V$ is a bounded linear map with $L(K) \subseteq K$. Independently Hopf [90, 91] proved that

$$N(L; K, K) \leq \tanh(\frac{1}{4}\Delta(L; K, K)),$$

though he restricted himself to Banach spaces V of measurable functions, e.g., L^∞, and the closed cone of measurable functions that are almost everywhere nonnegative. Moreover, Hopf never explicitly defined $N(L)$ and never used Hilbert's metric. Hopf's work was generalized by Bauer [17] and Ostrowksi [175]. Bushell [44] proved that $\kappa(L; K, K) = N(L; K, K)$, where K is a closed cone in a Banach space.

Before we start discussing the proof of the Birkhoff–Hopf theorem, we point out that (A.12) need not hold if K and K' are not almost Archimedean, and we define $\Delta(L; K, K')$ by $\Delta(L; K, K') = \text{diam}_1(L(K); K)$. Indeed, consider $V = W = \mathbb{R}^n$ and let $K = K' = K_{\text{lex}}$. If L is the identity, then it is easy to verify that $\omega(x/y) = 0$ and $d_H(x, y) = 0$ for all $x \sim_{K_{\text{lex}}} y$ in K_{lex}, so that $N(L) = k(L) = 0$. However, $\text{diam}_1(L(K_{\text{lex}}); K_{\text{lex}}) = \infty$ and $\text{diam}_2(L(K_{\text{lex}}); K_{\text{lex}}) = 0$.

The quantity $N(L; K, K')$ is defined in terms of $x, y \in K$ with $x \sim_K y$. In practice it is convenient to weaken the restrictions on x and y in K.

Lemma A.4.2 *The following equality holds:*

$$N(L; K, K') = \inf\{\mu \geq 0\colon \omega(Lx/Ly; K') \leq \mu\omega(x/y; K) \text{ for all } y \in K \setminus \{0\} \text{ and } x \in V \text{ such that } y \text{ dominates } x\}. \tag{A.13}$$

Proof Clearly $N(L; K, K')$ does not exceed the right-hand side. To obtain the opposite inequality we show that if $x \in V$ is dominated by $y \in K \setminus \{0\}$, then there exists $z \in K$ with $y \sim_K z$ such that $\omega(Lx/Ly) = \omega(Lz/Ly)$ and $\omega(x/y) = \omega(z/y)$. By definition there exist $a, b \in \mathbb{R}$ such that $ay \leq x \leq by$. Define $z = x + (1 - a)y$ and note that $y \leq z = y + x - ay \leq (1 + b - a)y$. Thus, $y \sim_K z$ and Equation (A.2) gives $\omega(z/y) = \omega(x/y)$ and $\omega(Lz/Ly) = \omega(Lx/Ly)$. □

The next lemma, while relatively straightforward, is a crucial step in the proof of the Birkhoff–Hopf theorem, and shows that the general case can be reduced to the two-dimensional case.

Lemma A.4.3 *To prove the Birkhoff–Hopf theorem A.4.1 it suffices to prove (A.12) when* $\dim(K) = \dim(K') = 2$, $\dim(V) = \dim(W) = 2$, *and* $L\colon V \to W$ *is one-to-one.*

Proof Let us first eliminate some trivial cases. If $\dim(K) \leq 1$ or $\dim(K') \leq 1$, one can easily verify that $N(L) = k(L) = \Delta(L) = 0$, so that (A.12) holds. If $\dim(K) = \dim(K') = 2$ and $\dim(V) = \dim(W) = 2$, but L is not one-to-one, it follows that $\dim(L(K)) \leq 1$ and again $N(L) = k(L) = \Delta(L) = 0$.

Now let us consider the general case, so K is a cone in a real vector space V and K' is a cone in a real vector space W. Let $x, y \in K$ and define $C = K \cap V(x, y)$ and $C' = K' \cap W(Lx, Ly)$, where $V(x, y)$ denotes the linear span of x and y in V, and $W(Lx, Ly)$ is the linear span of Lx and Ly in W.

For $u, v \in C$ we have that $u \leq_C v$ if and only if $u \leq_K v$. Likewise $Lu \leq_{C'} Lv$ if and only if $Lu \leq_{K'} Lv$. Thus, if $u, v \in C$ and $u \sim_K v$, then

$$M(u/v; C) = M(u/v; K) \quad \text{and} \quad m(u/v; C) = m(u/v; K). \tag{A.14}$$

Analogous equalities hold for Lu and Lv in C'. Obviously, $L(V(x, y)) \subseteq W(Lx, Ly)$ and $L(C) \subseteq C'$. Define $N(x, y) = N(L; C, C')$, $k(x, y) = k(L; C, C')$, and $\Delta(x, y) = \Delta(L; C, C')$. Technically we should distinguish $L: V \to W$ and the restriction of L to $V(x, y)$, but no confusion should arise. By assumption

$$N(x, y) = k(x, y) = \tanh(\frac{1}{4}\Delta(x, y)).$$

Equation (A.14) implies that

$$N(x, y) = \inf\{\mu \geq 0: \omega(Lu/Lv; K') \leq \mu\omega(u/v; K) \text{ for all } u, v \in C \text{ with } u \sim_K v\},$$

$$k(x, y) = \inf\{\lambda \geq 0: d_H(Lu, Lv; K') \leq \lambda d_H(u, v; K) \text{ for all } u, v \in C \text{ with } u \sim_K v\},$$

and

$$\Delta(x, y) = \sup\{d_H(Lu, Lv): u, v \in C \text{ and } Lu \sim_{K'} Lv\}.$$

This implies that

$$\begin{aligned}\sup\{k(x, y): x, y \in K\} &= \sup\{N(x, y): x, y \in K\} \\ &= \sup\{\tanh(\frac{1}{4}\Delta(x, y)): x, y \in K\} \\ &= \tanh(\sup\{\frac{1}{4}\Delta(x, y): x, y \in K\}),\end{aligned}$$

so that $N(L) = k(L) = \tanh(\frac{1}{4}\Delta(L))$, and we are done. □

The following theorem proves a special two-dimensional case which lies at the heart of the proof of the Birkhoff–Hopf theorem. In fact, most of the subsequent reasoning is aimed at deriving the general two-dimensional case from this special case.

Theorem A.4.4 *Let $V = W = \mathbb{R}^2$ and $K = K' = \mathbb{R}^2_+$. If $L: \mathbb{R}^2 \to \mathbb{R}^2$ is such that $L(\mathbb{R}^2_+) \subseteq \mathbb{R}^2_+$ and L is represented by the nonnegative matrix*

$$\begin{bmatrix} \alpha & 1 \\ 1 & \alpha \end{bmatrix},$$

where $\alpha > 1$, *then*

$$N(L) = k(L) = \tanh(\frac{1}{4}\Delta(L)) = \frac{\alpha - 1}{\alpha + 1}.$$

Proof We have that

$$N(L) = \sup\{\frac{\omega(Lx/Ly)}{\omega(x/y)} : x \sim y \text{ in } \mathbb{R}^2_+ \text{ and } \omega(x/y) \neq 0\}. \tag{A.15}$$

As $K = \mathbb{R}^2_+$, we can restrict x and y in (A.15) to be in $\text{int}(\mathbb{R}^2_+)$. Moreover, as

$$\frac{\omega(L(\lambda x)/L(\mu y))}{\omega(\lambda x/\mu y)} = \frac{\omega(Lx/Ly)}{\omega(x/y)}$$

for all $\lambda, \mu > 0$, we may assume that $x = (1, s)$ and $y = (1, t)$, with $s, t > 0$ and $s \neq t$ in (A.15).

A simple calculation shows that

$$N(L) = \sup\{\left|\frac{\alpha + t}{\alpha + s} - \frac{1 + \alpha t}{1 + \alpha s}\right| \Big/ \left|1 - \frac{t}{s}\right| : s, t > 0 \text{ and } s \neq t\} = \sup\{\vartheta(s) : s > 0\},$$

where

$$\vartheta(s) = \frac{(\alpha^2 - 1)s}{(\alpha + s)(\alpha s + 1)}.$$

It is an easy calculus exercise to show that

$$\sup_{s>0} \vartheta(s) = \vartheta(1) = (\alpha - 1)/(\alpha + 1).$$

A similar calculation yields the formula

$$k(L) = \sup\{\log \frac{(\alpha + s)(1 + \alpha t)}{(\alpha + t)(1 + \alpha s)} \Big/ \log \frac{t}{s} : 0 < s < t\}. \tag{A.16}$$

Define $f(t) = \log\big((1 + \alpha t)/(\alpha + t)\big)$; so, (A.16) gives

$$k(L) = \sup\{\frac{f(t) - f(s)}{\log t - \log s} : 0 < s < t\}.$$

Then the generalized mean value theorem implies that there exists $s < \tau < t$ with

$$\frac{f(t) - f(s)}{\log t - \log s} = f'(\tau)\tau = \frac{(\alpha^2 - 1)\tau}{(\alpha + \tau)(1 + \alpha\tau)} = \vartheta(\tau).$$

Since

$$\lim_{t \to s} \frac{f(t) - f(s)}{\log t - \log s} = \vartheta(s),$$

we conclude that $k(L) = \sup\{\vartheta(s) : s > 0\} = (\alpha - 1)/(\alpha + 1)$.

Finally we show that $\tanh(\frac{1}{4}\Delta(L)) = (\alpha - 1)/(\alpha + 1)$. Let $e_1 = (1, 0)$ and $e_2 = (0, 1)$ be the standard basis vectors in $\mathbb{R}^2$. As $\mathbb{R}^2_+$ is almost Archimedean, $\mathrm{diam}_1(A; \mathbb{R}^2_+) = \mathrm{diam}_2(A; \mathbb{R}^2_+)$. Therefore it follows from Lemma A.3.4 that

$$\begin{aligned}\Delta(L) &= \mathrm{diam}_1(\mathrm{co}(\{Le_1, Le_2\}); \mathbb{R}^2_+)\\ &= \mathrm{diam}_1(\{Le_1, Le_2\}; \mathbb{R}^2_+)\\ &= d_H(Le_1, Le_2)\\ &= \log(\alpha^2).\end{aligned}$$

This implies that

$$\tanh(\frac{1}{4}\Delta(L)) = \frac{e^{\Delta(L)/2} - 1}{e^{\Delta(L)/2} + 1} = \frac{\alpha - 1}{\alpha + 1}.$$

□

A useful observation to show that the general two-dimensional case of the Birkhoff–Hopf theorem can be reduced to Theorem A.4.4 is given in the following lemma.

Lemma A.4.5 *Let V and W be real vector spaces and $K \subseteq V$ and $K' \subseteq W$ be cones. Suppose that $L : V \to W$ is a linear map such that $L(K) \subseteq K'$. Let V_1 and W_1 be real vector spaces and suppose that $S : V_1 \to V$ and $T : W \to W_1$ are linear maps, which are one-to-one and onto. Denote $K_1 = S^{-1}(K)$ and $K_1' = T(K')$. If $L_1 : V_1 \to W_1$ is given by $L_1 = T \circ L \circ S$, then we have that*

$$N(L; K, K') = N(L_1; K_1, K_1'),$$

$$k(L; K, K') = k(L_1; K_1, K_1'),$$

and

$$\Delta(L; K, K') = \Delta(L_1; K_1, K_1').$$

Proof As S is one-to-one and onto, we know from (A.9) that $M(Sx/Sy; K) = M(x/y; K_1)$ and $m(Sx/Sy; K) = m(x/y; K_1)$ for all $x, y \in K_1$ with $x \sim_{K_1} y$. Thus, $\omega(Sx/Sy; K) = \omega(x/y; K_1)$ and $d_H(Sx, Sy; K) = d_H(x, y; K_1)$ for all $x, y \in K_1$ with $x \sim_{K_1} y$. Similarly $\omega(Tx/Ty; K_1') = \omega(x/y; K')$ and $d_H(Tx, Ty; K_1') = d_H(x, y; K')$ for all $x, y \in K'$ with $x \sim_{K'} y$. From these equalities the lemma follows. □

A.5 Two-dimensional cones

To show that the general two-dimensional case of the Birkhoff–Hopf theorem can be reduced to Theorem A.4.4, we need to better understand the geometry of two-dimensional cones. It turns out that there are essentially two types of two-dimensional cones: those that are almost Archimedean and those that are not. The following theorem gives a precise characterization. Although the results are geometrically obvious, the proof seems to demand some care.

Theorem A.5.1 *If K is a two-dimensional cone in a two-dimensional vector space E, where E is equipped with the standard topology, then either*

(i) $\mathrm{cl}(K)$ *is a cone, in which case there exist linearly independent vectors $u, v \in E$ such that*

$$\mathrm{cl}(K) = \{\lambda u + \mu v \colon \lambda \geq 0 \textit{ and } \mu \geq 0\} \tag{A.17}$$

and

$$\mathrm{int}(K) = \{\lambda u + \mu v \colon \lambda > 0 \textit{ and } \mu > 0\}, \tag{A.18}$$

or

(ii) $\mathrm{cl}(K)$ *is not a cone, in which case there exist linearly independent vectors $u, v \in E$ such that*

$$\mathrm{cl}(K) = \{\lambda u + \mu v \colon \lambda \in \mathbb{R} \textit{ and } \mu \geq 0\} \tag{A.19}$$

and

$$\mathrm{int}(K) = \{\lambda u + \mu v \colon \lambda \in \mathbb{R} \textit{ and } \mu > 0\}. \tag{A.20}$$

Proof We begin by recalling the following basic fact, which is easily proved with the aid of Lemma A.1.3: if A is a convex subset of a Hausdorff topological vector space and $\mathrm{int}(A) \neq \emptyset$, then $\mathrm{int}(\mathrm{cl}(A)) = \mathrm{int}(A)$. In particular, we have that $\mathrm{int}(\mathrm{cl}(K)) = \mathrm{int}(K)$.

As K is a two-dimensional cone, there exist linearly independent vectors $a, b \in K$ such that $\lambda a + \mu b \in K$ for all $\lambda, \mu \geq 0$. Note that $\{\lambda a + \mu b \colon \lambda, \mu > 0\}$ is an open subset of K, so that $\lambda a + \mu b \in \mathrm{int}(K)$ for all $\lambda, \mu > 0$. In particular, $a + b \in \mathrm{int}(K)$. If $B \subseteq K$ denotes an open ball about $a + b$, then $-B$ is an open ball about $-(a + b)$ in the complement of K. Therefore $-(a + b) \notin \mathrm{cl}(K)$. Define $x = 2a + b \in \mathrm{int}(K)$ and $y = -(a + b) \notin \mathrm{cl}(K)$. Note that x and y are linearly independent. Let

$$\tau = \sup\{t \in (0, 1) \colon tx + (1 - t)y \notin \mathrm{cl}(K)\}.$$

By definition of x and y, we know that $0 < \tau < 1$. As $\mathrm{cl}(K)$ is convex, $tx + (1 - t)y \notin \mathrm{cl}(K)$ for all $0 \leq t < \tau$ and $\tau x + (1 - \tau)y \in \partial K \setminus \{0\}$.

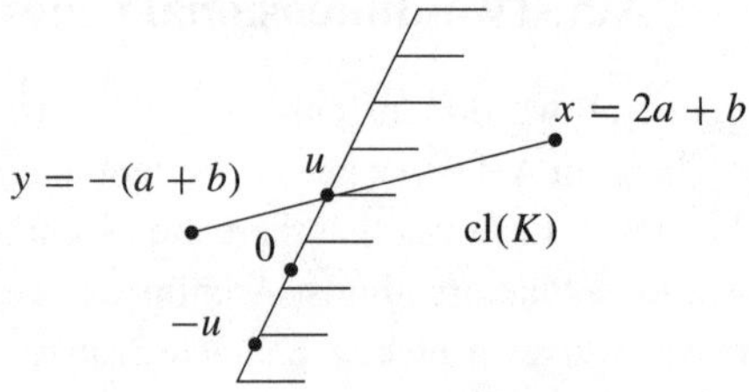

Figure A.2 cl(K) is not a cone.

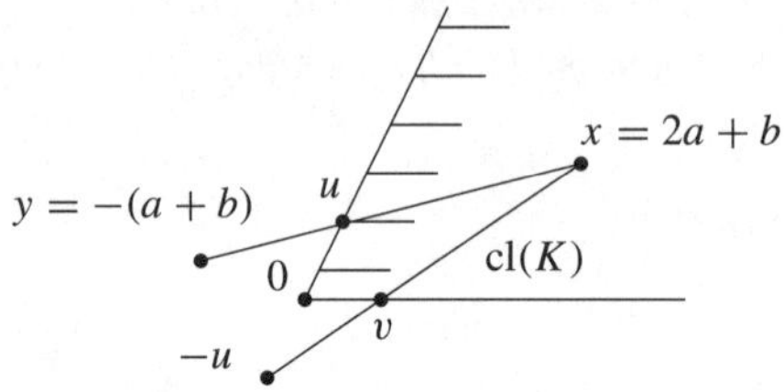

Figure A.3 cl(K) is a cone.

Put $u = \tau x + (1-\tau)y$. If $-u \in \mathrm{cl}(K)$, then $\mathrm{cl}(K)$ is not a cone and we are in the second case; see Figure A.2. We claim that $-u \in \partial K$ in that case. Indeed, otherwise $-u \in \mathrm{int}(\mathrm{cl}(K)) = \mathrm{int}(K)$, in which case Lemma A.1.3 gives $0 = u/2 + (-u/2) \in \mathrm{int}(K)$, which is impossible. Thus, $\lambda u \in \partial K$ for all $\lambda \in \mathbb{R}$, and u is linearly independent of any $v \in \mathrm{int}(K)$. Let $v \in \mathrm{int}(K)$. By construction,

$$\mathrm{cl}(K) \supseteq \{\lambda u + \mu v \colon \lambda \in \mathbb{R} \text{ and } \mu \geq 0\}$$

and

$$\mathrm{int}(K) = \mathrm{int}(\mathrm{cl}(K)) \supseteq \{\lambda u + \mu v \colon \lambda \in \mathbb{R} \text{ and } \mu > 0\}.$$

To prove (A.19) and (A.20), we consider $\lambda u + \rho v \in \mathrm{cl}(K)$. Suppose by way of contradiction that $\rho < 0$. Then Lemma A.1.3 implies that $\lambda u = (\lambda u + \rho v) + |\rho| v \in \mathrm{int}(K)$, which contradicts the fact that $\lambda u \in \partial K$ for all $\lambda \in \mathbb{R}$.

Now suppose that $-u \notin \partial K$. Then we shall show that we are in the first case; see Figure A.3.

Let $\sigma = \sup\{s \in (0,1) \colon sx + (1-s)(-u) \notin \mathrm{cl}(K)\}$. As before we know that $0 < \sigma < 1$. Put $v = \sigma x + (1-\sigma)(-u)$. Note that $v \in \partial K$ and, as x and y are linearly independent, $v \neq 0$. We claim that u and v are linearly independent. Indeed, if $u = \lambda v$ for some $\lambda \in \mathbb{R}$, then a simple calculation shows that

$$\big(\lambda\sigma - \lambda(1-\sigma)\tau - \tau\big)x + \big(-\lambda(1-\sigma)(1-\tau) - (1-\tau)\big)y = 0.$$

As x and y are linearly independent and $\sigma < 1$ and $\tau < 1$, we deduce that $\lambda = -1/(1-\sigma)$ and $\sigma = 0$, which contradicts the fact that $\sigma > 0$.

Since u and v are linearly independent and $u, v \in \partial K \setminus \{0\}$, we conclude that

$$\mathrm{cl}(K) \supseteq \{\lambda u + \mu v \colon \lambda \geq 0 \text{ and } \mu \geq 0\}$$

and

$$\mathrm{int}(K) = \mathrm{int}(\mathrm{cl}(K)) \supseteq \{\lambda u + \mu v \colon \lambda > 0 \text{ and } \mu > 0\}.$$

If $\lambda u + \mu v \in \mathrm{cl}(K)$ for some $\lambda < 0$ and $\mu \in \mathbb{R}$, then $|\lambda| u + (|\mu| + 1)v \in \mathrm{int}(K)$, so that

$$(\lambda u + \mu v) + |\lambda| u + (|\mu| + 1)v = (\mu + |\mu| + 1)v \in \mathrm{int}(K)$$

by Lemma A.1.3, which is impossible. Thus, $\lambda u + \mu v \in \mathrm{cl}(K)$ implies that $\lambda \geq 0$. In the same way it can be shown that $\mu \geq 0$, when $\lambda u + \mu v \in \mathrm{cl}(K)$, and therefore

$$\mathrm{cl}(K) = \{\lambda u + \mu v \colon \lambda \geq 0 \text{ and } \mu \geq 0\}.$$

Since $u, v \in \partial K \setminus \{0\}$, we also find that

$$\mathrm{int}(K) = \{\lambda u + \mu v \colon \lambda > 0 \text{ and } \mu > 0\}.$$

It follows from (A.17) that $\mathrm{cl}(K)$ is a cone. □

Theorem A.5.1 has a number of consequences that are used in the proof of the Birkhoff–Hopf theorem.

Corollary A.5.2 *Let K be a two-dimensional cone in a two-dimensional vector space E. Assume that $\mathrm{cl}(K)$ is a cone and let $u, v \in \partial K$ be such that*

$$\mathrm{cl}(K) = \{\lambda u + \mu v \colon \lambda \geq 0 \textit{ and } \mu \geq 0\}.$$

If $x, y \in K \setminus \{0\}$ and $x \sim_K y$, then

$$M(x/y; \mathrm{cl}(K)) = M(x/y; K) \quad \textit{and} \quad m(x/y; \mathrm{cl}(K)) = m(x/y; K).$$

Moreover, if $x, y \in \mathrm{int}(K)$ and $x = \lambda_1 u + \mu_1 v$ and $y = \lambda_2 u + \mu_2 v$, where $\lambda_1, \lambda_2, \mu_1, \mu_2 > 0$, then

$$M(x/y; K) = \max\{\lambda_1/\lambda_2, \mu_1/\mu_2\} \quad \textit{and} \quad m(x/y; K) = \min\{\lambda_1/\lambda_2, \mu_1/\mu_2\}.$$

In that case we also have that

$$\omega(x/y; K) = \omega(x/y; \mathrm{cl}(K)) = |\lambda_1/\lambda_2 - \mu_1/\mu_2|$$

and

$$d_H(x, y; K) = d_H(x, y; \text{cl}(K)) = \left|\log\left(\frac{\lambda_1\mu_2}{\lambda_2\mu_1}\right)\right|.$$

Proof If $x, y \in K \setminus \{0\}$ and $x \sim_K y$, then either $x = \lambda y$ for some $\lambda > 0$, or x and y are in $\text{int}(K)$. In the latter case, the equalities can be easily derived from (A.17) and (A.18). On the other hand, if $x = \lambda y$, then $M(x/y; \text{cl}(K)) = M(x/y; K) = \lambda$ and $m(x/y; \text{cl}(K)) = m(x/y; K) = \lambda$. □

Corollary A.5.3 *Let K be a two-dimensional cone in a two-dimensional real vector space E. If $\text{cl}(K)$ is not a cone, then for each $x, y \in K$ with $x \sim_K y$ we have that $\omega(x/y; K) = 0$ and $d_H(x, y) = 0$.*

Proof It follows from Theorem A.5.1 that if $x \sim_K y$, then either $x = y = 0$, or $x, y \in K \setminus \{0\}$ and $y = \lambda x$ for some $\lambda > 0$, or $x, y \in \text{int}(K)$. The statement is clear in the first two cases. So suppose that $x, y \in \text{int}(K)$. Then we can write $x = \lambda_1 u + \mu_1 v$ and $y = \lambda_2 u + \mu_2 v$, where $u, v \in E$ are as in Theorem A.5.1, $\lambda_1, \lambda_2 \in \mathbb{R}$ and $\mu_1, \mu_2 > 0$. It follows from (A.19) and (A.20) that $M(x/y; K) = \mu_1/\mu_2$ and $m(x/y; K) = \mu_1/\mu_2$, so that $\omega(x, y; K) = 0$ and $d_H(x, y) = 0$. □

We conclude this section with the following corollary.

Corollary A.5.4 *Let V and W be two-dimensional real vector spaces and let $K \subseteq V$ and $K' \subseteq W$ be almost Archimedean cones. If $L\colon V \to W$ is a linear map such that $L(K) \subseteq K'$, then*

$$N(L; K, K') = N(L; \text{cl}(K), \text{cl}(K')),$$

$$k(L; K, K') = k(L; \text{cl}(K), \text{cl}(K')),$$

and

$$\Delta(L; K, K') = \Delta(L; \text{cl}(K), \text{cl}(K')).$$

Proof We may assume that $\dim(K) = \dim(K') = 2$ and L is one-to-one, otherwise the left- and right-hand sides are zero in the equations. As K is almost Archimedean it follows from Theorem A.5.1 that there exist $u, v \in \partial K$ with $\text{cl}(K) = \{\lambda u + \mu v\colon \lambda \geq 0 \text{ and } \mu \geq 0\}$ and $\text{int}(K) = \{\lambda u + \mu v\colon \lambda > 0 \text{ and } \mu > 0\}$. Similarly, there exist $u', v' \in \partial K'$ such that $\text{cl}(K') = \{\lambda u' + \mu v'\colon \lambda \geq 0 \text{ and } \mu \geq 0\}$ and $\text{int}(K') = \{\lambda u' + \mu v'\colon \lambda > 0 \text{ and } \mu > 0\}$. We shall only prove that $k(L; K, K') = k(L; \text{cl}K, \text{cl}(K'))$ and leave the proofs of other quantities in the equations as an exercise to the reader.

By definition

$$k(L; K, K') = \inf\{\lambda \geq 0\colon d_H(Lx, Ly; K') \leq \lambda d_H(x, y; K) \text{ for all } x \sim_K y \text{ in } K\} \tag{A.21}$$

and

$$k(L; \mathrm{cl}(K), \mathrm{cl}(K')) = \inf\{\lambda \geq 0\colon d_H(Lx, Ly; \mathrm{cl}(K')) \leq \lambda d_H(x, y; \mathrm{cl}(K)) \text{ for all } x \sim_{\mathrm{cl}(K)} y \text{ in } \mathrm{cl}(K)\}. \tag{A.22}$$

If $x, y \in \mathrm{int}(K) = \mathrm{int}(\mathrm{cl}(K))$, then $x \sim_K y$ and $x \sim_{\mathrm{cl}(K)} y$. Since L is one-to-one, Lx and Ly are in $\mathrm{int}(K') = \mathrm{int}(\mathrm{cl}(K'))$. It follows from Corollary A.5.2 that

$$d_H(x, y; K) = d_H(x, y; \mathrm{cl}(K)) \quad \text{and} \quad d_H(Lx, Ly; K') = d_H(Lx, Ly; \mathrm{cl}(K')). \tag{A.23}$$

If $x \sim_K y$ and $x \notin \mathrm{int}(K)$, then $x \sim_{\mathrm{cl}(K)} y$, and either $x = y = 0$ or $x = \alpha y = \lambda u$ for some $\alpha, \lambda > 0$, or $x = \alpha y = \mu v$ for some $\alpha, \mu > 0$. In either case $d_H(x, y; K) = d_H(x, y; \mathrm{cl}(K)) = 0$, so that $d_H(Lx, Ly; K') = d_H(Lx, Ly; \mathrm{cl}(K')) = 0$. This implies that

$$k(L; K, K') = \inf\{\lambda \geq 0\colon d_H(Lx, Ly; K') \leq \lambda d_H(x, y; K) \text{ for all } x, y \in \mathrm{int}(K)\} \tag{A.24}$$

and

$$k(L; \mathrm{cl}(K), \mathrm{cl}(K')) = \inf\{\lambda \geq 0\colon d_H(Lx, Ly; \mathrm{cl}(K')) \leq \lambda d_H(x, y; \mathrm{cl}(K)) \text{ for all } x, y \in \mathrm{int}(K)\}. \tag{A.25}$$

It thus follows from (A.23) that $k(L; K, K') = k(L; \mathrm{cl}(K), \mathrm{cl}(K'))$. □

A.6 Completion of the proof of the Birkhoff–Hopf theorem

Before we prove the Birkhoff–Hopf theorem we show that the case $\Delta(L) = \infty$ can be obtained via a limiting argument from the case where $\Delta(L) < \infty$.

Lemma A.6.1 *Suppose that the Birkhoff–Hopf Theorem A.4.1 is true whenever $\Delta(L) < \infty$. Then it is also true if $\Delta(L) = \infty$.*

Proof If $\Delta(L) = \infty$, it follows from the definition that for every $R > 0$ there exist $x, y \in K \setminus \{0\}$ with $Lx \sim_{K'} Ly$ and $d_H(Lx, Ly; K') > R$. We may assume that $x \sim_K y$, otherwise we replace x by $x+\varepsilon y$ and y by $y+\varepsilon x$ for some small $\varepsilon > 0$. As $d_H(Lx, Ly; K') > R > 0$, x and y are linearly independent. Consider $V(x, y) \subseteq V$ and $W(Lx, Ly) \subseteq W$. Let $K(x, y) = K \cap V(x, y)$ and $K'(Lx, Ly) = K' \cap W(Lx, Ly)$. Write $L' \colon V(x, y) \to W(Lx, Ly)$ to denote the restriction of L to $V(x, y)$. Define a cone $C \subseteq V(x, y)$ by

$$C = \{\lambda x + \mu y \colon \lambda \geq 0 \text{ and } \mu \geq 0\} \subseteq K(x, y).$$

Note that $L'(C) \subseteq K'(Lx, Ly)$. For convenience we write $C' = K'(Lx, Ly)$.

Recall that for $u, v \in C'$ we have that $u \sim_{C'} v$ if and only if $u \sim_{K'} v$. In that case $d_H(u, v; C') = d_H(u, v; K')$. As $Lx \sim_{K'} Ly$, we have that $Lx \sim_{C'} Ly$. Moreover,

$$\begin{aligned} L(C \setminus \{0\}) &= \{\lambda Lx + \mu Ly \colon \lambda \geq 0, \mu \geq 0 \text{ and } \lambda + \mu > 0\} \\ &= \{sz \colon s > 0 \text{ and } z \in \mathrm{co}(\{Lx, Ly\})\}. \end{aligned}$$

Thus, $L(C \setminus \{0\})$ is contained in a part of C' and $L(C \setminus \{0\}) \neq \{0\}$. It follows from Definitions A.3.1 and A.3.2 and Lemma A.3.4 that

$$\begin{aligned} \Delta(L; C; C') &= \mathrm{diam}_2(L'(C); C') \\ &= \mathrm{diam}_1(L'(C); C') \\ &= \mathrm{diam}_1(\{Lx, Ly\}; C') \\ &= d_H(Lx, Ly; K') > R. \end{aligned}$$

As $\Delta(L'; C, C') < \infty$, we can apply (A.12) and find $u, v \in C$, with $u \sim_C v$ and $d_H(u, v; C) > 0$, such that

$$\frac{d_H(Lu, Lv; C')}{d_H(u, v; C)} > \tanh\left(\frac{R}{4}\right).$$

Since $C \subseteq K(x, y)$, we have that

$$d_H(u, v; C) \geq d_H(u, v; K(x, y)) = d_H(u, v; K)$$

for all $u, v \in C$ with $u \sim_C v$. We already know that $d_H(Lu, Lv; C') = d_H(Lu, Lv; K')$, so that

$$\frac{d_H(Lu, Lv; K')}{d_H(u, v; K)} > \tanh\left(\frac{R}{4}\right).$$

Since $R > 0$ was arbitrary, we conclude that $k(L; K, K') = 1$.

Note that (A.12) also implies that there exist $u', v' \in C$ with $\omega(u'/v'; C) > 0$ and

$$\frac{\omega(Lu'/Lv'; C')}{\omega(u'/v'; C)} > \tanh\left(\frac{R}{4}\right).$$

Since $\omega(u'/v'; C) \geq \omega(u'/v'; K(x, y)) = \omega(u'/v'; K)$ and $\omega(Lu'/Lv'; C') = \omega(Lu'/Lv'; K')$, we find that

$$\frac{\omega(Lu'/Lv'; K')}{\omega(u'/v'; K)} > \tanh\left(\frac{R}{4}\right),$$

so that $N(L; K, K') = 1$. □

Let us now prove the Birkhoff–Hopf Theorem.

Proof of Birkhoff–Hopf Theorem A.4.1 By Lemma A.4.3 we know that it suffices to prove (A.12) when $\dim(V) = \dim(W) = 2$ and L is one-to-one. We may also assume that $\Delta(L) < \infty$ by Lemma A.6.1. We distinguish two cases: (1) $\mathrm{cl}(K)$ or $\mathrm{cl}(K')$ is not a cone, and (2) $\mathrm{cl}(K)$ and $\mathrm{cl}(K')$ are both cones.

If we are in the first case, we claim that

$$N(L) = k(L) = \tanh(\frac{1}{4}\Delta(L)) = 0, \tag{A.26}$$

which proves (A.12). To prove these equalities we first assume that $\mathrm{cl}(K')$ is not a cone. If $x, y \in K$ and $Lx \sim_{K'} Ly$, then $\omega(Lx/Ly; K') = 0$ and $d_H(Lx, Ly) = 0$ by Corollary A.5.3, so that (A.26) holds. Now suppose that $\mathrm{cl}(K')$ is a cone, but $\mathrm{cl}(K)$ is not. If $x \sim_K y$, then it follows from Corollary A.5.3 that $d_H(x, y) = 0$ and $\omega(x/y; K) = 0$. This implies that $d_H(Lx, Ly) = 0$ and $\omega(Lx/Ly; K') = 0$, and hence $N(L) = k(L) = 0$. By Corollary A.5.2 wc know that

$$d_H(u, v; K') = d_H(u, v; \mathrm{cl}(K'))$$

for all $u \sim_{K'} v$ in K'. Therefore

$$\Delta(L; K, K') = \mathrm{diam}_2(L(K); K') = \mathrm{diam}_2(L(K); \mathrm{cl}(K')).$$

As $L(\mathrm{int}(K)) \subseteq L(K) \subseteq \mathrm{cl}(L(\mathrm{int}(K)))$, we get that

$$\begin{aligned} \mathrm{diam}_2(L(\mathrm{int}(K)); \mathrm{cl}(K')) &\leq \mathrm{diam}_2(L(K); \mathrm{cl}(K')) \\ &\leq \mathrm{diam}_2(\mathrm{cl}(L(\mathrm{int}(K))); \mathrm{cl}(K')). \end{aligned} \tag{A.27}$$

By applying Lemma A.3.4(iii) we find that each inequality in (A.27) is an equality. Therefore

$$\mathrm{diam}_2(L(\mathrm{int}(K)); \mathrm{cl}(K')) = \mathrm{diam}_2(L(K); \mathrm{cl}(K)') = \Delta(L; K, K'). \tag{A.28}$$

On the other hand, Corollary A.5.3 implies that $d_H(x, y; K) = 0$ for all $x, y \in \mathrm{int}(K)$, so that $d_H(Lx, Ly; \mathrm{cl}(K')) = d_H(Lx, Ly; K') = 0$ for all $x, y \in \mathrm{int}(K)$. Thus, we find that $\Delta(L; K, K') = 0$ by (A.28), which shows (A.26).

Let us now consider the case where $\mathrm{cl}(K)$ and $\mathrm{cl}(K')$ are cones. Then we may assume that K and K' are closed by Corollary A.5.4. It follows from Theorem A.5.1 that there exist $u, v \in V$ linearly independent such that

$$K = \{\lambda u + \mu v : \lambda \geq 0 \text{ and } \mu \geq 0\}.$$

Likewise there exist $u', v' \in W$ linearly independent such that

$$K' = \{\lambda u' + \mu v' : \lambda \geq 0 \text{ and } \mu \geq 0\}.$$

Define $S : \mathbb{R}^2 \to V$ by $S(\lambda, \mu) = \lambda u + \mu v$, and define $T : W \to \mathbb{R}^2$ by $T(\lambda u' + \mu v') = (\lambda, \mu)$. Clearly $S^{-1}(K) = \mathbb{R}^2_+$ and $T(K') = \mathbb{R}^2_+$.

The Birkhoff–Hopf Theorem A.4.1 holds for L if it holds for the linear map $L_1 : \mathbb{R}^2 \to \mathbb{R}^2$ given by $L_1 = T \circ L \circ S$ by Lemma A.4.5. Note that $L_1(\mathbb{R}^2_+) = \mathbb{R}^2_+$. If we identify $\mathbb{R}^2$ with 2×1 column vectors, then L_1 is represented with respect to the standard basis vectors $e_1 = (1, 0)$ and $e_2 = (0, 1)$ by the matrix

$$A = \begin{bmatrix} a & b \\ c & d \end{bmatrix},$$

where $L_1(e_1) = ae_1 + ce_2$ and $L_1(e_2) = be_1 + de_2$. As $L_1(\mathbb{R}^2_+) = \mathbb{R}^2_+$, we have that $a, b, c, d \geq 0$ and $\det(A) \neq 0$, as L_1 is one-to-one.

To finish the proof we further transform L_1 so that Theorem A.4.4 can be applied. Let D_1 and D_2 be diagonal matrices given by

$$D_1 = \begin{bmatrix} x_1 & 0 \\ 0 & x_2 \end{bmatrix} \quad \text{and} \quad D_2 = \begin{bmatrix} y_1 & 0 \\ 0 & y_2 \end{bmatrix}.$$

We claim that there exist $x_1, x_2, y_1, y_2 > 0$ such that

$$D_1 A D_2 = \begin{bmatrix} \beta & 1-\beta \\ 1-\beta & \beta \end{bmatrix},$$

where $0 < \beta < 1$ and $\beta \neq 1/2$. Indeed,

$$D_1 A D_2 = \begin{bmatrix} ax_1y_1 & bx_1y_2 \\ cx_2y_1 & dx_2y_2 \end{bmatrix}.$$

For $0 < x_1 < 1$, let $x_2 = 1 - x_1$, $y_1 = 1/(ax_1 + cx_2)$, and $y_2 = 1/(bx_1 + dx_2)$. This ensures that $D_1 A D_2$ is a column stochastic matrix. Let us now show that $0 < x_1 < 1$ can be chosen such that $D_1 A D_2$ is row stochastic. The sum of the entries of the first row of $D_1 A D_2$ is given by

$$\sigma(x_1) = \frac{ax_1}{ax_1 + c(1 - x_1)} + \frac{bx_1}{bx_1 + d(1 - x_1)}.$$

As $\lim_{x_1 \to 0^+} \sigma(x_1) = 0$ and $\lim_{x_1 \to 1^-} \sigma(x_1) = 2$, there exists $0 < x_1 < 1$ such that $\sigma(x_1) = 1$. In fact, x_1 is unique. As $D_1 A D_2$ is a 2×2 column stochastic matrix, it follows that D_1AD_2 is also row stochastic for that value of x_1. Hence

$$D_1AD_2 = \begin{bmatrix} \beta & 1-\beta \\ 1-\beta & \beta \end{bmatrix},$$

where $0 < \beta < 1$. Moreover, $\beta \neq 1/2$, as D_1AD_2 is one-to-one. If $\beta < 1/2$, we let

$$P = \begin{bmatrix} 0 & 1 \\ 1 & 0 \end{bmatrix}$$

and, if $\beta > 1/2$, we let P be the 2×2 identity matrix. Thus,

$$PD_1AD_2 = \begin{bmatrix} \gamma & 1-\gamma \\ 1-\gamma & \gamma \end{bmatrix},$$

where $1/2 < \gamma < 1$. Finally, let

$$D = \begin{bmatrix} 1/(1-\gamma) & 0 \\ 0 & 1/(1-\gamma) \end{bmatrix}$$

and put $L_2 = PD_1AD_2D$. Note that

$$L_2 = \begin{bmatrix} \alpha & 1 \\ 1 & \alpha \end{bmatrix},$$

where $\alpha > 1$. Moreover, $L_2(\mathbb{R}^2_+) = \mathbb{R}^2_+$ and L_2 is one-to-one. It now follows from Theorem A.4.4 that (A.12) holds for L_2. We conclude from Lemma A.4.5 that (A.12) also holds for L. □

The Birkhoff–Hopf theorem has a simple explicit form when $K = \mathbb{R}^n_+$ and $K' = \mathbb{R}^m_+$. If $A = (a_{ij})$ is an $m \times n$ matrix, then $A(\mathbb{R}^n_+) \subseteq \mathbb{R}^m_+$ if and only if $a_{ij} \geq 0$ for all $1 \leq i \leq m$ and $1 \leq j \leq n$. Moreover, for each $x, y \in \mathbb{R}^n_+$ we have that $x \sim_{\mathbb{R}^n_+} y$ is equivalent to

$$I_x = \{i : x_i > 0\} = \{i : y_i > 0\} = I_y.$$

In that case

$$d_H(x, y) = \log\left(\max_{i,j \in I_x} \frac{x_i y_j}{x_j y_i}\right). \tag{A.29}$$

These observations yield the following theorem.

Theorem A.6.2 *Let $A = (a_{ij})$ be an $m \times n$ nonnegative matrix, and let $e_1, \ldots, e_n$ denote the standard basis vectors in $\mathbb{R}^n$. If there exists $J \subseteq \{1, \ldots, n\}$ such that $Ae_i \sim_{\mathbb{R}^m_+} Ae_j$ for all $i, j \in J$ and $Ae_i = 0$ for all $i \notin J$, then*

$$\Delta(A) = \max_{i,j \in J} d_H(Ae_i, Ae_j) < \infty. \tag{A.30}$$

In particular, if A is positive, so $J = \{1, \ldots, n\}$, then

$$\Delta(A) = \max_{1 \leq i,j \leq n} d_H(Ae_i, Ae_j) = \log\Big(\max_{i,jp,q} \frac{a_{pi}a_{qj}}{a_{pj}a_{qi}}\Big). \tag{A.31}$$

Proof Equation (A.30) follows from Lemma A.3.4, since

$$A(\mathbb{R}^n_+) = \{\sum_{j \in J} \lambda_j Ae_j : \lambda_j > 0 \text{ for all } j \in J\}.$$

The second equality is a consequence of (A.29). □

Explicit formulas like those in Theorem A.6.2 can also be given when K and K' are polyhedral cones by using the facet-defining functionals. Analogues of Theorem A.6.2 for integral operators with positive kernels were obtained by Hopf [90]; see also [62].

In the infinite-dimensional case it useful to note that we have proved slightly more than is stated in Theorem A.4.1. To see this we need the notion of a cone-linear map.

Definition A.6.3 If K is a cone in a real vector space V, K' is a cone in a real vector space W, and $f: K \to K'$ is a map such that

$$f(\alpha x + \beta y) = \alpha f(x) + \beta f(y)$$

for all $x, y \in K$ and $\alpha, \beta \geq 0$, then f is said to be *cone-linear*.

It is known (see [32]) that if K is a closed cone in a Banach space X and K' is a closed cone in a Banach space Y, then a cone-linear map $f: K \to K'$ may not have a continuous linear extension $F: X \to Y$, even if $X = \mathrm{cl}(K - K)$. As continuity does not enter the statement of the Birkhoff–Hopf Theorem A.4.1, we easily obtain the following extension.

Theorem A.6.4 *If K is a cone in a real vector space V, K' is a cone in a real vector space W, and $f: K \to K'$ is a cone-linear map, then*

$$N(f) = k(f) = \tanh(\frac{1}{4}\Delta(f)).$$

Proof By Theorem A.4.1 it suffices to note that f has a linear extension $F: V_0 \to W$, where $V_0 = \{u - v : u, v \in K\}$ is a linear subspace of V. Indeed, if $x = u - v \in V_0$, then we can define $F(x) = f(u) - f(v)$. To see that F is well defined we remark that if $x = u - v \in V_0$ and $x = u' - v' \in V_0$ for some $u, u', v, v' \in K$, then $u + v' = u' + v$, so that $f(u) + f(v') = f(u + v') = f(u' + v) = f(u') + f(v)$. This implies that F is well defined. It is easy to verify that F is linear and $F(K) \subseteq K'$. □

A.7 Eigenvectors of cone-linear maps

As in finite-dimensional Perron–Frobenius theory one typically wishes to prove the existence of a (unique) eigenvector $v \in K \setminus \{0\}$ of a cone-linear map $f: K \to K$ with $f(v) = rv$, where $r = r_K(f)$ is the cone spectral radius; see [32, 62, 110, 111, 117, 138, 139, 155, 166, 167, 173, 193, 194, 196]. If K is a closed cone in Banach space X, then for each $x \in K$, we define, as in Section 5.3,

$$\mu(x) = \limsup_{k\to\infty} \|f^k(x)\|_X^{1/k},$$

where $f: K \to K$ is a cone-linear map which is continuous at 0. The *cone-spectral radius* of f is defined by

$$r_K(f) = \sup\{\mu(x): x \in K\}.$$

Alternatively, we can define $\|f^k\|_K = \sup\{\|f^k(x)\|_X : x \in K \text{ and } \|x\|_X \leq 1\}$ and consider the *Bonsall cone spectral radius*,

$$\hat{r}_K(f) = \lim_{k\to\infty} \|f^k\|_K^{1/k} = \inf_{k\geq 1} \|f^k\|_K^{1/k}.$$

It is known [138] that $\hat{r}_K(f) = r_K(f)$ when $f: K \to K$ is a cone-linear map that is continuous at 0.

To find conditions under which a cone-linear map $f: K \to K$ has a unique eigenvector in $K \setminus \{0\}$ with eigenvalue equal to the cone spectral radius, we can combine the Birkhoff–Hopf Theorem A.4.1 and Banach's contraction Theorem 3.2.1. Indeed, if K is a closed normal cone in a Banach space X and $x \in K$, then the same argument as in Lemma 2.7.2 shows that (Σ_x, d_H) is a complete metric space, where $\Sigma_x = \{y \in K : x \sim_K y \text{ and } \|y\| = 1\}$. Recall that a cone K in a Banach space is normal if there exists $\delta \geq 1$ such that $\|x\| \leq \delta\|y\|$ for all $x, y \in K$ with $0 \leq x \leq y$. This idea leads to the following theorem.

Theorem A.7.1 *Let K be a closed normal cone in a Banach space X and let $f: K \to K$ be a cone-linear map which is continuous at* 0. *If there exists an integer $p \geq 1$ such that $\Delta(f^p) < \infty$ and $f^{p+1}(K) \neq \{0\}$, then f has a unique normalized eigenvector $v \in K \setminus \{0\}$ such that $f(v) = r_K(f)v$. Moreover, if we let $c = \tanh(\Delta(f^p)/4) < 1$, then*

$$d_H(f^{kp}(x), v) \leq c^k d_H(x, v)$$

for all $x \in K \setminus \{0\}$.

Proof As $\Delta(f^p) < \infty$, it follows from Lemma A.3.3 that $f^p(K) \setminus \{0\}$ is contained in a part of K. Since $f^{p+1}(K) \neq \{0\}$, there exists $u \in f^p(K)$ such that $f(u) \neq 0$. Let K_u denote the part of K containing u. Note that $u = f^p(x)$ for some $x \in K \setminus \{0\}$, so that $f(u) = f(f^p(x)) = f^p(f(x)) \in K_u$. This implies that $f(K_u) \subseteq K_u$.

Let $\Sigma_u = \{x \in K_u : \|x\| = 1\}$ and define $g\colon \Sigma_u \to \Sigma_u$ by

$$g(x) = \frac{f(x)}{\|f(x)\|} \quad \text{for } x \in \Sigma_u.$$

It is easy to see that $g^k(x) = f^k(x)/\|f^k(x)\|$ for all $k \geq 1$. As $f(K_u) \subseteq K_u$, we have that $g^k(\Sigma_u) \subseteq \Sigma_u$ for all $k \geq 1$. It follows from Theorem A.6.4 that

$$d_H(g^p(x), g^p(y)) = d_H(f^p(x), f^p(y)) \leq c d_H(x, y)$$

for all $x, y \in \Sigma_u$, where $c = \tanh(\Delta(f^p)/4) < 1$. As K is normal, (Σ_u, d_H) is a complete metric space, so that g^p has a unique fixed point, say $v \in \Sigma_u$, by Banach's contraction theorem (see Theorem 3.2.1). Note that $g(v) = g^p(g(v)) \in \Sigma_u$ is also a fixed point of g^p and hence $g(v) = v$.

Let $\lambda = \|f(v)\| > 0$ and remark that $f(v) = \lambda v$. By Theorem A.6.4 we get that

$$\begin{aligned} c^k d_H(x, v) &\geq d_H(f^{kp}(x), f^{kp}(v)) \\ &= d_H(f^{kp}(x), \lambda^{kp} v) \\ &= d_H(f^{kp}(x), v). \end{aligned}$$

It remains to show that $\lambda = r_K(f)$. Note that $\mu(v) = \lambda$, so that $\lambda \leq r_K(f)$. Let $x \in K$ be such that $f^p(x) \neq 0$. Thus, $f^p(x) \sim_K u$. As $v \in \Sigma_u$, $u \sim_K v$, so that $f^p(x) \sim_K v$. Hence there exist $a, b > 0$ such that

$$a\lambda^p v \leq_K f^p(x) \leq_K b\lambda^p v.$$

Applying f^m to these inequalities yields

$$a\lambda^{p+m} v \leq_K f^{p+m}(x) \leq_K b\lambda^{p+m} v.$$

As K is normal, there exists $\delta \geq 1$ with $\|f^{p+m}(x)\| \leq \delta b \lambda^{p+m}$. Therefore

$$\mu(x) = \limsup_{m\to\infty} \|f^{p+m}(x)\|^{1/(p+m)} \leq \limsup_{m\to\infty} (\delta b)^{1/(p+m)} \lambda = \lambda,$$

so that $r_K(f) \leq \lambda$, which completes the proof. □

If all conditions of Theorem A.7.1 are satisfied except for the normality of K, the original Kreĭn–Rutman theorem and subsequent generalizations give further conditions which ensure the existence of a normalized eigenvector $v \in K$ of f with eigenvalue $r_K(f)$. Once v is known to exist, the arguments of Theorem A.7.1 prove that v is unique and that the geometric-rate convergence estimates given in Theorem A.7.1 remain valid.

Even if $\Delta(f) = \infty$ in the Birkhoff–Hopf theorem, the Hilbert metric can still provide useful information about f. In particular, it may happen that $d_H(f(x), f(y)) < d_H(x, y)$ for all $x, y \in K$ with $d_H(x, y) > 0$. For nonnegative matrices this leads to the notion of a "scrambling matrix."

An $m \times n$ nonnegative matrix $A = (a_{ij})$ is said to be *row-allowable* if $A(\mathrm{int}(\mathbb{R}^n_+)) \subseteq \mathrm{int}(\mathbb{R}^m_+)$. Equivalently, A is row-allowable if each row of A contains at least one positive entry. A row-allowable nonnegative matrix $A = (a_{ij})$ is called *scrambling* if for each $1 \le i_1 < i_2 \le m$ there exists $1 \le j \le n$ such that $a_{i_1 j} a_{i_2 j} > 0$. It is known (see, for example, [202, pp. 80–3]) that if A is a row-allowable scrambling matrix, then

$$d_H(Ax, Ay) < d_H(x, y) \tag{A.32}$$

for all $x, y \in \mathrm{int}(\mathbb{R}^n_+)$ with $d_H(x, y) > 0$. For $m = n$, Equation (A.32) implies that A has at most one normalized eigenvector in $\mathrm{int}(\mathbb{R}^n_+)$. It may, however, happen that a row-allowable scrambling matrix has no eigenvector in $\mathrm{int}(\mathbb{R}^n_+)$. For example, the matrix

$$A = \begin{bmatrix} a & b \\ 0 & c \end{bmatrix},$$

where $a, b, c > 0$, has an eigenvector in $\mathrm{int}(\mathbb{R}^n_+)$ if and only if $c > a$.

More generally there exists the following result for cone-linear maps.

Theorem A.7.2 *Let V and W be Hausdorff topological vector spaces and $K \subseteq V$ and $K' \subseteq W$ be closed cones with non-empty interiors. Let $f\colon K \to K'$ be a cone-linear map with $f(K) \subseteq K'$. If for all linearly independent $u, v \in \partial K$, with $(u + v)/2 \in \mathrm{int}(K)$, it is true that either $f(u) \in \mathrm{int}(K')$ or $f(v) \in \mathrm{int}(K')$, then*

$$d_H(f(x), f(y)) < d_H(x, y)$$

for all $x, y \in \mathrm{int}(K)$ with $d_H(x, y) > 0$.

Proof Let $x, y \in \mathrm{int}(K)$ with $d_H(x, y) > 0$, so x and y are linearly independent. As usual we write $K(x, y) = K \cap V(x, y)$ and $K'(f(x), f(y)) = K' \cap W(f(x), f(y))$. Thus, $K(x, y)$ and $K'(f(x), f(y))$ are closed cones in the two-dimensional spaces $V(x, y)$ and $W(f(x), f(y))$, respectively. If $f(x)$ and $f(y)$ are linearly dependent, then $d_H(f(x), f(y)) = 0$, and we are done. So, suppose that $f(x)$ and $f(y)$ are linearly independent. Let L denote the linear extension to $V(x, y)$ of the restriction of f to $K(x, y)$, and note that L is a one-to-one linear map that maps $V(x, y)$ onto $W(f(x), f(y))$. As $x, y \in \mathrm{int}(K)$, we know that x and y are also in $\mathrm{int}(K(x, y))$ in $V(x, y)$.

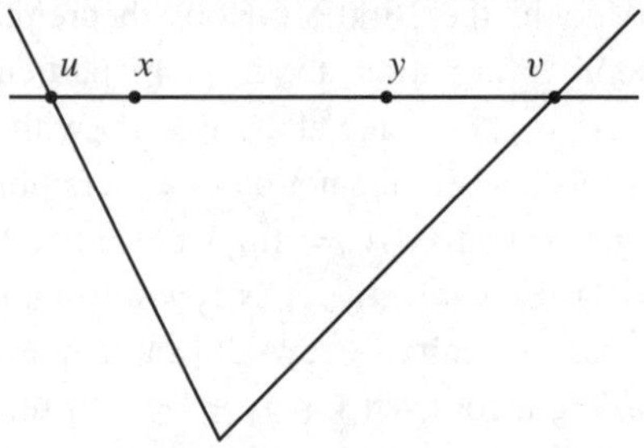

Figure A.4 $K(x, y)$.

This implies that $f(x)$ and $f(y)$ lie in the interior of $K'(f(x), f(y))$ in $W(f(x), f(y))$, as L is one-to-one.

By the Hahn–Banach separation theorem there exists a continuous linear functional $\vartheta\colon V(x, y) \to \mathbb{R}$ such that $\vartheta(w) > 0$ for all $w \in K(x, y) \setminus \{0\}$. Similarly, there exists a continuous linear functional $\psi\colon W(f(x), f(y)) \to \mathbb{R}$ such that $\psi(z) > 0$ for all $z \in K'(f(x), f(y)) \setminus \{0\}$. By rescaling x and y we may assume that $\vartheta(x) = \vartheta(y) = 1$. Let l be the straight line through x and y. As $K(x, y)$ is a closed cone, there exist u and v in $l \cap \partial K(x, y)$ as in Figure A.4.

Clearly $(u + v)/2 \in \text{int}(K(x, y))$; so, either $f(u) \in \text{int}(K')$ or $f(v) \in \text{int}(K')$ by assumption. For definiteness we suppose that $f(u) \in \text{int}(K')$. By Theorem 2.1.2,

$$d_H(x, y; K) = \log[u, x, y, v].$$

As L is one-to-one, we have that

$$[u, x, y, v] = [Lu, Lx, Ly, Lv].$$

Write $u' = Lu/\psi(Lu)$ and $v' = Lv/\psi(Lv)$, and let $S\colon W(f(x), f(y)) \to W(f(x), f(y))$ be defined by

$$S(af(u) + bf(v)) = au' + bv'$$

for $a, b \in \mathbb{R}$. Note that S is a one-to-one linear map that maps $f(u)$, $f(x)$, and $f(y)$ into the interior of $K'(f(x), f(y))$ and $S(f(v)) \in K'(f(x), f(y))$. Since S is one-to-one, $f(x) = Lx$, $f(y) = Ly$, $f(u) = Lu$, and $f(v) = Lv$, we have that

$$[u', S(Lx), S(Ly), v'] = [Lu, Lx, Ly, Lv].$$

Now note that u', $S(Lx)$, $S(Ly)$, and v' lie on the line

$$l' = \{z \in K'(f(x), f(y))\colon \psi(z) = 1\}.$$

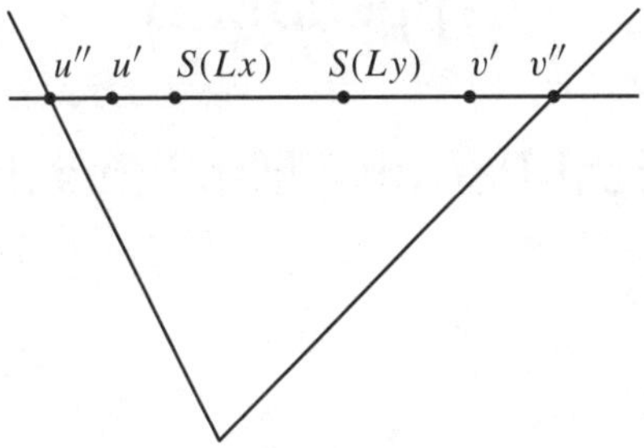

Figure A.5 $K'(f(x), f(y))$.

As $u' \in \text{int}(K')$, we know that there exist u'' and v'' in $l' \cap \partial K'(f(x), f(y))$ as in Figure A.5. (Obviously, $u'' \neq u'$, but v'' could be equal to v'.) As $u' \in \text{int}(K')$ we find that

$$\begin{aligned}[u'', f(x), f(y), v''] &= [u'', S(Lx), S(Ly), v''] \\ &< [u', S(Lx), S(Ly), v'] = [u, x, y, v].\end{aligned}$$

This implies that $d_H(f(x), f(y)) < d_H(x, y)$, and we are done. □

The condition in Theorem A.7.2 can be shown to be optimal. We conclude the appendix by showing that row-allowable scrambling matrices are contractions under Hilbert's metric.

Corollary A.7.3 *Let $A = (a_{ij})$ be an $m \times n$ nonnegative matrix. If A is row-allowable and scrambling, then*

$$d_H(Ax, Ay) < d_H(x, y)$$

for all $x, y \in \text{int}(\mathbb{R}^n_+)$ with $d_H(x, y) > 0$.

Proof We will use Theorem A.7.2. So, suppose that u and v are linearly independent in $\partial\mathbb{R}^n_+$ and $(u + v)/2 \in \text{int}(\mathbb{R}^n_+)$. We need to show that either $Au \in \text{int}(\mathbb{R}^m_+)$ or $Av \in \text{int}(\mathbb{R}^m_+)$. Assume that $Au \in \partial\mathbb{R}^m_+$. Then there exists $1 \leq i \leq m$ such that

$$(Au)_i = \sum_{j=1}^{n} a_{ij}u_j = 0. \tag{A.33}$$

Let $J = \{j : u_j = 0\}$. As $(u + v)/2 \in \text{int}(\mathbb{R}^n_+)$, $v_j > 0$ for all $j \in J$. It follows from (A.33) that $a_{ij} = 0$ for all $j \notin J$. Now for each $1 \leq k \leq m$ there exists $1 \leq j_k \leq n$ such that $a_{kj_k}a_{ij_k} > 0$, since A is scrambling. So, each $j_k \in J$ and $a_{kj_k}v_{j_k} > 0$. Thus, $\sum_j a_{kj}v_j > 0$ for each $1 \leq k \leq m$, so that $Av \in \text{int}(\mathbb{R}^m_+)$. □

Appendix B

Classical Perron–Frobenius theory

In this appendix we provide proofs of most of the results from Section 1.1 concerning classical linear Perron–Frobenius theory. We begin (see Theorem B.1.1) by proving a generalization, valid for general cones, of Perron's theorem which is stated in Theorem 1.1.1. From this result we then derive the finite-dimensional Kreĭn–Rutman theorem (Theorem 1.1.6). We also show that many of the results in the general version of Perron's theorem remain valid for irreducible linear maps, and this yields Theorem 1.1.7. We subsequently give a complete proof of the third assertion in the classical Perron–Frobenius Theorem 1.1.2, which depends on special properties of the cone $\mathbb{R}^n_+$ and is of a qualitatively different nature from the other two assertions (see Proposition B.4.3). We next use this part of the Perron–Frobenius theorem to prove Theorems 1.1.8 and 1.1.9 concerning the peripheral spectrum and iterative behavior of linear maps on polyhedral cones.

Our treatment here is concise and meant only as an introduction to the linear theory. The reader should consult the books by Bapat and Raghavan [15], Berman and Plemmons [22], Minc [148], and Seneta [202], or the survey paper by Tam [214], for a more thorough discussion of linear Perron–Frobenius theory.

B.1 A general version of Perron's theorem

It is well known that a linear map $L\colon V \to V$ on a real n-dimensional vector space V can be identified with a real $n \times n$ matrix A by selecting a basis $v_1, \ldots, v_n$ for V. More precisely, the basis induces a linear isomorphism, $T\colon x_1v_1 + \cdots + x_nv_n \mapsto (x_1, \ldots, x_n)$, from V onto $\mathbb{R}^n$, and TLT^{-1} corresponds to a real $n \times n$ matrix A. The *spectrum*, $\sigma(L)$, of L is defined by

$$\sigma(L) = \{\lambda \in \mathbb{C}\colon Ax = \lambda x \text{ for some } x \in \mathbb{C}^n \setminus \{0\}\},$$

i.e., the set of eigenvalues of L. From basic linear algebra we know that $\sigma(L)$ is independent of the choice of the basis of V. If $\lambda \in \mathbb{R}$ is an eigenvalue of L, then there exists $x \in V \setminus \{0\}$ such that $L(v) = \lambda v$. Furthermore, if $\lambda = \rho e^{i\vartheta} \in \mathbb{C} \setminus \mathbb{R}$ is an eigenvalue of L, then there exist $x, y \in V \setminus \{0\}$ and, when suitably interpreted, $x + iy$ is an eigenvector of L with eigenvalue $\rho e^{i\vartheta}$. This implies that

$$L(x) = \rho\cos(\vartheta)x - \rho\sin(\vartheta)y \quad\text{and}\quad L(y) = \rho\sin(\vartheta)x + \rho\cos(\vartheta)y. \tag{B.1}$$

It is easy to verify that x and y in (B.1) are linearly independent over $\mathbb{R}$. We shall need this elementary fact later.

A linear map $L\colon V \to V$ which leaves a solid closed cone K in V invariant is said to be *primitive* if there exists an integer $m \geq 1$ such that

$$L^m(K \setminus \{0\}) \subseteq \operatorname{int}(K).$$

Note that an $n \times n$ nonnegative matrix $A = (a_{ij})$ maps $\mathbb{R}^n_+ \setminus \{0\}$ into $\operatorname{int}(\mathbb{R}^n_+)$ if and only if $a_{ij} > 0$ for all $1 \leq i, j \leq n$. So, an $n \times n$ nonnegative matrix A is primitive if and only if there exists an integer $m \geq 1$ such that A^m is positive.

In the following theorem, recall that if V is a finite-dimensional real vector space and $L\colon V \to V$ is a linear map, then $r(L)$, the spectral radius of L, satisfies $r(L) = \sup\{|z|\colon z \in \sigma(L)\}$. It is known that $r(L) = \lim_{k\to\infty} \|L^k\|^{1/k}$, where $\|L^k\| = \sup\{\|L^k x\| : x \in V \text{ with } \|x\| \leq 1\}$ and $\|\cdot\|$ is a norm on the vector space V.

Theorem B.1.1 *Let K be a solid closed cone in a finite-dimensional real normed space V with $\dim(V) \geq 1$, and let $L\colon V \to V$ be a linear map with $L(K) \subseteq K$ and spectral radius $r(L)$. If L is primitive, then the following assertions hold:*

(i) $r(L) > 0$ and $r(L)$ is an algebraically simple eigenvalue of L with an eigenvector $v \in \operatorname{int}(K)$ with $\|v\| = 1$.
(ii) For every $x \in K \setminus \{0\}$ we have that

$$\lim_{k\to\infty} \frac{L^k(x)}{\|L^k(x)\|} = v.$$

In particular, every eigenvector $w \in K$ of L is a scalar multiple of v.
(iii) If $\lambda \in \sigma(L)$ and $\lambda \neq r(L)$, then $|\lambda| < r(L)$.

Proof Recall that if $x \in K$ and $y \in \operatorname{int}(K)$, then $x + y \in \operatorname{int}(K)$. Moreover, if $x, y \in K$ with $y \leq x$ and $x \notin \operatorname{int}(K)$, then $y \notin \operatorname{int}(K)$. Indeed, if $y \in \operatorname{int}(K)$, then $x + y \in \operatorname{int}(K)$, and since we assume that $x - y \in K$, $(x - y) + (x + y) = 2x \in \operatorname{int}(K)$, contrary to our assumption.

We claim that $L(\mathrm{int}(K)) \subseteq \mathrm{int}(K)$. To show this we assume, by way of contradiction, that $u \in \mathrm{int}(K)$ with $L(u) \in \partial K$. Since $u \in \mathrm{int}(K)$, there exists $M > 0$ such that $L(u) \leq Mu$, so that $L^2(u) \leq ML(u)$. Since $L(u) \notin \mathrm{int}(K)$, the previous remarks imply that $L^2(u) \notin \mathrm{int}(K)$. Continuing in this way, we see that $L^k(u) \notin \mathrm{int}(K)$ for all $k \geq 1$, which is impossible, as $L^m(K \setminus \{0\}) \subseteq \mathrm{int}(K)$. Let $\Sigma = \{x \in K : \|x\| = 1\}$ and $\Sigma^\circ = \{x \in \mathrm{int}(K) : \|x\| = 1\}$. The assumptions imply that $L^m(\Sigma)$ is a compact subset of $\mathrm{int}(K)$. Select $u \in \Sigma^\circ$ and let d_H be Hilbert's metric on $\mathrm{int}(K)$. As $L^m(\Sigma)$ is compact, there exists $R > 0$ such that

$$L^m(\Sigma) \subseteq B_R(u), \tag{B.2}$$

where $B_R(u) = \{x \in \mathrm{int}(K) : d_H(x, u) \leq R\}$. In particular, it follows from (B.2) that

$$\sup\{d_H(L^m(x), L^m(y)) : x, y \in \mathrm{int}(K)\} \leq 2R.$$

Now define $f : \mathrm{int}(K) \to \mathrm{int}(K)$ by

$$f(x) = \frac{L(x)}{\|L(x)\|} \quad \text{for } x \in \mathrm{int}(K).$$

Note that $f^j(x) = L^j(x)/\|L^j(x)\|$ for all $j \geq 1$, and define $g(x) = f^m(x)$ for all $x \in \mathrm{int}(K)$. It follows from the fact that $g(\Sigma) \subseteq \Sigma^\circ$ and from the Birkhoff–Hopf Theorem A.4.1 that there exists a constant $0 \leq c < 1$ such that, for each $x, y \in \Sigma^\circ$,

$$d_H(g(x), g(y)) \leq c d_H(x, y).$$

Thus, it follows from Banach's contraction Theorem 3.2.1 that g has a unique fixed point $v \in \Sigma^\circ$ and $g^k(x) \to v$ as $k \to \infty$, for all $x \in \Sigma^\circ$. Furthermore, since $g(f(v)) = f(g(v)) = f(v)$, $f(v)$ is also a fixed point of g. So, $f(v) = v$, and we see that $L(v) = \|L(v)\|v$. Because $f^m(x) \in \Sigma^\circ$ for all $x \in K \setminus \{0\}$, we also find that

$$\lim_{k\to\infty} \frac{L^{mk}(x)}{\|L^{mk}(x)\|} = v \tag{B.3}$$

for all $x \in K \setminus \{0\}$. If we replace x in (B.3) by $L^j(x)$ for $x \in K \setminus \{0\}$ and $0 \leq j < m$, and if we note that $L^j(x)$ is unequal to 0, we obtain from (B.3) that $\lim_{k\to\infty} L^{km+j}(x)/\|L^{km+j}(x)\| = v$. We deduce from this that

$$\lim_{k\to\infty} \frac{L^k(x)}{\|L^k(x)\|} = v \quad \text{for all } x \in K \setminus \{0\}.$$

Let $r = \|L(v)\| > 0$, so $L(v) = rv$ and $v \in \mathrm{int}(K)$. We now show that r is equal to the spectral radius of L. To begin we note that there exists a constant $C > 0$ such that

$$-Cv \leq x \leq Cv \quad \text{for all } x \in V \text{ with } \|x\| \leq 1.$$

It follows that

$$0 \leq L^j(x) + Cr^j v = L^j(x + Cv) \leq L^j(2Cv) = 2Cr^j v.$$

Dividing by r^j we get that

$$0 \leq r^{-j} L^j(x) + Cv \leq 2Cv \quad \text{for all } x \in V \text{ with } \|x\| \leq 1. \tag{B.4}$$

From Lemma 1.2.5 we know that every closed finite-dimensional cone is normal. So, we can use Equation (B.4) to find a constant $C_1 > 0$ such that $r^{-j}\|L^j(x)\| \leq C_1 C$, which shows that r is the spectral radius of L.

If $L(w) = rw$ for some $w \in V$ and $w \neq \lambda v$ for all $\lambda \in \mathbb{R}$, then for $\varepsilon > 0$ sufficiently small $\xi = v + \varepsilon w \in \text{int}(K)$, $L(\xi) = r\xi$, and ξ is not a real multiple of v. But then $L^k(\xi)/\|L^k(\xi)\| = \xi/\|\xi\| \neq v$, which contradicts (B.3), so we conclude that v is the unique, up to scalar multiples, eigenvector of L with eigenvalue r.

To show that r has algebraic multiplicity 1, we must prove that there does not exist $w \in V$ with $(L - rI)(w) = v$. If $L(w) = rw + v$, then a simple induction argument shows that

$$L^k(w) = r^k w + kr^{k-1} v \quad \text{for all } k \geq 1. \tag{B.5}$$

Note that (B.4) implies that there exists a constant $C_2 > 0$ such that

$$r^{-k}\|L^k(w)\| \leq C_2\|w\|.$$

So, if we divide both sides in (B.5) by r^k, we get that

$$C_2\|w\| \geq r^{-k}\|L^k(w)\| = \|w + k(r^{-1}v)\| \geq kr^{-1} - \|w\|,$$

which is impossible for k large.

To complete the proof of the theorem, it remains to show that if $\lambda \in \sigma(L)$ and $|\lambda| = r$, then $\lambda = r$. By replacing L by $r^{-1}L$ we may as well assume that $r = 1$. If $L(x) = -x$ for some $x \in V$ with $\|x\| = 1$, then an elementary linear algebra argument shows that x and v are linearly independent. For $\varepsilon > 0$ sufficiently small, $v + \varepsilon x \in \text{int}(K)$. So, $L^k(v + \varepsilon x) = v + (-1)^k \varepsilon x$ and x is not a multiple of v, which contradicts Equation (B.3).

Now suppose that $\lambda \in \sigma(L)$ with $|\lambda| = 1$ and $\lambda \in \mathbb{C} \setminus \mathbb{R}$. Then $\lambda = e^{i\vartheta}$ for some $0 < \vartheta < 2\pi$ and $\vartheta \neq \pi$. In that case there exist linearly independent vectors $x, y \in V$ satisfying (B.1) with $\rho = 1$. We claim that $\{v, x, y\}$ is a linearly independent subset of V. The proof is an elementary exercise in linear algebra, but we include it for completeness.

Suppose by way of contradiction that there exist $\alpha, \beta, \gamma \in \mathbb{R}$, not all zero, such that

$$\alpha v + \beta x + \gamma y = 0. \tag{B.6}$$

Applying L to both sides of (B.6) gives

$$\alpha v + (\beta \cos(\vartheta) + \gamma \sin(\vartheta))x + (\gamma \cos(\vartheta) - \beta \sin(\vartheta))y = 0. \tag{B.7}$$

Subtracting (B.6) from (B.7), and using the fact that x and y are linearly independent, we find that

$$(\cos(\vartheta) - 1)\beta + \sin(\vartheta)\gamma = 0 \quad \text{and} \quad (-\sin(\vartheta)\beta + (\cos(\vartheta) - 1)\gamma = 0.$$

Since $(\cos(\vartheta) - 1)^2 + \sin^2(\vartheta) = 2(1 - \cos(\vartheta))$ and $1 - \cos(\vartheta) > 0$ when $\sin(\vartheta) \neq 0$, we conclude that $\beta = \gamma = 0$. This implies that $\alpha = \beta = \gamma = 0$, which is a contradiction.

Now select $\delta > 0$ such that $\{u \in V : \|u - v\| < \delta\} \subseteq \operatorname{int}(K)$. For x, y, and ϑ as above, select $\varepsilon > 0$ such that $\varepsilon\sqrt{\|x\|^2 + \|y\|^2} < \delta$. Because

$$L^k(v + \varepsilon x) = v + \varepsilon(\cos(k\vartheta)x - \sin(k\vartheta)y),$$

the Cauchy–Schwarz inequality implies that $\|L^k(v + \varepsilon x) - v\| < \delta$. For $0 \leq \psi \leq 2\pi$, define

$$g(\psi) = v + \varepsilon(\cos(\psi)x - \sin(\psi)y),$$

so $g(k\vartheta \bmod 2\pi) = L^k(v + \varepsilon x)$. Since $\{v, x, y\}$ is a linearly independent set in V,

$$\frac{g(\psi)}{\|g(\psi)\|} \neq v.$$

The map $\psi \mapsto \frac{g(\psi)}{\|g(\psi)\|} - v$ is continuous, and hence there exists $\mu > 0$ such that

$$\left\| \frac{g(\psi)}{\|g(\psi)\|} - v \right\| \geq \mu$$

for all $0 \leq \psi \leq 2\pi$. In particular, we find for each $k \geq 1$ that

$$\left\| \frac{L^k(v + \varepsilon x)}{\|L^k(v + \varepsilon x)\|} - v \right\| \geq \mu,$$

which contradicts (B.3). We conclude that if $\lambda \in \sigma(L)$ with $\lambda \neq r$, then $|\lambda| < r$, which completes the proof. □

Notice that Perron's Theorem 1.1.1 is a special case of Theorem B.1.1 where $m = 1$, $V = \mathbb{R}^n$, $K = \mathbb{R}^n_+$, and L is an $n \times n$ matrix with all entries positive.

B.2 The finite-dimensional Kreĭn–Rutman theorem

Theorem B.1.1 has a number of important consequences. For example, one can use it to prove a finite-dimensional version of the Kreĭn–Rutman Theorem 1.1.6. To get started we need to make a few preliminary observations.

Let B be an $n \times n$ real matrix and suppose that $(B_k)_k$ is a sequence of real $n \times n$ matrices with $\|B_k - B\| \to 0$ as $k \to \infty$. For $k \geq 1$ let r_k denote the spectral radius of B_k, and let r denote the spectral radius of the matrix B. Assume that the characteristic polynomial $p(z)$ of B has distinct roots $\zeta_1, \ldots, \zeta_p \in \mathbb{C}$, where ζ_j has algebraic multiplicity m_j; so, $\sum_{j=1}^{p} m_j = n$. Now if $p_k(z)$ denotes the characteristic polynomial of B_k, then we can use Rouché's theorem [191, p. 218] from complex analysis to show that for each $\delta > 0$ and $1 \leq j \leq p$ there exists an integer $k(\delta, j) \geq 1$ such that for each $k \geq k(\delta, j)$ the polynomial $p_k(z)$ has precisely m_j roots $z \in \mathbb{C}$, counting multiplicity, satisfying $|z - \zeta_j| \leq \delta$. From this fact it is easy to deduce that

$$\lim_{k \to \infty} r_k = r.$$

We also see that the corresponding result holds for sequences of bounded linear maps $(B_k)_k$ on finite-dimensional normed spaces V with $\|B_k - B\| \to 0$ as $k \to \infty$. In infinite-dimensional Banach spaces, however, it may happen that $\|B_k - B\| \to 0$ but r_k does not converge to r. S. Kakutani constructed an interesting counterexample, which can be found in [185, pp. 282–3]).

Theorem B.2.1 (Kreĭn–Rutman) *Let K be a solid closed cone in a finite-dimensional real normed space V with* $\dim(V) \geq 1$. *If $B : V \to V$ is a linear map with $B(K) \subseteq K$ and spectral radius r, then there exists $v \in K$ with $B(v) = rv$ and $\|v\| = 1$.*

Proof By Lemma 1.2.4 we know there exists $\varphi \in K^*$ such that $\varphi(x) > 0$ for all $x \in K \setminus \{0\}$. Let $(\varepsilon_k)_k$ be a sequence of positive reals with $\varepsilon_k \to 0$ as $k \to \infty$, and fix $u \in \operatorname{int}(K)$. For $k \geq 1$ and $x \in V$, define

$$B_k(x) = B(x) + \varepsilon_k \varphi(x) u.$$

So, B_k is a primitive linear map, as $B_k(K \setminus \{0\}) \subseteq \operatorname{int}(K)$.

Let r_k denote the spectral radius of B_k for $k \geq 1$. Theorem B.1.1 implies that $r_k > 0$ and there exists $v_k \in \operatorname{int}(K)$ with $B_k(v_k) = r_k v_k$ and $\|v_k\| = 1$ for all $k \geq 1$. By the remarks preceding the theorem, we know that

$$\lim_{k \to \infty} r_k = r,$$

since $\|B_k - B\| \to 0$ as $k \to \infty$. By taking a subsequence we may assume that $v_k \to v$, where $v \in K$ and $\|v\| = 1$. It follows that

$$B(v) = \lim_{k\to\infty} B(v_k) = \lim_{k\to\infty} B_k(v_k) = r_k v_k = rv,$$

which completes the proof. □

B.3 Irreducible linear maps

As we shall see in this section, much of Theorem B.1.1 remains true if L in Theorem B.1.1 is only assumed to be irreducible in the sense of Definition 1.1.4. See also Proposition 1.1.5.

Theorem B.3.1 *Let K be a solid closed cone in a finite-dimensional real normed space V with* $\dim(V) \geq 1$. *If $L: V \to V$ is a linear map with $L(K) \subseteq K$ and spectral radius $r(L)$, and L is irreducible, then the following assertions hold:*

(i) $r(L)$ is an algebraically simple eigenvalue of L with an eigenvector $v \in$ $\mathrm{int}(K)$. *Moreover, $r(L) > 0$ if $L \neq 0$.*
(ii) Every eigenvector $w \in K$ of L is a scalar multiple of v.

Proof First we remark that the assertions are trivial if $\dim(V) = 1$. So, assume that $\dim(V) \geq 2$. For simplicity write $r = r(L)$. It follows from Proposition 1.1.5 that there exists $\lambda > r$ such that $(\lambda I - L)^{-1}(K \setminus \{0\}) \subseteq \mathrm{int}(K)$. Let $B = (\lambda I - L)^{-1}$ and note that Theorem B.1.1 applies to B.

From Theorem B.2.1 we know that there exists $v \in K$, with $\|v\| = 1$, such that $L(v) = rv$, and hence $(\lambda I - L)(v) = (\lambda - r)v$. This implies that

$$(\lambda - r)^{-1}v = (\lambda I - L)^{-1}(v) = B(v),$$

so we see that $v \in \mathrm{int}(K)$. Because $v \in \mathrm{int}(K)$, the same argument as in the proof of Theorem B.1.1 implies that $(\lambda - r)^{-1}$ is the spectral radius of B. It also follows from Theorem B.1.1 that $(\lambda - r)^{-1} > 0$ and $(\lambda - r)^{-1}$ is an algebraically simple eigenvalue of B. Moreover, each eigenvector $w \in K$ of B must be a scalar multiple of v. We shall use these facts to derive the corresponding results for L.

We first show that $r > 0$, for which it essential to assume that $\dim(V) \geq 2$, since in our definition the zero 1×1 matrix is irreducible. If $r = 0$, then $L(v) = 0$, and since $v \in \mathrm{int}(K)$, there exists $\delta > 0$ such that $v + x \in K$ for all $x \in V$ with $\|x\| < \delta$. It follows that if $\|x\| < \delta$, then $L(v + x) = L(x) \in K$ and $L(v - x) = -L(x) \in K$, from which we deduce that $L(x) = 0$ for all

$x \in V$. Thus, $(\lambda I - L)^{-1}(x) = \lambda^{-1}x$ for all $x \in V$. This contradicts the fact that $(\lambda I - L)^{-1}(K \setminus \{0\}) \subseteq \text{int}(K)$, since K contains at least one proper face if $\dim(K) \geq 2$.

If $L(w) = rw$ for some $w \in V \setminus \{0\}$, and $w \neq \mu v$ for all $\mu \in \mathbb{R}$, then

$$(\lambda - r)^{-1}w = (\lambda I - L)^{-1}(w) = B(w),$$

which is impossible, as each eigenvector of B with eigenvalue $(\lambda - r)^{-1}$ is a scalar multiple of v.

To prove that r is an algebraically simple eigenvalue, we need to show that there does not exist $u \in V$ with $(L - rI)(u) = v$. Arguing by contradiction, assume that $(L - rI)(u) = v$. Using the fact that $B(v) = (\lambda - r)^{-1}v$, a simple calculation shows that

$$v = ((\lambda - r)I - B^{-1})(u) = (B - (\lambda - r)^{-1}I)((\lambda - r)^2 u),$$

which is impossible, as $(\lambda - r)^{-1}$ is an algebraically simple eigenvalue of B.

Finally we show that each eigenvector $w \in K$ of L is a scalar multiple of v. Assume that $L(w) = \rho w$ for some $w \in K \setminus \{0\}$ and $\rho \geq 0$. This implies that $(L - \lambda I)(w) = (\rho - \lambda)w$, so that $B(w) = (\rho - \lambda)^{-1}w$. Since the only eigenvectors of B in K are scalar multiples of v, we conclude that w must be a scalar multiple of v. □

Notice that the first two statements in the Perron–Frobenius theorem as stated in Theorem 1.1.2 are a direct consequence of Theorem B.3.1. In the next section we prove the third statement in Theorem 1.1.2, concerning the peripheral spectrum. This statement is of a qualitatively different nature and depends, to a large extent, on special properties of $\mathbb{R}^n_+$.

B.4 The peripheral spectrum

We begin the analysis of the peripheral spectrum of an irreducible nonnegative matrix with an observation that holds for general linear maps that leave a solid closed cone invariant.

Lemma B.4.1 *Let K be a solid closed cone in a finite-dimensional real normed space V. If $L\colon V \to V$ is a linear map with $L(K) \subseteq K$, and there exist $u \in K \setminus \{0\}$ and $\alpha > 0$ such that*

$$L(u) - \alpha u \in K,$$

then the spectral radius $r(L)$ of L satisfies $r(L) \geq \alpha$. If, in addition, L is irreducible and $L(u) - \alpha u \neq 0$, then $r(L) > \alpha$.

Proof For simplicity write $r = r(L)$. By assumption $\alpha u \leq L(u)$, so that

$$\alpha^k u \leq L^k(u) \quad \text{for all } k \geq 1.$$

If $\alpha > r$, we find that

$$\left(\frac{1}{\alpha}\right)^k L^k(u) \to 0 \quad \text{as } k \to \infty,$$

and

$$\left(\frac{1}{\alpha}\right)^k L^k(u) - u \in K \quad \text{for all } k \geq 1.$$

This implies that $-u \in K$, which is impossible, and hence $\alpha \leq r$.

If, in addition, L is irreducible, and $L(u) \neq \alpha u$, then we know from Proposition 1.1.5 that there exists $\lambda > r$ such that

$$(\lambda I - L)^{-1}(K \setminus \{0\}) \subseteq \mathrm{int}(K).$$

Writing $w = (\lambda I - L)^{-1}(u)$, we find that

$$(\lambda I - L)^{-1}(L(w) - \alpha u) = L(w) - \alpha w \in \mathrm{int}(K).$$

Since $L(w) - \alpha w \in \mathrm{int}(K)$, there exists $\delta > 0$ such that $L(w) - \alpha w \geq \delta w$, and now the first part of the lemma tells us that $r \geq \alpha + \delta > \alpha$. □

To proceed further we shall need an elegant result by Wielandt [225] concerning nonnegative irreducible matrices. (Here we mean irreducible with respect to $\mathbb{R}^n_+$.) Before we state it, we introduce the following notation. Given a matrix $C = (c_{ij})$ with entries from $\mathbb{C}$ we write $|C| = (|c_{ij}|)$. Similarly, for $z = (z_1, \ldots, z_n)$ in $\mathbb{C}^n$ we let $|z| = (|z_1|, \ldots, |z_n|) \in \mathbb{R}^n_+$.

Theorem B.4.2 (Wielandt) *Let $A = (a_{ij})$ be an $n \times n$ nonnegative irreducible matrix. If $C = (c_{ij})$ is an $n \times n$ matrix with entries from $\mathbb{C}$ and*

$$|c_{ij}| \leq a_{ij} \quad \textit{for all } 1 \leq i, j \leq n, \tag{B.8}$$

then $r(C) \leq r(A)$. Moreover, if C has an eigenvector $z \in \mathbb{C}^n \setminus \{0\}$ with eigenvalue $r(A)e^{i\vartheta}$, then $|z| \in \mathrm{int}(\mathbb{R}^n_+)$, $|C| = A$, and $A|z| = r(A)|z|$. Furthermore, the diagonal matrix

$$D = \mathrm{diag}(z_1/|z_1|, \ldots, z_n/|z_n|)$$

satisfies

$$C = e^{i\vartheta} DAD^{-1}. \tag{B.9}$$

Proof If $Cw = \rho w$ for some $w \in \mathbb{C}^n$, then by (B.8) we find that

$$|\rho||w| = |Cw| \leq |C||w| \leq A|w|.$$

So, by Lemma B.4.1 we know that $\rho \leq r(A)$, and hence $r(C) \leq r(A)$.

For simplicity write $r = r(A)$ and suppose that there exists $z \in \mathbb{C}^n \setminus \{0\}$ with

$$Cz = re^{i\vartheta} z,$$

so $r|z| = |Cz| \leq |C||z| \leq A|z|$. If $A|z| \neq |C||z|$ or $|Cz| \neq |C||z|$, then $A|z| \neq r|z|$, and because A is irreducible Lemma B.4.1 implies that the spectral radius of A is strictly greater than r, which is impossible. It follows that $A|z| = |C||z| = r|z|$; and since A is irreducible, $|z|$ is the unique, up to scalar multiples, eigenvector of A in $\mathbb{R}^n_+$, and $|z| \in \text{int}(\mathbb{R}^n_+)$.

To show that $|C| = A$, remark that

$$0 = r|z_i| - r|z_i| = \sum_{j=1}^{n} (a_{ij} - |c_{ij}|)|z_j|$$

for $1 \leq i \leq n$. Note that $|z_j| > 0$ and $a_{ij} - |c_{ij}| \geq 0$ for all i and j. Thus, we must have that $a_{ij} = |c_{ij}|$ for all i and j.

Now let $D = \text{diag}(z_1/|z_1|, \ldots, z_n/|z_n|)$ and let $E = e^{-i\vartheta} D^{-1} C D = (e_{ij})$. Using the equality $D|z| = z$, we see that $CD|z| = re^{i\vartheta} z$ and

$$E|z| = re^{-i\vartheta} D^{-1}(e^{i\vartheta} z) = r|z| = A|z|.$$

As $|e_{ij}| = |c_{ij}| = a_{ij}$ for $1 \leq i, j \leq n$, we also have that $|E||z| = r|z|$. Thus,

$$0 = r|z_i| - r|z_i| = \sum_{j=1}^{n} (|e_{ij}| - e_{ij})|z_j|,$$

and hence

$$\sum_{j=1}^{n} (|e_{ij}| - \text{Re}(e_{ij}))|z_j| = 0,$$

for all $1 \leq i \leq n$. Here $\text{Re}(\xi)$ denotes the real part of $\xi \in \mathbb{C}$. Obviously, $|e_{ij}| \geq \text{Re}(e_{ij})$, so $|e_{ij}| = \text{Re}(e_{ij})$, as $|z_j| > 0$ for all j. Thus, $e_{ij} = |e_{ij}|$ for all i and j, which shows that $e_{ij} = |c_{ij}| = a_{ij}$ for all i and j, and hence (B.9) holds. □

We now prove the third part of the Perron–Frobenius Theorem 1.1.2.

Proposition B.4.3 *If A is a nonnegative irreducible $n \times n$ matrix, and A has exactly q eigenvalues $\lambda_1, \dots, \lambda_q \in \mathbb{C}$ with modulus $r(A)$, then these eigenvalues are precisely the roots of the equation*

$$z^q - r(A)^q = 0,$$

and each λ_i has algebraic multiplicity 1.

Proof By replacing A with $r(A)^{-1}A$ we may as well assume that $r(A) = 1$. Let $G = \{\lambda_1, \dots, \lambda_q\}$ be the set of eigenvalues of A with modulus 1. If $e^{i\vartheta} \in G$, then it follows from Theorem B.4.2 that there exists a diagonal matrix D with $|D| = I$, where I is the identity matrix, such that

$$A = D(e^{i\vartheta}A)D^{-1}. \tag{B.10}$$

This equation implies that $\sigma(A) = \sigma(e^{i\vartheta}A)$, as $\sigma(e^{i\vartheta}A) = e^{i\vartheta}\sigma(A)$. We conclude that $\mu \in \sigma(A)$ if and only if $\mu e^{i\vartheta} \in \sigma(A)$.

Now if $e^{i\vartheta_1}$ and $e^{i\vartheta_2}$ are in $\sigma(A)$, we see that $e^{i(\vartheta_1+\vartheta_2)}\mu \in \sigma(A)$ for all $\mu \in \sigma(A)$. In particular, since $1 \in \sigma(A)$, we find that $e^{i(\vartheta_1+\vartheta_2)} \in \sigma(A)$. Because A has real entries, if $e^{i\vartheta} \in \sigma(A)$, we must have that $e^{-i\vartheta} \in \sigma(A)$. So, G forms an abelian group with at most n elements.

For $\mu \in G$ we can write $\mu = e^{i\vartheta}$, where $0 \leq \vartheta < 2\pi$. If G contains at least two elements, we let $\vartheta_0 = \min\{\vartheta \in (0, 2\pi) : e^{i\vartheta} \in G\}$. So, $0 < \vartheta_0 \leq \pi$, since $e^{i(2\pi-\vartheta_0)} = e^{-i\vartheta_0} \in G$. As G is a group, $e^{ki\vartheta_0} \in G$ for all $k \in \mathbb{Z}$. We claim that, for each $e^{i\psi} \in G$, we have that $\psi = k\vartheta_0$ for some integer k. Indeed, if $\psi \neq m\vartheta_0$ for all integers $m \geq 1$, then there exists m_0 such that $0 < \psi - m_0\vartheta_0 < \vartheta_0$. But $e^{i(\psi - m_0\vartheta_0)} \in G$, which contradicts the choice of ϑ_0. Thus, G is a the cyclic group of order q generated by $e^{i\vartheta_0}$, where $\vartheta_0 = 2\pi/q$.

Note that if $e^{i\vartheta} \in G$, then it follows from (B.9) that

$$p(\lambda) = \det(\lambda I - A) = e^{in\vartheta}\det((\lambda e^{-i\vartheta})I - A) = e^{in\vartheta}p(\lambda e^{-i\vartheta}).$$

We see that

$$p'(\lambda) = e^{in\vartheta}e^{-i\vartheta}p'(\lambda e^{-i\vartheta})$$

and, more generally,

$$p^{(m)}(\lambda) = e^{in\vartheta}e^{-im\vartheta}p^{(m)}(\lambda e^{-i\vartheta})$$

for all $m \geq 1$. To see that $\lambda_k = e^{i\vartheta_k} \in G$ has algebraic multiplicity 1, remark that, for each $m \geq 1$,

$$p^{(m)}(\lambda_k) = e^{in\vartheta_k}e^{-im\vartheta_k}p^{(m)}(\lambda_k e^{-i\vartheta_k}) = e^{i(n-m)\vartheta_k}p^{(m)}(1) \neq 0,$$

as $1 = r(A) \in G$ has algebraic multiplicity 1 by Theorem B.3.1. Thus, each $\lambda_k \in G$ has algebraic multiplicity 1, and we are done. □

The number of eigenvalues $\lambda \in \mathbb{C}$ of an irreducible linear map $L: V \to V$ with $|\lambda| = r(L)$ is usually called the *index of cyclicity*. It is known that if $L: V \to V$ is an irreducible linear map, then L is primitive if and only if its index of cyclicity is 1; see [117, theorem 6.3] or [22, p. 18].

Proposition B.4.3 can be extended to linear maps that leave a polyhedral cone invariant. This observation goes back to Kreĭn and Rutman [117, pp. 303–5]. It can also be found as Exercise 5.25 in the book by Berman and Plemmons [22, p. 23]. For a solution they refer to a paper by Barker and Turner [16]. This paper, however, contains a mistake, according to Tam [214, p. 244], who gave a correct proof in [213]. In our presentation we follow the original proof by Kreĭn and Rutman. The main idea is to reduce the general case to the case where the linear map leaves the standard positive cone invariant. We start with an auxiliary lemma.

Lemma B.4.4 *Let $K \subseteq V$ be a solid polyhedral cone with N facets and facet-defining functionals $\psi_1, \ldots, \psi_N \in V^*$. If $Q: \mathbb{R}^N \to V^*$ is defined by*

$$Q(w_1, \ldots, w_N) = \sum_{j-1}^{N} w_j \psi_j \quad \textit{for } (w_1, \ldots, w_N) \in \mathbb{R}^N,$$

then

(i) Q is surjective,
(ii) $Q^*: V^{**} = V \to (\mathbb{R}^N)^*$ is injective, and
(iii) $Qw \neq 0$ for all $w \in \mathbb{R}^N_+ \setminus \{0\}$.

Proof Suppose that Q is not surjective. By the Hahn–Banach theorem there exists a non-zero functional $\vartheta: V^* \to \mathbb{R}$ such that $\vartheta(\psi_j) = 0$ for all $1 \leq j \leq N$. We know that there exists $x \in V \setminus \{0\}$ such that $\vartheta(\varphi) = \varphi(x)$ for all $\varphi \in V^*$. In particular, $\psi_j(x) = 0$ for all $1 \leq j \leq N$. It follows from Lemma 1.2.1 that $x, -x \in K$, which is impossible, as $x \neq 0$.

The second assertion is a simple consequence of the first one. Simply note that $Q^*: V^{**} = V \to (\mathbb{R}^N)^*$ is given by $(Q^*v)(w) = (Qw)(v) = \sum_j w_j \psi_j(v)$ for all $w \in \mathbb{R}^N$. So, $Q^*v = 0$ implies that $0 = (Qw)(v)$ for all $w \in \mathbb{R}^N$. As Q is surjective, $v = 0$.

To prove the final assertion we note that if we take $x \in \mathrm{int}(K)$, then for each $w \in \mathbb{R}^N_+ \setminus \{0\}$ we have that $(Qw)(x) = \sum_j w_j \psi_j(x) > 0$. □

It is easy to verify that the first two assertions in Lemma B.4.4 also hold for Q and Q^* on the complexifications of $\mathbb{R}^N$, $(\mathbb{R}^N)^*$, V, and V^*.

Theorem B.4.5 *If $A\colon V \to V$ is a linear map that leaves a solid polyhedral cone K with N facets invariant, then for each $\lambda \in \sigma(A)$ with $|\lambda| = r(A)$ there exists an integer $1 \le q \le N$ such that $\lambda^q = r(A)^q$.*

Proof By rescaling A we may assume that $r(A) = 1$. Let $\psi_1, \dots, \psi_N \in V^*$ be the facet-defining functionals of K. Define $C = \{\sum_{j=1}^N a_j\psi_j \in V^* : a_j \ge 0$ for all $1 \le j \le N\}$. We first show that $K^* = C$. Obviously $C \subseteq K^*$. Suppose that $\varphi \in K^* \setminus C$. By the Hahn–Banach separation theorem (see [186, theorem 11.4]), there exists $\rho \in V^{**}$ such that $\rho(\varphi) < 0$ and $\rho(\psi_j) \ge 0$ for all $1 \le j \le N$. So there exists a non-zero $x \in V$ with $\psi_j(x) \ge 0$ for all $1 \le j \le N$, and hence $x \in K$ by Lemma 1.2.1. Thus, $\rho(\varphi) = \varphi(x) \ge 0$, which is impossible.

Now let $Q\colon \mathbb{R}^N \to V^*$ be defined as in Lemma B.4.4. We know that $Q(\mathbb{R}^N_+) \subseteq K^*$. It follows that for each $1 \le i \le N$ there exist $\alpha_{ij} \ge 0$, $1 \le j \le N$, such that

$$A^*\psi_i = \sum_{j=1}^N \alpha_{ij}\psi_j.$$

Moreover, for each $(w_1, \dots, w_N) \in \mathbb{R}^N$,

$$A^*\Big(\sum_{i=1}^N w_i\psi_i\Big) = \sum_{i=1}^N w_i\Big(\sum_{j=1}^N \alpha_{ij}\psi_j\Big) = \sum_{j=1}^N\Big(\sum_{i=1}^N w_i\alpha_{ij}\Big)\psi_j.$$

Now define $B\colon \mathbb{R}^N \to \mathbb{R}^N$ by $B(w_1, \dots, w_N) = (v_1, \dots, v_N)$, where

$$v_j = \sum_{i=1}^N w_i\alpha_{ij} \quad \text{for } 1 \le j \le N.$$

Note that $A^*Q = QB$ and $B(\mathbb{R}^N_+) \subseteq \mathbb{R}^N_+$. If $z \in \mathbb{C}^N$ and $Bz = \lambda z$, then $A^*Qz = \lambda Qz$. Now let $w \in \mathbb{R}^N_+ \setminus \{0\}$ be such that $Bw = r(B)w$; then $Qw \in K^*$ and $Qw \ne 0$ by Lemma B.4.4. Moreover, $A^*Qw = r(B)Qw$, which shows that $r(A) = r(A^*) \ge r(B)$.

We also have the relation

$$Q^*A = B^*Q^*. \tag{B.11}$$

If $Az = \lambda z$, where $z \in \tilde{V} \setminus \{0\}$ and $\tilde{V}$ is the complexification of V, then $B^*Q^*z = Q^*Az = \lambda Q^*z$ and $Q^*z \ne 0$, as Q^* is injective on $\tilde{V}$ by Lemma B.4.4. This implies that if $\lambda \in \sigma(A)$, then $\lambda \in \sigma(B^*) = \sigma(B)$ and hence $r(A) \le r(B)$. We conclude that $r(A) = r(B)$.

If $Az = e^{i\vartheta} z$, where $z \in \tilde{V} \setminus \{0\}$, then Q^*z is an eigenvector of B^* with eigenvalue $e^{i\vartheta}$. The linear map B^* maps $(\mathbb{R}^N_+)^*$ into itself, and can be represented by an $N \times N$ nonnegative matrix T. We know that there exists an $N \times N$ permutation matrix P such that $P^T T P$ is in Frobenius normal form,

$$P^T T P = \begin{bmatrix} D^{11} & 0 & \dots & 0 \\ D^{21} & D^{22} & \dots & 0 \\ \vdots & \vdots & \ddots & 0 \\ D^{m1} & D^{m2} & \dots & D^{mm} \end{bmatrix}, \tag{B.12}$$

where each diagonal block D^{ii} is irreducible; see [148, p. 142]. So, $e^{i\vartheta}$ is an eigenvalue of one of the matrices D^{ii}. It now follows from Proposition B.4.3 that there exists $1 \leq q \leq N$ such that $(e^{i\vartheta})^q = 1$, and we are done. □

If we assume in Theorem B.4.5, in addition, that A is irreducible, then one may be tempted to believe that one can sharpen the assertion to say that there exists $1 \leq q \leq N$ such that for each $\lambda \in \sigma(A)$ with $|\lambda| = r(A)$ we have that $\lambda^q = r(A)^q$, as in Proposition B.4.3. This turns out to be false. Examples showing that such an integer $1 \leq q \leq N$ need not exist for irreducible linear maps that leave a general solid polyhedral cone with N facets invariant were communicated to the authors by Tam and Lins.

We will now use Theorem B.4.5 to prove the following result concerning the iterative behavior of linear maps on polyhedral cones.

Theorem B.4.6 *If $A\colon V \to V$ is a linear map that leaves a solid polyhedral cone $K \subseteq V$ with N facets invariant, then there exists an integer $p \geq 1$, which is the order of a permutation on N letters, such that $\lim_{k\to\infty} A^{kp}x$ exists for all $x \in K$ with $(\|A^k x\|)_k$ bounded.*

The idea is again to reduce it to the case where $K = \mathbb{R}^N_+$, and to apply [171, theorem 9.1], which says the following.

Theorem B.4.7 *If $A\colon \mathbb{R}^n \to \mathbb{R}^n$ is a linear map which leaves $\mathbb{R}^n_+$ invariant and $r(A) \leq 1$, then there exists an integer $p \geq 1$, which is the order of a permutation on n letters, such that $\lim_{k\to\infty} A^{kp}x$ exists for all $x \in \mathbb{R}^n$ with $(\|A^k x\|)_k$ bounded.*

The additional assumption, $r(A) \leq 1$, is required to ensure that the limit exists not only for points in $\mathbb{R}^n_+$, but for all points in $\mathbb{R}^n$ with a norm-bounded orbit. In fact, Theorem B.4.7 has the following consequence.

Corollary B.4.8 *If $A\colon \mathbb{R}^n \to \mathbb{R}^n$ is a linear map that leaves $\mathbb{R}^n_+$ invariant, then there exists an integer $p \geq 1$, which is the order of a permutation on n letters, such that $\lim_{k\to\infty} A^{kp}x$ exists for all $x \in \mathbb{R}^n_+$ with $(\|A^k x\|)_k$ bounded.*

Proof Let C be the $n \times n$ nonnegative matrix representing the linear map A. Then there exists a permutation matrix P such that $P^T C P$ is in Frobenius normal form, so

$$P^T C P = \begin{bmatrix} D^{11} & 0 & \dots & 0 \\ D^{21} & D^{22} & \dots & 0 \\ \vdots & \vdots & \ddots & 0 \\ D^{m1} & D^{m2} & \dots & D^{mm} \end{bmatrix}$$

and each diagonal block D^{ii} is irreducible; see [148, p. 142]. Denote the right-hand side by D, and the spectral radius of the D^{ii} by r_i for $1 \leq i \leq m$. Furthermore, let $S_i \subseteq \{1, \dots, n\}$ be the set of indices corresponding to the diagonal block D^{ii}.

We claim that, for each $y \in \mathbb{R}^n_+$ with $(\|C^k y\|)_k$ bounded, the point $P^T y$ has no support on S_i for all i with $r_i > 1$, i.e., there exists no $j \in S_j$ with $(P^T y)_j > 0$, if $r_i > 1$. To prove this we argue by contradiction. Let $z = P^T y$ and let $z^i \in \mathbb{R}^{S_i}_+$ denote the restriction of z to its coordinates in S_i. Suppose that z has a support on S_i and $r_i > 1$. Because $z^i \neq 0$ and D^{ii} is irreducible, we know from Proposition 1.1.5 that there exists an integer $q \geq 1$ such that

$$v = z^i + D^{ii} z^i + \cdots + (D^{ii})^q z^i \in \operatorname{int}(\mathbb{R}^{S_i}_+).$$

Now let $\|\cdot\|_1$ be the ℓ_1-norm, so $\|x\|_1 = \sum_i |x_i|$. We have that

$$\begin{aligned} \Big(\sum_{j=0}^{q} \|C^{k+j} y\|_1\Big)^{1/k} &\geq \Big\| \sum_{j=0}^{q} C^{k+j} y \Big\|_1^{1/k} = \Big\| \sum_{j=0}^{q} P^T C^{k+j} P z \Big\|_1^{1/k} \\ &= \Big\| \sum_{j=0}^{q} D^{k+j} z \Big\|_1^{1/k} \geq \Big\| \sum_{j=0}^{q} (D^{ii})^{k+j} z^i \Big\|_1^{1/k} = \Big\| (D^{ii})^k v \Big\|_1^{1/k}. \end{aligned}$$

Note that the left-hand side converges to 1 as $k \to \infty$, since y has a norm-bounded orbit under C. However, the right-hand side converges to $r_i > 1$, by Proposition 5.3.6, which is impossible.

Let $\hat{S} = \bigcup_{i:\, r_i > 1} S_i$ and denote by $\hat{D}$ the $n \times n$ nonnegative matrix obtained from D by replacing the entry d_{ij} in D by 0 if and only if $i \in \hat{S}$ or $j \in \hat{S}$. Obviously, $r(\hat{D}) \leq 1$. So we can apply Theorem B.4.7 to deduce that there exists an integer $p \geq 1$ which is the order of a permutation on n letters, such that $\lim_{k \to \infty} \hat{D}^{kp} w$ exists for all $w \in \mathbb{R}^n$ with $(\|\hat{D}^k w\|)_k$ bounded.

Now let $x \in \mathbb{R}^n_+$ with $(\|C^k x\|)_k$ bounded and let $u = P^T x$. As x and $C^p x$ have a norm-bounded orbit under C, we know from the claim that u and $D^p u$ have no support on $\hat{S}$. So, $\hat{D}^p u = D^p u$. More generally, $C^{(k-1)p} x$ and $C^{kp} x$ have a norm-bounded orbit under C. So, it follows from the claim that $\hat{D}^{kp}u = D^{kp}u$ for all $k \geq 1$. This implies that $\lim_{k\to\infty} D^{kp}u$ exists. As $C^{kp} = PD^{kp}P^T$ for all $k \geq 1$, we conclude that $\lim_{k\to\infty} C^{kp}x$ exists. □

We can now prove Theorem B.4.6.

Proof of Theorem B.4.6 Recall from the proof of Theorem B.4.5 that $Q^*A = B^*Q^*$; see Equation (B.11). From Lemma B.4.4 we know that Q^* is injective. Moreover, for $v \in K$ we have that $(Q^*v)(w) = \sum_j w_j\psi_j(v) \geq 0$ for all $w \in \mathbb{R}^N_+$. Thus, $Q^*(K) \subseteq (\mathbb{R}^N_+)^*$. Note that if $x \in K$ has a norm-bounded orbit under A, then $Q^*x \in (\mathbb{R}^N_+)^*$ has a norm-bounded orbit under B^*, as $\|(B^*)^k Q^*x\| = \|Q^*A^kx\|$ for all $k \geq 1$. We know from Corollary B.4.8 that there exists an integer $p \geq 1$ which is the order of a permutation on N letters, such that $\lim_{k\to\infty}(B^*)^{kp}w$ exists for all $w \in (\mathbb{R}^N_+)^*$ with $(\|(B^*)^k w\|)_k$ bounded. Using the relation $A^k = (Q^*)^{-1}(B^*)^kQ^*$ we conclude that $\lim_{k\to\infty} A^{kp}x$ exists for all $x \in K$ with $(\|A^kx\|)_k$ bounded. □

Notes and comments

Chapter 1

The theory of nonnegative matrices was pioneered by Perron [179, 180] and Frobenius [70–72] in the early twentieth century. An excellent historical account of the Perron–Frobenius theorem is given by Hawkins [84]. A thorough overview of the spectral theory of linear maps that leave a cone in a finite-dimensional vector space invariant can be seen in the survey paper by Tam [214].

The generalization of the notion of irreducibility of nonnegative matrices to linear maps that leave a cone in a finite-dimensional vector space invariant as given in Definition 1.1.4 goes back to Vandergraft [219]. The equivalent notion of irreducibility given in Proposition 1.1.5 appears in the work of Schaefer [195].

Early extensions of the general infinite-dimensional Kreĭn–Rutman theorem are due to Bonsall [29–32] and Schaefer [193, 195], and have been further developed by many authors. An extensive list of references can by found in Mallet-Paret and Nussbaum [139].

The earliest proof of Theorem 1.1.8 is due to Kreĭn and Rutman [117, pp. 303–5]. It also appears as an exercise in the book by Berman and Plemmons [22, exercise 5.25, p. 23]. They refer the reader to Barker and Turner [16]. Tam [214, p. 244], however, has noted that there is a mistake in [16] and he has given a correct proof in [213, theorem 7.6].

Results related to Theorem 1.1.9 can be found in Friedland and Schneider [69, theorem 3.6], Lins and Nussbaum [133, theorem 2], and in Nussbaum and Verduyn Lunel [171, theorem 9.1].

Chapter 2

Hilbert's metric spaces are also known in the literature as Hilbert's geometries and naturally generalize Klein's model of the hyperbolic plane. They were introduced by Hilbert in a letter to Klein and play a role in the solution of Hilbert's fourth problem; see Álvarez Paiva [9]. Hilbert later published the contents of the letter [86].

Proposition 2.2.4 stating that Hilbert's metric on an n-dimensional simplex is isometric to an n-dimensional normed space was proved for $n = 2$ by Phadke [181], and independently by de la Harpe [83]. The general case is due to Nussbaum [160]. It was shown by Foertsch and Karlsson [66] that the n-dimensional simplices are the only domains for which the Hilbert metric space is isometric to an n-dimensional normed space.

The standard positive cone, the Lorentz cone, and the cone of symmetric positive semi-definite $n \times n$ matrices are all examples of symmetric cones, which means that the cone is self-dual, and its group of linear automorphisms acts transitively on the interior. Symmetric cones carry an additional algebraic structure known as a Euclidean Jordan algebra. In fact, there is a one-to-one correspondence between symmetric cones and Euclidean Jordan algebras. Euclidean Jordan algebras were classified by Jordan, von Neumann, and Wigner [94]. A detailed exposition of the rich theory of symmetric cones can be found in the book by Faraut and Korányi [63]. A formula for Hilbert's metric on general symmetric cones was obtained by Koufany [108].

The idea that one can combine Hilbert's metric and Banach's contraction theorem to analyze the spectral properties of linear operators that leave a cone in a Banach space invariant goes back to Birkhoff [25] and Samelson [192], and was further developed in the setting of nonlinear cone maps by Bushell [43, 44, 46], Kohlberg [105], Kohlberg and Pratt [107], Krasnosel'skii [109], Krause [112], Morishima [150], Nussbaum [158, 159], Oshime [174], Potter [183], Thompson [216–218], and many others.

Sections 2.5 and 2.6 on completeness, convexity, and geodesics in Hilbert's and Thompson's metric spaces is based on the treatment by Nussbaum [158].

Sections 2.7 and 2.8 are based on Crandall and Tartar [54].

Chapter 3

Fixed-point problems for non-expansive maps on infinite-dimensional Banach spaces are much more subtle than in finite-dimensional spaces, and are not fully understood. For example, it is unknown if every non-expansive map $f\colon Y \to Y$ on a convex bounded closed subset Y of a reflexive Banach space V

has a fixed point. An extensive survey of this problem and many other related problems has been given by Goebel and Kirk [76]. There exists an extensive literature concerning non-expansive maps and non-expansive retractions in infinite-dimensional Banach spaces. The interested reader can also consult Bruck [40, 54, 55, 60, 106, 114, 127] and the further authors and articles in these references.

The construction of the horofunction boundary $C(\infty)$ as given in Section 3.3 is usually attributed to Gromov [78]. Our presentation follows Ballmann [12]. A detailed description of the horofunction boundary for finite-dimensional normed spaces and Hilbert's metric spaces can be found in the work of Walsh [220, 221].

The Denjoy–Wolff-type theorems discussed in Sections 3.3 and 3.4 originated with Beardon [18, 19] and Karlsson [99]. Generalizations of the Denjoy–Wolf theorem to fixed-point free holomorphic maps on convex domains $D \subseteq \mathbb{C}^n$ exist; see Abate [1] and Hervé [85]. Further generalizations for holomorphic maps on complex Banach spaces are due to Kapeluszky, Kuczumow, and Reich [96, 97] and to Mellon [142]; see also the book by Goebel and Reich [77].

The material in Section 3.5 is based on Lemmens and van Gaans [126], who use earlier ideas by Edelstein [60].

Chapter 4

It was first shown by Akcoglu and Krengel [2] that the orbits of ℓ_1-norm non-expansive maps $f : X \to X$, where X is a compact subset of $\mathbb{R}^n$, converge to periodic orbits. Their methods were of a topological nature and did not yield an a-priori upper bound for the possible periods of periodic points in terms of the dimension of the underlying space. In his thesis, Weller [222] generalized the result of Akcoglu and Krengel to maps that are non-expansive under a polyhedral norm, but he also did not obtain an a-priori upper bound. The first upper bound for the largest possible period of a periodic point of an ℓ_1-norm non-expansive map goes back to Misiurewicz [149]. Martus derived upper bounds for maps that are non-expansive under a polyhedral norm in his thesis [140]. Other upper bounds were obtained in [27, 121, 123, 137, 160, 204].

For strictly convex norms, Lemmens and van Gaans [127] proved that if $f : \mathbb{R}^n \to \mathbb{R}^n$, with $f(0) = 0$, is non-expansive under a strictly convex norm $\|\cdot\|$, and $(\mathbb{R}^n, \|\cdot\|)$ contains no two-dimensional plane H such that $(H.\|\cdot\|)$ is Hilbertian and the range of a linear norm-1 projection, then every orbit of f converges to a periodic orbit. Similar results exist for non-expansive maps on strictly convex Hilbert's metric spaces; see Lemmens [119].

Additional supporting evidence for Nussbaum's Conjecture 4.2.2 can be found in [125]. In particular, it was conjectured there that the points in a periodic orbit $\mathcal{O}$ of a sup-norm non-expansive map $f: \mathbb{R}^n \to \mathbb{R}^n$ can be perturbed over an arbitrarily small positive distance so that the set of perturbed points, $\mathcal{O}'$, has no sup-norm triangle equality; that is to say, there exist no distinct $x', y', z' \in \mathcal{O}'$ such that $\|x' - z'\|_\infty = \|x' - y'\|_\infty + \|y' - z'\|_\infty$. Sets S in $\mathbb{R}^n$ with no sup-norm triangle equalities have at most 2^n points by a result of Blokhuis and Wilbrink [27]. Thus the perturbation conjecture in [125] implies Nussbaum's Conjecture 4.2.2. At present there is no counterexample to this stronger conjecture. In [125] several related conjectures were given. All these conjectures turn out to be equivalent, which was not proved in [125].

Theorem 4.3.7 and Corollary 4.3.8 were proved by Lemmens and Scheutzow [123], and confirmed a conjecture by Gunawardena [79, 80].

Chapter 5

The results in Section 5.1 were found by Burbanks, Nussbaum, and Sparrow [41]. The results in Sections 5.2 and 5.3 were obtained by Lemmens and Nussbaum [120].

The material in Sections 5.3 and 5.4 concerning the cone spectral radius and the corresponding eigenvector is based on Mallet-Paret and Nussbaum [138], and Nussbaum [156]. For closed cones K in an infinite-dimensional Banach space V, the problem of finding (close to) optimal conditions on the continuous order-preserving homogeneous map $f: K \to K$ so that f has an eigenvector $v \in K \setminus \{0\}$ with $f(v) = r_K(f)v$ is open. Further analysis of this problem can be found in Mallet-Paret and Nussbaum [139] and in [155]. In particular, it follows from [155, theorem 2.1] that it is sufficient to assume that $r_K(f) > 0$ and f is compact, but this compactness condition is unnecessarily strong.

The question of whether $r_K(f) = \hat{r}_K(f)$ for general continuous order-preserving homogeneous maps $f: K \to K$ on a closed cone in an infinite-dimensional Banach space is open. The equality holds in numerous important special cases; see [138, theorems 2.2 and 2.3] and [139, theorem 3.4]. In particular, $r_C(f) = \hat{r}_C(f)$ if f^m is compact for some integer m. At present no counterexample to the equality is known

The results in Section 5.6 can be found in Nussbaum [156]; see also [6].

Chapter 6

Semi-differentiability is an intermediate concept, which is stronger than Gateaux differentiability, but weaker than Fréchet differentiability. The interest in semi-differentiability stems from the fact that most nonlinear cone maps arising in applications in game theory and control theory (see Akian, Gaubert, and Nussbaum [7]; Shapley [203]; and Sorin [208]), in mathematical economics (see Morishima [150] and Oshime [174]), and in mathematical biology (see chapter 3 of Nussbaum [159] and section 4 of [156]) are not Fréchet differentiable at some point(s) of the closed cone where they are defined. However, it is frequently the case that the maps are semi-differentiable at points where Fréchet differentiability fails.

A systematic treatment of semi-differentiability for maps defined on subsets of $\mathbb{R}^n$ is given by Rockafellar and Wets [187]. A treatment of semi-differentiability, concentrating on the class $\mathcal{M}_-$ of maps $f: \mathbb{R}^n_+ \to \mathbb{R}^n_+$, is given in chapter 3 of [159], though semi-differentiability is not formally defined in [159]. Semi-differentiability is used there to determine exactly when certain order-preserving homogeneous maps $f: \mathbb{R}^4_+ \to \mathbb{R}^4_+$ which arise as models in population biology have eigenvectors in $\text{int}(\mathbb{R}^4_+)$. A generalization of the Rockafellar–Wets results to the context of normed linear spaces has been given by Akian, Gaubert, and Nussbaum [7], and this is the basis for most of our exposition in Sections 6.1 and 6.4.

If K is a closed cone with non-empty interior in a Banach space X, and $f: \text{int}(K) \to \text{int}(K)$ is continuous, homogeneous, and order-preserving, the question of whether f has an eigenvector in $\text{int}(K)$ is often quite difficult, even in finite dimensions. A generalized perturbation method approach, in the spirit of Section 6.2, is sometimes a useful way to attack the problem and has been applied, for example, by Nussbaum [163] to prove scaling theorems for integral kernels. If $K = \mathbb{R}^n_+$, Gaubert and Gunawardena [74] have introduced the idea of associating to f a "digraph" G_f and a "recession map" $\hat{f}$, and Theorems 6.2.3 and 6.3.2 are due to them. The digraph G_f is a direct generalization of the notion of an incidence matrix defined in Section 6.6, an idea introduced by Nussbaum [156, section 4].

Generally speaking, if one can establish the existence of a normalized eigenvector u in $\text{int}(K)$ for $f: K \to K$, existing results concerning uniqueness of u and convergence of various normalized iterates to u are satisfactory, at least for the classes of maps of interest here. Theorem 6.4.1 is a finite-dimensional version of Theorem 7.5 and Corollary 7.7 in [7]. Theorem 6.4.6 is essentially a finite-dimensional version of Theorem 2.5 in [158, p. 47], although some unnecessarily restrictive hypotheses in Theorem 2.5 have been removed.

Variants of the idea of a map being strongly order-preserving play an important role in Section 6.5. This is an old concept and has been studied by many authors: see Brualdi, Parter, and Schneider [39] and, much more recently, Kloeden and Rubinov [103]. In a different context, strongly order-preserving maps also play a central role in work of Hirsch and Smith [87–89, 207], Krause and Ranft [113, 115] (see also [114]), and Takáč [211, 212].

Chapter 7

There is a large literature concerning DAD theorems and their generalizations to various kinds of "matrix scaling problems." Early important work concerning the classic DAD problem, where $m = n$ and $\alpha_i = \beta_i = 1$ for all $1 \leq i \leq n$, can be found in Sinkhorn [205], Sinkhorn and Knopp [206], Brualdi, Parter, and Schneider [39], and Menon [143, 144]. In particular, Sinkhorn and Knopp [206] and Brualdi, Parter, and Schneider [39] independently, and with different methods, obtained necessary and sufficient conditions for the classic DAD problem to have a solution. Menon [143] introduced the important observation that solving the classic DAD problem is equivalent to a fixed-point problem. Further work in this direction can be found in Menon's paper [144], a joint paper by Menon and Schneider [145], and in the previously mentioned paper by Brualdi, Parter, and Schneider [39]. Somewhat more recent related work can be found in [14, 59, 67, 199, 200]. There is also considerable interest (see [65, 95, 129, 178]) in finding efficient methods to compute D, E, and DAE, and there is an extensive literature of generalizations of DAD problems to more general "matrix scaling problems," see, e.g., [188, 189].

The problem of extending some of the results on the matrix scaling problem to corresponding problems about integral kernels $k(s, t)$ involves many difficulties which are not present in the finite-dimensional case. The interested reader is referred to [34, 35, 50, 98, 153, 163].

It was apparently first noted in [157] that the fixed-point approach pioneered by Menon, combined with facts about Hilbert's metric and the Birkhoff–Hopf theorem (see Appendix A), provides a powerful tool in studying DAD problems. In fact, the arguments presented in Chapter 7 can be extended to the problem of scaling integral kernels; see [35, 163]. The reader should, however, bear in mind that a variety of other methods exist, notably so-called entropy minimization methods; see [35, 199, 200] and references therein.

Brualdi [38] has found necessary and sufficient conditions for the row scaling problem, discussed in Section 7.5, to have a solution when the matrix is symmetric.

Chapter 8

The material in Sections 8.1 and 8.2 is based on Akian, Gaubert, Lemmens, and Nussbaum [5]. Theorem 8.2.7 was proved by Lemmens and Sparrow [124].

Related results on the iterative behavior of convex topical maps can be found in [3,4].

Denjoy–Wolff type theorems for fixed-point free non-expansive maps on Hilbert's metric spaces were first studied by Beardon [19,20], and were further developed by Karlsson [99]; see also [101]. Applications of this work to the iterative behavior of order-preserving homogeneous maps on cones go back to Gunawardena and Walsh [81], and were further investigated by Lins in his thesis [130] (see also [131]) and by Nussbaum [168] (see also [133, 134]). Section 8.3 is based on Beardon [19], Karlsson [99], Karlsson and Noskov [101], and Nussbaum [168]. The results discussed in Section 8.4 were found by Lins [131]. Related results on the asymptotic behavior of fixed-point free non-expansive maps can be found in [75, 104, 106, 132].

Chapter 9

The study of the dynamics of ℓ_1-norm non-expansive maps was initiated by Akcoglu and Krengel [2]. The connection with lower semi-lattice homomorphisms goes back to Scheutzow [197, 198], and was further developed by Nussbaum [161]. The notion of an admissible array was introduced by Nussbaum and Scheutzow [169]. The proof of Theorem 9.3.8 is due to Nussbaum, Scheutzow, and Verduyn Lunel [170]. An extensive computational study of the possible periods of admissible arrays can be found in [171], and detailed number-theoretic analysis of the largest possible period of an admissible array on n symbols is given in [172].

The results on the periodic points of ℓ_1-norm non-expansive maps on $\mathbb{R}^n$ are due to Lemmens and Scheutzow [122]. It should be noted, however, that the finite set $\tilde{R}(n)$ consisting of periods of periodic points of ℓ_1-norm non-expansive maps $f\colon X \to X$, where X can be an arbitrary subset of $\mathbb{R}^n$, is not well understood. We know from Corollary 4.2.5 that $\tilde{\psi}(n) = \max\{p\colon p \in \tilde{R}(n)\}$ satisfies

$$\tilde{\psi}(n) \leq \max_k 2^k \binom{m}{k},$$

where $m = 2^{n-1}$. In [121] it was shown that $\tilde{\psi}(n) \geq 3 \cdot 2^{n-1}$ for all $n \geq 3$. It is conjectured that there exists $c > 2$ such that $\tilde{\psi}(n) \leq c^n$ for all $n \geq 1$, but as yet no proof exists.

Extensions of the results in Chapter 9 to order-preserving maps $f\colon \mathbb{R}^n_+ \to \mathbb{R}^n_+$ that are non-expansive under a strictly monotone norm and satisfy $f(0) = 0$ can be found in [126, 165].

References

[1] M. Abate, *Iteration Theory of Holomorphic Maps on Taut Manifolds*. Research and Lecture Notes in Mathematics, Rende, Mediterranean Press, 1989.

[2] M.A. Akcoglu and U. Krengel, Nonlinear models of diffusion on a finite space. *Probab. Theory Related Fields* **76**(4), (1987), 411–20.

[3] M. Akian and S. Gaubert, Spectral theorem for convex monotone homogeneous maps, and ergodic control. *Nonlinear Anal.* **52**(2), (2003), 637–79.

[4] M. Akian, S. Gaubert, and B. Lemmens, Stability and convergence in discrete convex monotone dynamical systems. *J. Fixed Point Theory Appl.* **9**(2), (2011), 295–325.

[5] M. Akian, S. Gaubert, B. Lemmens, and R.D. Nussbaum, Iteration of order preserving subhomogeneous maps on a cone. *Math. Proc. Cambridge Philos. Soc.* **140**(1), (2006), 157–76.

[6] M. Akian, S. Gaubert, and R. Nussbaum, A Collatz–Wielandt characterization of the spectral radius of order-preserving homogeneous maps on cones, preprint. arXiv:1112.5968, 2012.

[7] M. Akian, S. Gaubert, and R. Nussbaum, Uniqueness of the fixed point of nonexpansive semidifferentiable maps, preprint. arXiv:1201.1536, 2012.

[8] P. Alexandroff and H. Hopf, *Topologie*. Berlin, Springer Verlag, 1935.

[9] J.C. Álvarez Paiva, Hilbert's fourth problem in two dimensions. In *MASS Selecta: Teaching and Learning Advanced Undergraduate Mathematics*. S. Katok, A. Sossinsky, and S. Tabachnikov, eds., Providence, RI, American Mathematical Society, 2003, pp. 165–83.

[10] N. Aronszajn and P. Panitchpakdi, Extension of uniformly continuous transformations and hyperconvex metric spaces. *Pacific J. Math.* **6**, (1956), 405Ð439.

[11] F. Baccelli, G. Cohen, G.J. Olsder, and J.P. Quadrat, *Synchronization and Linearity: An Algebra for Discrete Event Systems*. Wiley Series in Probability and Statistics, Chichester, Wiley, 1992.

[12] W. Ballmann, *Lectures on Spaces of Nonpositive Curvature*. DMV Seminar, 25, Basel, Birkhäuser Verlag, 1995.

[13] S. Banach, Sur les opérations dans les ensembles abstraits et leurs applications aux équations intégrales. *Fund. Math.* **3**, (1922), 133–80.

[14] R. Bapat, D_1AD_2 theorems for multidimensional matrices. *Linear Algebra Appl.* **48**, (1982), 437–42.

[15] R.B. Bapat and T.E.S. Raghavan, *Nonnegative Matrices and Applications*. Encyclopedia of Mathematics and its Applications, 64, Cambridge, Cambridge University Press, 1997.

[16] G.P. Barker and R.E.L. Turner, Some observations on the spectra of cone-preserving maps. *Linear Algebra Appl.* **6**, (1973), 149–53.

[17] F.L. Bauer, An elementary proof of the Hopf inequality for positive operators. *Numer. Math.* **7**, (1965), 331–7.

[18] A.F. Beardon, Iteration of contractions and analytic maps. *J. Lond. Math. Soc. (2)* **41**(1), (1990), 141–50.

[19] A.F. Beardon, The dynamics of contractions. *Ergodic Theory Dynam. Systems* **17**(6), (1997), 1257–66.

[20] A.F. Beardon, The Klein, Hilbert and Poincaré metrics of a domain. *J. Comput. Appl. Math.* **105**(1–2), (1999), 155–62.

[21] R. Bellman, *Dynamic Programming*. Princeton, NJ, Princeton University Press, 1957.

[22] A. Berman and R.J. Plemmons, *Nonnegative Matrices in the Mathematical Sciences*. Computer Science and Applied Mathematics, New York, Academic Press, 1979.

[23] T. Bewley and E. Kohlberg, The asymptotic theory of stochastic games. *Math. Oper. Res.* **1**(3), (1976), 197–208.

[24] R. Bhatia, *Matrix Analysis*. Graduate Texts in Mathematics, 169, New York, Springer Verlag, 1996.

[25] G. Birkhoff, Extensions of Jentzsch's theorem. *Trans. Amer. Math. Soc.* **85**(1), (1957), 219–27.

[26] G. Birkhoff, Uniformly semi-primitive multiplicative processes. *Trans. Amer. Math. Soc.* **104**, (1962), 37–51.

[27] A. Blokhuis and H.A. Wilbrink, Alternative proof of Sine's theorem on the size of a regular polygon in $\mathbb{R}^k$ with the ℓ_∞-metric. *Discrete Comput. Geom.* **7**(4), (1992), 433–4.

[28] B. Bollobás, *Combinatorics*. Cambridge, Cambridge University Press, 1989.

[29] F.F. Bonsall, Sublinear functionals and ideals in partially ordered vector spaces. *Proc. Lond. Math. Soc.* **4**, (1954). 402–18.

[30] F.F. Bonsall, Endomorphisms of partially ordered vector spaces. *J. Lond. Math. Soc.* **30**, (1955), 133–44.

[31] F.F. Bonsall, Endomorphisms of partially ordered vector spaces without order unit. *J. Lond. Math. Soc.* **30**, (1955), 145–53.

[32] F.F. Bonsall, Linear operators in complete positive cones. *Proc. Lond. Math. Soc.* **8**(3), (1958), 53–75.

[33] J.M. Borwein and P.B. Borwein, The arithmetic-geometric mean and fast computation of elementary functions. *SIAM Rev.* **26**(3), (1984), 351–66.

[34] J.M. Borwein and A.S. Lewis, Decomposition of multivariate functions. *Canad. J. Math.* **44**(3), (1992), 463–82.

[35] J.M. Borwein, A.S. Lewis, and R.D. Nussbaum, Entropy minimization, *DAD* problems, and doubly stochastic kernels. *J. Funct. Anal.* **123**(2), (1994), 264–307.

[36] T. Bousch and J. Mairesse, Finite-range topical functions and uniformly topical functions. *Dyn. Syst.* **21**(1), (2006), 73–114.

[37] M.R. Bridson and A. Haefliger, *Metric Spaces of Non-positive Curvature*. Grundlehren der Mathematischen Wissenschaften, 319, Berlin, Springer Verlag, 1999.

[38] R.A. Brualdi, The DAD theorem for arbitrary row sums. *Proc. Amer. Math. Soc.* **45**(2), (1974), 189–94.

[39] R.A. Brualdi, S.V. Parter, and H. Schneider, The diagonal equivalence of a nonnegative matrix to a stochastic matrix. *J. Math. Anal. Appl.* **16**, (1966), 31–50.

[40] R.E. Bruck, Jr., Properties of fixed-point sets of nonexpansive mappings in Banach spaces. *Trans. Amer. Math. Soc.* **179**, (1973), 251–62.

[41] A.D. Burbanks, R.D. Nussbaum, and C. Sparrow, Extension of order-preserving maps on a cone. *Proc. Roy. Soc. Edinburgh Sect. A* **133**A, (2003), 35–59.

[42] H. Busemann, *The Geometry of Geodesics*. Pure and Applied Mathematics, 6, New York, Academic Press, 1955.

[43] P.J. Bushell, Hilbert's projective metric and positive contraction mappings in a Banach space. *Arch. Ration. Mech. Anal.* **52**, (1973), 330–8.

[44] P.J. Bushell, On the projective contraction ratio for positive linear mappings. *J. Lond. Math. Soc.* **6**, (1973), 256–8.

[45] P.J. Bushell, On solutions of the matrix equation $T'AT = A^2$. *Linear Algebra Appl.* **8**, (1974), 465–9.

[46] P.J. Bushell, The Cayley-Hilbert metric and positive operators. *Linear Algebra Appl.* **84**, (1986), 271–80.

[47] P. Butkovič, *Max-linear Systems: Theory and Algorithms*. Springer Monographs in Mathematics, Berlin, Springer Verlag, 2010.

[48] A. Całka, On a condition under which isometries have bounded orbits. *Colloq. Math.* **48**, (1984), 219–27.

[49] B.C. Carlson, Algorithms involving arithmetic and geometric means. *Amer. Math. Monthly* **78**, (1971), 496–505.

[50] I. Csiszár, I-divergence geometry of probability distributions and minimization problems. *Ann. Probability* **3**, (1975), 146–58.

[51] J. Cochet-Terrasson, S. Gaubert, and J. Gunawardena, A constructive fixed-point theorem for min-max functions. *Dynam. Stability Systems* **14**(4), (1999), 407–33.

[52] L.Collatz, Einschliessungs satz für die charakteristischen Zahlen von Matrizen. *Math. Z.* **48**, (1942), 221–6.

[53] D.A. Cox, The arithmetic-geometric mean of Gauss. *Enseign. Math. (2)* **30**(3–4), (1984), 275–330.

[54] M.G. Crandall and L. Tartar, Some relations between nonexpansive and order preserving mappings. *Proc. Amer. Math. Soc.* **78**(3), (1980), 385–90.

[55] C.M. Dafermos and M. Slemrod, Asymptotic behavior of nonlinear contraction semigroups. *J. Funct. Anal.* **13**, (1973), 97–106.

[56] A. Denjoy, Sur l'itération des fonctions analytiques. *C. R. Acad. Sc. Paris, Serie I* **182**, (1926), 255–7.

[57] W.F. Donoghue, Jr., *Monotone Matrix Functions and Analytic Continuation*. Grundlehren der Mathematischen Wissenschaften, 207, New York, Springer Verlag, 1974.

[58] N. Dunford and J.T. Schwartz. *Linear Operators.* Wiley Classics Library, New York, John Wiley & Sons, 1988.

[59] B.C. Eaves, A.J. Hoffman, U.G. Rothblum, and H. Schneider, Line-sum-symmetric scalings of square nonnegative matrices. *Math. Programming Stud.* **25**, (1985), 124–41.

[60] M. Edelstein, On non-expansive mappings of Banach spaces. *Proc. Cambridge Philos. Soc.* **60**, (1964), 439–47.

[61] S.P. Eveson and R.D. Nussbaum, An elementary proof of the Birkhoff–Hopf theorem. *Math. Proc. Cambridge Philos. Soc.* **117**, (1995), 31–55.

[62] S.P. Eveson and R.D. Nussbaum, Applications of the Birkhoff–Hopf theorem to the spectral theory of positive linear operators. *Math. Proc. Cambridge Philos. Soc.* **117**, (1995), 491–512.

[63] J. Faraut and A. Korányi, *Analysis on Symmetric Cones*. Oxford, Clarendon Press, 1994.

[64] M. Fekete, Über die Verteilung der Wurzeln bei gewissen algebraischen Gleichungen mit ganzzahligen Koeffizienten. *Math. Z.* **17**(1), (1923), 228–49.

[65] S.E. Fienberg, An iterative procedure for estimation in contingency tables. *Ann. Math. Statist.* **41**, (1970), 907–17.

[66] T. Foertsch and A. Karlsson, Hilbert metrics and Minkowski norms. *J. Geom.* **83**(1-2), (2005), 22–31.

[67] J. Franklin and J. Lorenz, On the scaling of multidimensional matrices. *Linear Algebra Appl.* **114/115**, (1989), 717–35.

[68] H. Freudenthal and W. Hurewicz, Dehnungen, Verkürzungen, Isometrien. *Fund. Math.* **26** (1936), 120–2.

[69] S. Friedland and H. Schneider, The growth of powers of a nonnegative matrix. *SIAM J. Algebraic Discrete Methods* **1**, (1980), 185–200.

[70] G. Frobenius, Über Matrizen aus positiven Elementen. *S.-B. Preuss. Akad. Wiss (Berlin)*, (1908), 471–6.

[71] G. Frobenius, Über Matrizen aus positiven Elementen, II. *S.-B. Preuss. Akad. Wiss (Berlin)*, (1908), 514–18.

[72] G. Frobenius, Über Matrizen aus nicht negativen Elementen. *S.-B. Preuss. Akad. Wiss (Berlin)*, (1912), 456–77.

[73] F.R. Gantmacher, *The Theory of Matrices*, Vols. 1, 2. New York, Chelsea, 1959.

[74] S. Gaubert and J. Gunawardena, The Perron–Frobenius theory for homogeneous, monotone functions. *Trans. Amer. Math. Soc.* **356**(12), (2004), 4931–50.

[75] S. Gaubert and G. Vigeral, A maximin characterization of the escape rate of nonexpansive mappings in metrically convex spaces, *Math. Proc. Cambridge Phil. Soc.*, in press. `arXiv: 1012.4765`, 2010. DOI: 10.1017/S0305004111000673, 2011.

[76] K. Goebel and W.A. Kirk, Some problems in metric fixed point theory. *J. Fixed Point Theory Appl.* **4**(1), (2008), 13–25.

[77] K. Goebel and S. Reich, *Uniform Convexity, Hyperbolic Geometry, and Nonexpansive Mappings.* Monographs and Textbooks in Pure and Applied Mathematics, 83, New York, Marcel Dekker, 1984.

[78] M. Gromov, Hyperbolic manifolds, groups and actions. In *Riemann Surfaces and Related Topics: Proceedings of the 1978 Stony Brook Conference (State*

Univ. New York, Stony Brook, NY, 1978), Annals of Mathematics Studies, 97, Princeton, NJ, Princeton University Press, 1981, pp. 183–213.

[79] J. Gunawardena, An introduction to idempotency. In *Idempotency (Bristol, 1994)*, J. Gunawardena, ed., Publications of the Newton Institute, 11, Cambridge, Cambridge University Press, 1998, pp. 1–49.

[80] J. Gunawardena, From max-plus algebra to nonexpansive mappings: A nonlinear theory for discrete event systems. *Theoret. Comput. Sci.* **293**(1), (2003), 141–67.

[81] J. Gunawardena and C. Walsh, Iterates of maps which are non-expansive in Hilbert's projective metric. *Kybernetika (Prague)* **39**(2), (2003), 193–204.

[82] G.H. Hardy, J.E. Littlewood, and G. Pólya, *Inequalities*. Cambridge Mathematical Library, Cambridge, Cambridge University Press, 1988.

[83] P. de la Harpe, On Hilbert's metric for simplices. In *Geometric Group Theory*, Vol. 1, Niblo, G.A., *et al.*, eds., Lecture Note Series, 181, London Mathematical Society, 1993, pp. 97–119.

[84] T. Hawkins, Continued fractions and the origins of the Perron–Frobenius theorem. *Arch. Hist. Exact Sci.* **62**(6), (2008), 655–717.

[85] M. Hervé, Quelques proprietés des applications analytiques d'une boule à *m* dimensions dans elle-même. *J. Math. Pures Appl.* **42**, (1963), 117–47.

[86] D. Hilbert, Über die gerade Linie als kürzeste Verbindung zweier Punkte. *Math. Ann.* **46**, (1895), 91–6

[87] M.W. Hirsch, Stability and convergence in strongly monotone dynamical systems. *J. Reine Angew. Math.* **383**, (1988), 1–53.

[88] M.W. Hirsch, Positive equilibria and convergence in subhomogeneous monotone dynamics. In *Comparison Methods and Stability Theory*, X. Liu and D. Siegel, eds., Lecture Notes in Pure and Applied Mathematics, 162, New York, Marcel Dekker, 1994, pp. 169–88.

[89] M.W. Hirsch and H. Smith, Monotone dynamical systems. In *Handbook of Differential Equations: Ordinary Differential Equations*, Vol. II, Amsterdam, Elsevier, 2005, pp. 239–357.

[90] E. Hopf, An inequality for positive integral operators. *J. Math. Mech* **12**, (1963), 683–92.

[91] E. Hopf, Remarks on my paper "An inequality for positive integral operators." *J. Math. Mech.* **12**, (1963), 889–92.

[92] R.A. Howard, *Dynamic Programming and Markov Processes*. New York, John Wiley & Sons, 1960.

[93] G. Jameson, *Ordered Linear Spaces*. Lecture Notes in Mathematics, 141, Berlin, Springer Verlag, 1970.

[94] P. Jordan, J. von Neumann, and E. Wigner, On an algebraic generalization of the quantum mechanical formalism. *Ann. of Math. (2)* **35**(1), (1934), 29–64.

[95] B. Kalantari and L. Khachiyan, On the complexity of nonnegative-matrix scaling. *Linear Algebra Appl.* **240**, (1996), 87–103.

[96] J. Kapeluszny, T. Kuczumow, and S. Reich, The Denjoy–Wolff theorem in the open unit ball of a strictly convex Banach space. *Adv. Math.* **143**(1), (1999), 111–23.

[97] J. Kapeluszny, T. Kuczumow, and S. Reich, The Denjoy–Wolff theorem for condensing holomorphic mappings. *J. Funct. Anal.* **167**(1), (1999), 79–93.

[98] S. Karlin and L. Nirenberg, On a theorem of P. Nowosad. *J. Math. Anal. Appl.* **17**, (1967), 61–7.

[99] A. Karlsson, Nonexpanding maps and Busemann functions. *Ergodic Theory Dynam. Systems* **21**(5), (2001), 1447–57.

[100] A. Karlsson, V. Metz, and G.A. Noskov, Horoballs in simplices and Minkowski spaces. *Int. J. Math. Math. Sci.* **2006**, Article ID 23656, (2006), 20 pages.

[101] A. Karlsson and G.A. Noskov, The Hilbert metric and Gromov hyperbolicity. *Enseign. Math. (2)* **48**(1-2), (2002), 73–89.

[102] T. Kato, *Perturbation Theory for Linear Operators*. Grundlehren der Mathematischen Wissenschaften, 132, Berlin, Springer Verlag, 1976.

[103] P.E. Kloeden and A.M. Rubinov, A generalization of the Perron–Frobenius theorem. *Nonlinear Anal.* **41**(1–2), (2000), 97–115.

[104] E. Kohlberg, Invariant half-lines of nonexpansive piecewise-linear transformations. *Math. Oper. Res.* **5**(3), (1980), 366–72.

[105] E. Kohlberg, The Perron–Frobenius theorem without additivity. *J. Math. Econom.* **10**, (1982), 299–303.

[106] E. Kohlberg and A. Neyman, Asymptotic behavior of nonexpansive mappings in normed linear spaces. *Israel J. Math.* **38**(4), (1981), 269–75.

[107] E. Kohlberg and J.W. Pratt, The contraction mapping approach to the Perron–Frobenius theory: Why Hilbert's metric? *Math. Oper. Res.* **7**(2), (1982), 198–210.

[108] K. Koufany, Application of Hilbert's projective metric on symmetric cones. *Acta Math. Sin. (Engl. Ser.)* **22**(5), (2006), 1467–72.

[109] M.A. Krasnosel'skii, *Positive Solutions of Operator Equations.* Groningen, Noordhoff, 1964.

[110] M.A. Krasnosel'skii, J. A. Lifshits, and A.V. Sobolev, *Positive Linear Systems: The Method of Positive Operators.* Sigma Series in Applied Mathematics, 5, Lemgo, Heldermann Verlag, 1989.

[111] M.A. Krasnosl'skii and A.V. Sobolev, Spectral clearance of a focussing operator. *Funct. Anal. Appl.* **17**, (1983), 58–9.

[112] U. Krause, A nonlinear extension of the Birkhoff-Jentzsch theorem. *J. Math. Anal. Appl.* **114**(2), (1986), 552–68.

[113] U. Krause, Relative stability for ascending and positively homogeneous operators on Banach spaces. *J. Math. Anal. Appl.* **188**(1), (1994), 182–202.

[114] U. Krause and R.D. Nussbaum, A limit set trichotomy for self-mappings of normal cones in Banach spaces. *Nonlinear Anal.* **20**(7), (1993), 855–70.

[115] U. Krause and P. Ranft, A limit set trichotomy for monotone nonlinear dynamical systems. *Nonlinear Anal.* **19**(4), (1992), 375–92.

[116] M.G. Kreĭn and D. Milman, On the extreme points of regularly convex sets. *Studia Math.* **9**, (1940), 133–8.

[117] M.G. Kreĭn and M.A. Rutman, Linear operators leaving invariant a cone in a Banach space. *Amer. Math. Soc. Translation 1950* **26**, (1950), 199–325.

[118] E. Landau, *Handbuch der Lehre von der Verteilung der Primzahlen, I*, 2nd ed. New York, Chelsea, 1953.

[119] B. Lemmens, Nonexpansive mappings on Hilbert's metric spaces. *Topol. Methods Nonlinear Anal.* **38**(1), (2011), 45–58.

[120] B. Lemmens and R. Nussbaum, Continuity of the cone spectral radius. *Proc. Amer. Math. Soc.*, in press. `arXiv: 1107.4532`, 2011.

[121] B. Lemmens, R.D. Nussbaum, and S.M. Verduyn Lunel, Lower and upper bounds for ω-limit sets of nonexpansive maps. *Indag. Math. (N.S.)* **12**(2), (2001), 191–211.

[122] B. Lemmens and M. Scheutzow, A characterization of the periods of periodic points of 1-norm nonexpansive maps. *Selecta Math. (N.S.)* **9**(4), (2003), 557–78.

[123] B. Lemmens and M. Scheutzow, On the dynamics of sup-norm nonexpansive maps. *Ergodic Theory Dynam. Systems* **25**(3), (2005), 861–71.

[124] B. Lemmens and C. Sparrow, A note on periodic points of order preserving subhomogeneous maps. *Proc. Amer. Math. Soc.* **134**(5), (2006), 1513–17.

[125] B. Lemmens, C. Sparrow, and M. Scheutzow, Transitive actions of finite abelian groups of sup-norm isometries. *European J. Combin.* **28**(4), (2007), 1163–79.

[126] B. Lemmens and O.W. van Gaans, Periods of order-preserving nonexpansive maps on strictly convex normed spaces. *J. Nonlinear Convex Anal.* **4**(3), (2003), 353–63.

[127] B. Lemmens and O.W. van Gaans, On non-expansive maps dynamics on strictly convex Banach spaces. *Israel J. Math.* **171**, (2009), 425–42.

[128] Y. Lim, Hilbert's projective metric on Lorentz cones and Birkhoff formula for Lorentzian compressions. *Linear Algebra Appl.* **423**(2–3), (2007), 246–54.

[129] N. Linial, A. Samorodnitsky, and A. Wigderson, A deterministic strongly polynomial algorithm for matrix scaling and approximate permanents. *Combinatorica* **20**(4), (2000), 545–68.

[130] B. Lins, *Asymptotic Behavior and Denjoy–Wolff Theorems for Hilbert Metric Nonexpansive Maps*. PhD. thesis, Rutgers University, New Brunswick, NJ, 2007.

[131] B. Lins, A Denjoy–Wolff theorem for Hilbert metric nonexpansive maps on a polyhedral domain. *Math. Proc. Cambridge. Philos. Soc.* **143**(1), (2007), 157–64.

[132] B. Lins, Asymptotic behavior of nonexpansive mappings in finite-dimensional normed spaces. *Proc. Amer. Math. Soc.* **137**(7), (2009), 2387–92.

[133] B. Lins and R. Nussbaum, Iterated linear maps on a cone and Denjoy–Wolff theorems. *Linear Algebra Appl.* **416**(2–3), (2006), 615–26.

[134] B. Lins and R. Nussbaum, Denjoy–Wolff theorems, Hilbert metric nonexpansive maps and reproduction-decimation operators. *J. Funct. Anal.* **254**(9), (2008), 2365–86.

[135] K. Löwner, Über monotone Matrixfunktionen. *Math. Z.* **38**, (1934), 177–216.

[136] D. Lubell, A short proof of Sperner's lemma. *J. Combin. Theory* **1**, (1966), 299.

[137] R. Lyons and R.D. Nussbaum, On transitive and commutative finite groups of isometries. In *Fixed Point Theory and Applications*, K.-K. Tan, ed., Singapore, World Scientific, 1992, pp. 189–228.

[138] J. Mallet-Paret and R.D. Nussbaum, Eigenvalues for a class of homogeneous cone maps arising from max-plus operators. *Discrete Contin. Dyn. Syst.* **8** (2002), 519–63.

[139] J. Mallet-Paret and R.D. Nussbaum, Generalizing the Krein-Rutman theorem, measures of noncompactness and the fixed point index. *J. Fixed Point Theory Appl.* **7**(1), (2010), 103–43.

[140] P. Martus, *Asymptotic Properties of Nonstationary Operator Sequences in the Nonlinear Case*. PhD. thesis, Friedrich-Alexander Universität, Erlangen-Nürnberg, Germany, 1989.

[141] J.-P. Massias, Majoration explicite de l'ordre maximum d'un élément du groupe symétrique. *Ann. Fac. Sci. Toulouse Math. (5)* **6**(3–4), 269–81, 1984.

[142] P. Mellon, A general Wolff theorem for arbitrary Banach spaces. *Math. Proc. R. Ir. Acad.* **104A**(2), (2004), 127–42.

[143] M.V. Menon, Reduction of a matrix with positive elements to a doubly stochastic matrix. *Proc. Amer. Math. Soc.* **18**, (1967), 244–7.

[144] M.V. Menon, Some spectral properties of an operator associated with a pair of nonnegative matrices. *Trans. Amer. Math. Soc.* **132**, (1968), 369–75.

[145] M.V. Menon and H. Schneider, The spectrum of a nonlinear operator associated with a matrix. *Linear Algebra Appl.* **2**, (1969), 321–34.

[146] L.D. Meshalkin, Generalization of Sperner's theorem on the number of subsets of a finite set (in Russian). *Teor. Veroyatn. Primen.* **8**, (1963), 219–20.

[147] W. Miller, The maximum order of an element of a finite symmetric group. *Amer. Math. Monthly* **94**(6), (1987), 497–506.

[148] H. Minc, *Nonnegative Matrices*. Wiley-Interscience Series in Discrete Mathematics and Optimization, New York, John Wiley & Sons, 1988.

[149] M. Misiurewicz, Rigid sets in finite-dimensional l_1-spaces. Technical Report 45, Mathematica Göttingensis Schriftenreihe des Sonderforschungsbereichs Geometrie und Analysis, Göttingen, Mathematisches Institut, George-August-Universität Göttingen, 1987.

[150] M. Morishima. *Equilibrium, Stability and Growth*. Oxford, Clarendon Press, 1964

[151] B. de Sz.Nagy, Sur les lattis linéaires de dimension finie. *Comment. Math. Helv.* **17**, (1945), 209–13.

[152] A. Neyman, Stochastic games and nonexpansive maps. In *Stochastic Games and Applications (Stony Brook, NY, 1999)*, A. Neyman and S. Sorin, eds., NATO Science Series C: Mathematical and Physical Sciences, 570, Dordrecht, Kluwer Academic Publishers, 2003, pp. 397–415.

[153] P. Nowosad, On the integral equation $\kappa f = 1/f$ arising in a problem in communication. *J. Math. Anal. Appl.* **14**, (1966), 484–92.

[154] R.D. Nussbaum, A generalization of the Ascoli theorem and an application to functional differential equations. *J. Math. Anal. Appl.* **35**, (1971), 600–10.

[155] R.D. Nussbaum, Eigenvectors of nonlinear positive operators and the linear Krein-Rutman theorem. In *Fixed Point Theory*, E. Fadell and G. Fournier, eds., Lecture Notes in Mathematics, 886, Berlin, Springer Verlag, 1981, pp. 309–331.

[156] R.D. Nussbaum, Convexity and log convexity for the spectral radius. *Linear Algebra Appl.* **73**, (1986), 59–122.

[157] R.D. Nussbaum, Iterated nonlinear maps and Hilbert's projective metric: A summary. In *Dynamics of Infinite-Dimensional Systems (Lisbon, 1986)*, NATO Advanced Science Institutes Series F: Computer and Systems Science, 37, Berlin, Springer Verlag, 1987, pp. 231–248.

[158] R.D. Nussbaum, Hilbert's projective metric and iterated nonlinear maps. *Mem. Amer. Math. Soc.* **391**, (1988), 1–137.

[159] R.D. Nussbaum, Iterated nonlinear maps and Hilbert's projective metric, II, *Mem. Amer. Math. Soc.* **401**, (1989), 1–118.

[160] R. D. Nussbaum, Omega limit sets of nonexpansive maps: Finiteness and cardinality estimates. *Differential Integral Equations* **3**(3), (1990), 523–40.

[161] R.D. Nussbaum, Estimates of the periods of periodic points of nonexpansive operators. *Israel J. Math.* **76**(3), (1991), 345-80.

[162] R.D. Nussbaum, Convergence of iterates of a nonlinear operator arising in statistical mechanics. *Nonlinearity* **4**(4), (1991), 1223–40.

[163] R.D. Nussbaum, Entropy minimization, Hilbert's projective metric, and scaling integral kernels. *J. Funct. Anal.* **115**(1), (1993), 45–99.

[164] R.D. Nussbaum, Finsler structures for the part metric and Hilbert's projective metric and applications to ordinary differential equations. *Differential Integral Equations* **7**(5–6), (1994), 1649–1707.

[165] R.D. Nussbaum, Lattice isomorphisms and iterates of nonexpansive maps. *Nonlinear Anal.* **22**(8), (1994), 945–70.

[166] R.D. Nussbaum, Eigenvectors of order-preserving linear operators. *J. Lond. Math. Soc.* **58**, (1998), 480–96.

[167] R.D. Nussbaum, Periodic points of positive linear operators and Perron–Frobenius operators. *Integral Equations Operator Theory* **39**, (2001), 41–97.

[168] R.D. Nussbaum, Fixed Point Theorems and Denjoy–Wolff Theorems for Hilbert's Projective Metric in Infinite Dimensions. *Topol. Methods Nonlinear Anal.* **29**(2), (2007), 199–250.

[169] R.D. Nussbaum and M. Scheutzow, Admissible arrays and a generalization of Perron-Frobenius theory. *J. Lond. Math. Soc.* **58**(2), (1998), 526–44.

[170] R.D. Nussbaum, M. Scheutzow, and S.M. Verduyn Lunel, Periodic points of non-expansive maps and nonlinear generalizations of the Perron–Frobenius theory. *Selecta Math. (N.S.)* **4**(1), (1998), 1–41.

[171] R.D. Nussbaum and S.M. Verduyn Lunel, Generalizations of the Perron-Frobenius theorem for nonlinear maps. *Mem. Amer. Math. Soc.* **138**(659), (1999), 1–98.

[172] R.D. Nussbaum and S.M. Verduyn Lunel, Asymptotic estimates for the periods of periodic points of non-expansive maps. *Ergodic Theory Dynam. Systems* **23**, (2003), 1199–1226.

[173] R.D. Nussbaum and B.J. Walsh, Approximations by polynomials with nonnegative coefficients and the spectral theory of positive operators. *Trans. Amer. Math. Soc.* **350**, (1998), 2367–91.

[174] Y. Oshime, An extension of Morishima's nonlinear Perron–Frobenius theorem. *J. Math. Kyoto Univ.* **23**(4), (1983), 803–30.

[175] A.M. Ostrowski, On positive matrices. *Math. Ann.* **150**, (1963), 276–84.

[176] A.M. Ostrowski, Positive matrices and functional analysis. In *Recent Advances in Matrix Theory*, Madison, WI, University of Wisconsin Press, 1964, pp. 81–101.

[177] A. Papadopoulos, *Metric Spaces, Convexity and Nonpositive Curvature*. IRMA Lectures in Mathematics and Theoretical Physics, 6, Zürich, European Mathematical Society Publishing House, 2005.

[178] B.N. Parlett and T.L. Landis, Methods for scaling to doubly stochastic form. *Linear Algebra Appl.* **48**, (1982), 53–79.

[179] O. Perron, Grundlagen für eine Theorie des Jacobischen Kettenbruchalgorithmus. *Math. Ann.* **64**, (1907), 1–76.

[180] O. Perron, Zur Theorie der Über Matrizen. *Math. Ann.* **64**, (1907), 248–63.

[181] B.B. Phadke, A triangular world with hexagonal circles. *Geom. Dedicata* **3**, (1975), 511–20.

[182] P. Poláčik and I. Tereščák, Convergence to cycles as a typical asymptotic behavior in smooth strongly monotone discrete-time dynamical systems. *Arch. Ration. Mech. Anal.* **116**(4), (1992), 339–60.

[183] A.J.B. Potter, Applications of Hilbert's projective metric to certain classes of non- homogeneous operators. *Quart. J. Math. Oxford Ser. (2)* **28**, (1977), 93–9.

[184] H. Rademacher, Über partielle und totale Differenzierbarkeit von Funktionenmehrerer Variabeln und über die Transformation der Doppelintegrale. *Math. Ann.* **79**, 340–59.

[185] C.E. Rickart, *General Theory of Banach Algebras*. University Series in Higher Mathematics, Princeton, NJ, van Nostrand, 1960

[186] R.T. Rockafellar, *Convex Analysis*. Princeton Landmarks in Mathematics, Princeton, NJ, Princeton University Press, 1997.

[187] R.T. Rockafellar and R. J.-B. Wets, *Variational Analysis*. Grundlehren der Mathematischen Wissenschaften, 317, Berlin, Springer Verlag, 1998.

[188] U.G. Rothblum, Generalized scalings satisfying linear equations. *Linear Algebra Appl.* **114/115**, (1989), 765–83.

[189] U.G. Rothblum, H. Schneider, and M.H. Schneider, Scaling matrices to prescribed row and column maxima. *SIAM J. Matrix Anal. Appl.* **15**(1), (1994), 1–14.

[190] W. Rudin, *Functional Analysis*. International Series in Pure and Applied Mathematics, New York, McGraw-Hill, 1991.

[191] W. Rudin, *Real and Complex Analysis*. Series in Higher Mathematics, New York, McGraw-Hill, 1966.

[192] H. Samelson, On the Perron–Frobenius theorem. *Michigan Math. J.* **4**, (1957), 57–9.

[193] H.H. Schaefer, Positive Transformationen in lokalkonvexen halbgeordneten Vektorräumen. *Math. Ann.* **129**, (1955), 323–9

[194] H.H. Schaefer, On nonlinear positive operators. *Pacific J. Math.* **9**(3), (1959), 847–60.

[195] H.H. Schaefer, Some spectral properties of positive linear operators. *Pacific J. Math.* **10**, (1960), 1009–19.

[196] H.H. Schaefer, *Banach Lattices and Positive Operators*. Grundlehren der Mathematischen Wissenschaften, 215, New York, Springer Verlag, 1974.

[197] M. Scheutzow, Periods of nonexpansive operators on finite l_1-spaces. *European J. Combin.* **9**, (1988), 73–8.

[198] M. Scheutzow, Corrections to periods of nonexpansive operators on finite l_1-spaces. *European J. Combin.* **12**(2), (1991), 183.

[199] M.H. Schneider, Matrix scaling, entropy minimization, and conjugate duality, I: Existence conditions. *Linear Algebra Appl.* **114/115**, (1989), 785–813.

[200] M.H. Schneider, Matrix scaling, entropy minimization, and conjugate duality, II: The dual problem. *Math. Programming (Ser. B)* **48**(1), (1990), 103–24.

[201] A. Schrijver, *Theory of Linear and Integer Programming*. Chichester, John Wiley & Sons, 1986.

[202] E. Seneta, *Nonnegative Matrices and Markov Chains*, 2nd edn. Springer Series in Statistics, New York, Springer Verlag, 1981.

[203] L.S. Shapley, Stochastic games. *Proc. Natl. Acad. Sci. USA* **39**, (1953), 1095–1100.

[204] R. Sine, A nonlinear Perron–Frobenius theorem. *Proc. Amer. Math. Soc.* **109**(2), (1990), 331–6.

[205] R. Sinkhorn, A relationship between arbitrary positive matrices and doubly stochastic matrices. *Ann. Math. Statist.* **35**, (1964), 876–9.

[206] R. Sinkhorn and P. Knopp, Concerning nonnegative matrices and doubly stochastic matrices. *Pacific J. Math.* **21**. (1967), 343–8.

[207] H.L. Smith, *Monotone Dynamical Systems: An Introduction to the Theory of Competitive and Cooperative Systems*. Mathematical Surveys and Monographs, 41, Providence, RI, American Mathematical Society, 1995.

[208] S. Sorin, *A First Course on Zero-Sum Repeated Games*. Mathématiques et Applications, 37, Berlin, Springer Verlag, 2002.

[209] E. Sperner, Ein Satz über Untermengen einer endlichen Menge. *Math. Z.* **27**, (1928), 544–8.

[210] S. Straszewicz, Über exponierte Punkte abgeschlossener Punktmengen. *Fund. Math.* **24**, (1935), 139–43.

[211] P. Takáč, Asymptotic behavior of discrete-time semigroups of sublinear, strongly increasing mappings with applications to biology. *Nonlinear Anal.* **14**(1), (1990), 35–42.

[212] P. Takáč, Convergence in the part metric for discrete dynamical systems in ordered topological cones. *Nonlinear Anal.* **26**(11), (1996), 1753–77.

[213] B.-S Tam, On the distinguished eigenvalues of a cone-preserving map. *Linear Algebra Appl.* **131**, (1990), 17–37.

[214] B.-S. Tam, A cone-theoretic approach to the spectral theory of positive linear operators: The finite-dimensional case. *Taiwanese J. Math.* **5**(2), (2001), 207–77.

[215] A.E. Taylor, *Introduction to Functional Analysis*. London, John Wiley & Sons, 1958.

[216] A.C. Thompson, *Generalizations of the Perron–Frobenius Theorem to Operators Mapping a Cone into Itself*. PhD. thesis, Newcastle upon Tyne, UK, 1963.

[217] A.C. Thompson, On certain contraction mappings in a partially ordered vector space. *Proc. Amer. Math. Soc.* **14**, (1963), 438–43.

[218] A.C. Thompson, On the eigenvalues of some not-necessarily-linear transformations. *Proc. Lond. Math. Soc.* **15**(3), (1965), 577–98.

[219] J.S. Vandergraft, Spectral properties of matrices which have invariant cones. *SIAM J. Appl. Math.* **16**, (1968), 1208–22.

[220] C. Walsh, The horofunction boundary of finite-dimensional normed spaces. *Math. Proc. Cambridge Philos. Soc.* **142**(3), (2007), 497–507.

[221] C. Walsh, The horofunction boundary of the Hilbert geometry. *Adv. Geom.* **8**(4), (2008), 503–29.

[222] D. Weller, *Hilbert's Metric, Part Metric and Self-Mappings of a Cone*. PhD. thesis, Universität Bremen, Germany, 1987.

[223] J.H. Wells and L.R. Williams, *Embeddings and Extensions in Analysis*. Ergebnisse Mathematik und ihrer Grenzgebiete, 84, Berlin, Springer Verlag, 1975.

[224] P. Whittle, *Optimization Over Time. Vol. I. Dynamic Programming and Stochastic Control*. Wiley Series in Probability and Statistics, Chichester, John Wiley & Sons, 1982.

[225] H. Wielandt, Unzerlegbare, nicht negative Matrizen. *Math. Z.* **52**, (1950), 642–8.

[226] J. Wolff, Sur l'itération des fonctions bornées. *C. R. Acad. Sc. Paris, Serie 1,* **182**, (1926), 200–1.

[227] J. Wolff, Sur une généralisation d'un théoreme de Schwartz. *C. R. Acad. Sc. Paris, Serie 1,* **183**, (1926), 500–2.

[228] K. Wysocki, Behavior of directions of solutions of differential equations. *Differential Integral Equations* **5**(2), (1992), 281–305.

[229] K. Yamamoto, Logarithmic order of free distributive lattice. *J. Math. Soc. Japan* **6**, (1954), 343–53.

[230] P.P. Zabreiko, M.A. Krasnosel'skii, and Yu.V. Pokornyi, On a class of positive linear operators. *Funct. Anal. Appl.* **5**, (1972), 57–70.

Symbols

Index